Bacteria and Cancer

Abdul Arif Khan
Editor

Bacteria and Cancer

Springer

Editor
Abdul Arif Khan
Microbiology Unit
Department of Pharmaceutics
College of Pharmacy
King Saud University
Riyadh, Saudi Arabia
abdularifkhan@gmail.com

ISBN 978-94-007-2584-3 e-ISBN 978-94-007-2585-0
DOI 10.1007/978-94-007-2585-0
Springer Dordrecht Heidelberg London New York

Library of Congress Control Number: 2011944981

Printed on acid-free paper

Springer is part of Springer Science+Business Media (www.springer.com)

Contents

1 **Epidemiology of the Association Between Bacterial Infections and Cancer** 1
Christine P.J. Caygill and Piers A.C. Gatenby

2 **Gastric Cancer and *Helicobacter pylori*** 25
Amedeo Amedei and Mario M. D'Elios

3 ***Streptococcus bovis* and Colorectal Cancer** 61
Harold Tjalsma, Annemarie Boleij, and Ikuko Kato

4 **Chlamydial Disease: A Crossroad Between Chronic Infection and Development of Cancer** 79
Carlo Contini and Silva Seraceni

5 ***Salmonella typhi* and Gallbladder Cancer** 117
Catterina Ferreccio

6 **Ocular Adnexal Lymphoma of MALT-Type and Its Association with *Chlamydophila psittaci* Infection** 139
Andrés J.M. Ferreri, Riccardo Dolcetti, Silvia Govi, and Maurilio Ponzoni

7 **Possible Strategies of Bacterial Involvement in Cancer Development** 165
Puneet, Gopal Nath, and V.K. Shukla

8 **Bacteria as a Therapeutic Approach in Cancer Therapy** 185
Sazal Patyar, Ajay Prakash, and Bikash Medhi

9 **Targeting Cancer with Amino-Acid Auxotroph *Salmonella typhimurium* A1-R** 209
Robert M. Hoffman

10 Bacterial Asparaginase: A Potential Antineoplastic Agent for Treatment of Acute Lymphoblastic Leukemia 225
Abhinav Shrivastava, Abdul Arif Khan, S.K. Jain, and P.K. Singhal

11 Can Bacteria Evolve Anticancer Phenotypes? 245
Navya Devineni, Reshma Maredia, and Tao Weitao

12 Management of Bacterial Infectious Complications in Cancer Patients 259
Kenneth V.I. Rolston

Index 275

Chapter 1
Epidemiology of the Association Between Bacterial Infections and Cancer

Christine P.J. Caygill and Piers A.C. Gatenby

Abstract The role of infectious agents such as bacteria, viruses, fungi etc. has been of interest for many years. Many studies have linked chronic bacterial infection with subsequent development of cancer at a number of different sites in the body.

Most cancers have a multifactorial aetiology with a number of different steps between the normal and the malignant cell. One example of this is stomach cancer where it has been postulated that bacteria play a role at a number of stages but will also be true of cancers at other sites.

This chapter summarises those situations where cancers occur as a possible result of bacterial infection and covers oesophageal, stomach, colorectal, gallbladder, pancreatic, bladder and lung cancer.

Keywords Bacteria • Bacterial infections • Cancer • Epidemiology • Esophagus • Stomach • Colon • Rectum • Gallbladder • Pancreas • Bladder • Lung • Review • Cancer prevention • Infection

1.1 Introduction

It has been postulated that over 80% of cancers are caused by environmental factors (Higginson 1968) many of which factors are non-infectious such as diet and exposure to radiation. However the number of cancers caused by infectious agents is likely to rise with further research; for example until recently, it was thought that the acidic conditions of the stomach resulted in a sterile environment whereas in relatively

---

C.P.J. Caygill (✉) • P.A.C. Gatenby
UK Barrett's Oesophagus Registry, UCL Division of Surgery and Interventional Science, Royal Free Hospital, London NW3 2PF, UK
e-mail: ccaygill@medsch.ucl.ac.uk

A.A. Khan (ed.), *Bacteria and Cancer*, DOI 10.1007/978-94-007-2585-0_1,

recent times one of the most important infectious agents found to increase the risk of cancer, *Helicobacter pylori* was identified (Eslick 2010). Currently, more than 20% of cancer have been postulated to be linked to infectious agents (zur Hausen 2009). Of these, the majority of the causative agents are viruses, which make up nearly two thirds of the infectious causes (human papilloma virus linked to squamous cell carcinoma of the ano-genital region and nasopharynx, Epstein Barr virus linked to Burkitt's lymphoma and hepatitis B and C viruses linked to hepatocellular carcinoma) (zur Hausen 2009). Smaller numbers of tumours are related to infections from human herpes virus, liver flukes and schistosomes (Parkin 2006). Additionally, immuno-suppression caused iatrogenically, in patients with autoimmune disease and organ transplants, but also by HIV and HTLV results in higher rates of Kaposi's sarcoma, lip, vulval and penile cancers as well as non-Hodgkin's lymphoma compared to non-immuno-compromised subjects. Rates of salivary gland, eye, tongue, thyroid and cervical cancer are also higher than in non immuno-compromised controls (Ruprecht et al. 2008).

Overall, if the infectious causes of cancer were prevented there would be 26.3% fewer cancers in developing countries and 7.7% in developed countries (Parkin 2006).

The major bacterial cause of human cancer is *Helicobacter pylori*. This organism was classified as being carcinogenic for humans in 1994 (IARC Working Group 1994). It is causally associated with gastric carcinoma and gastric lymphoma as well as a number of other malignancies (Wu et al. 2009b). *Helicobacter pylori* infection is generally acquired during childhood, with a gradual increase in prevalence towards middle age (Parkin 2006; Robins et al. 2008). Its prevalence varies globally and in some countries is greater than 75% with overall prevalence of 74% in developing countries and 58% in developed countries (Parkin 2006). This organism has been implicated in one third of cancers caused by infective agents (including virus-caused cancers) and is found in 80% of patients with gastric cancer (zur Hausen 2009). In 2002, there were estimated to be 592,000 cases of gastric adenocarcinoma and 11,500 cases of gastric lymphoma attributable to *Helicobacter pylori* (Parkin 2006).

There are a huge number of bacteria living symbiotically with the human host (10^{15} in the alimentary tract flora (Ouwehand and Vaughan 2006)) and their presence is crucial for normal human physiological function.

The effects of bacteria are not ubiquitously harmful and the dichotomy of bacterial protection versus harm is illustrated by the relative protective effects of *Helicobacter pylori* infection of the stomach with regards to reduction of oesophageal cancer, but increased risk of gastric adenocarcinoma and lymphoma (Nakajima and Hattori 2004). Colonisation by bacterial species does not indicate a true infection and bacteria may colonise the abnormal host environment around a tumour. Additionally, some bacterial toxins have been used in anti-cancer therapy as chemotherapeutic agents (Patyar et al. 2010).

1.2 Oesophageal Cancer

The two major types of oesophageal cancer, squamous cell carcinoma and adenocarcinoma have different aetiologies. Squamous cell carcinoma develops most frequently in patients who smoke and have high alcohol intake or long standing achalasia. Adenocarcinoma is associated with gastro-oesophageal reflux and columnar metaplasia ("Barrett's oesophagus") (Allum et al. 2002).

The oesophageal mucosa is continuously bathed in swallowed saliva and food boluses have a rapid transit time due to the organ's coordinated peristalsis and appropriate lower oesophageal sphincter relaxation minimising the contact time of carcinogenic agents with the organ. In normal subjects a small volume of gastro-oesophageal reflux occurs with low frequency, however in patients with defective antireflux mechanisms and inadequate lower oesophageal muscular clearance, the lower oesophagus may be bathed in swallowed boluses and gastric contents for more prolonged periods (Gatenby and Bann 2009). The highest risk of oesophageal adenocarcinoma is seen in patients with the most frequent and prolonged reflux symptoms (Lagergren et al. 1999) and those with metaplastic columnar-lined oesophagus (Barrett's oesophagus) which has an annual incidence of adenocarcinoma of 0.69% per annum (Gatenby et al. 2008).

There has been a worldwide increase in the incidence of oesophageal cancers over the last 50 years, the oesophagus being the eighth commonest site of primary carcinoma in 2000 (Parkin 2001). This increase has been demonstrated specifically in the United Kingdom (Newnham et al. 2003; Kocher et al. 2001; Powell and McConkey 1992; Johnston and Reed 1991; McKinney et al. 1995) as well as in other countries (Ries et al. 2004; Daly et al. 1996; Liabeuf and Faivre 1997; Tuyns 1992; Moller 1992; Hansen et al. 1997). The histological type of these tumours has changed, from historically a strong predominance of squamous cell carcinomata (Bosch et al. 1979; Puestow et al. 1955; Turnbull and Goodner 1968; Webb and Busuttil 1978) to the present time, when adenocarcinomata comprise the majority of oesophageal tumours in the United States and United Kingdom (Gelfand et al. 1992; Putnam et al. 1994; Rahamim and Cham 1993; Chalasani et al. 1998; Johnston and Reed 1991; Devesa et al. 1998; Powell and McConkey 1992). Furthermore, current trends are predictive of a continued rise in oesophageal cancer in the UK (Gatenby et al. 2011; Moller et al. 2007) which is likely also to be seen in other countries, especially those with high proportions of adenocarcinoma (Curado et al. 2007). However globally, squamous cell carcinoma is still the predominant histological type (Curado et al. 2007).

Swallowed bacteria from normal oral flora include *Streptococcus*, *Neisseria*, *Veillonella*, *Fusobacterium*, *Bacteroides*, *Lactobacillus*, *Staphylococcus* and *Enterobacteriaceae* (Sjosted 1989). A difference has been noted in the oesophageal flora in patients with oesophageal cancer compared to the normal oesophagus (Eslick 2010) and Barrett's oesophagus compared to the normal oesophagus

(MacFarlane et al. 2007). However it is likely that the majority of the changes in microbiological flora occurs due to opportunistic colonisation of the altered host environment of the cancer rather than earlier in the process of carcinogenesis as causative agents, with the exception of *Campylobacter concisus* and *Campylobacter rectus* which have been associated with the development of adenocarcinoma in patients with columnar metaplasia of the oesophagus via mutagenic effects including nitrite, N-nitroso and nitrous oxide mediated damage (MacFarlane et al. 2007).

Streptococcus anginosus infection has been found in 44% of oesophageal cancer tissue samples (Morita et al. 2003), but a role in the development of cancer has not been demonstrated.

Treponema denticola, which is associated with gingivitis and periodontitis is frequently found in oesophageal cancer specimens. This was the most frequent organism found in resected oesophageal cancer specimens in one series (Narikiyo et al. 2004).

Helicobacter pylori infection results in stomach inflammation and reduced gastric acid production and its eradication has been shown to increase reflux oesophagitis and metaplastic columnar-lined oesophagus (Labenz et al. 1997; Corley et al. 2008). The EUROGAST group has demonstrated that the ratio of cases of squamous cell carcinoma of the oesophagus: adenocarcinoma of the oesophagus is higher in centres with higher population prevalence of *Helicobacter pylori* infection (14 centres total), but that the strain of *Helicobacter pylori* did not have a clear relationship with histological type (Robins et al. 2008). The FINBAR study demonstrated that the rate of *Helicobacter pylori* positivity was lower in patients with reflux oesophagitis (42.4% positive), Barrett's oesophagus (47.4% positive) and adenocarcinoma (51.9% positive) compared to control subjects (59.3% positive). Cag A positivity (the strain most strongly associated with peptic ulcer disease and development of gastric tumours) was lower in Barrett's oesophagus and oesophageal adenocarcinoma patients than in patients with reflux oesophagitis or control subjects. When the oesophageal cancer group was divided into those with true oesophageal tumours to tumours at the oesophagogastric junction, rates of *Helicobacter pylori* and the Cag A strain were similar in patients with junctional tumours and control subjects, but lower in true oesophageal tumours (Anderson et al. 2008).

Three meta-analyses have been published on the relationship between *Helicobacter pylori* infection and the Cag A strain in the last 4 years. Rokkas et al. (2007) demonstrated an odds ratio of 0.52 (95% confidence interval 0.37–0.73) for *Helicobacter* positive compared to negative patients in development of adenocarcinoma (with similar findings for *Helicobacter* positivity and Barrett's oesophagus). The odds ratio for Cag A positive *Helicobacter pylori* and development of adenocarcinoma was 0.51 (95% confidence limits 0.31–0.82). There was no significant relationship between *Helicobacter pylori* positivity and squamous cell carcinoma (odds ratio 0.85, 95% confidence limits 0.55–1.33). Zhuo et al. (2008) demonstrated that in 12 case-control studies, the odds ratio for development of oesophageal adenocarcinoma (9 studies, 684 cases oesophageal adenocarcinoma and 2,470 controls of which 259 cases and 1,287 controls were *Helicobacter pylori* positive) with

Helicobacter pylori infection was 0.58 (95% confidence interval 0.48–0.70) and for squamous cell carcinoma (5 studies, 644 cases squamous cell carcinoma and 2,021 controls of which 355 cases and 1,150 controls were *Helicobacter pylori* positive) was 0.80 (95% confidence interval 0.45–1.43). For the Cag A strain-infected subjects compared to non-Cag A strain-infected subjects the odds ratio for development of adenocarcinoma was 0.54 (95% confidence interval 0.40–0.73) and the odds ratio for development of squamous cell carcinoma was 1.20 (95% confidence interval 0.45–3.18) (Zhuo et al. 2008). Islami and Kamangar (2008) demonstrated that in their meta-analysis of 13 studies (840 cases and 2,890 controls) that infection with the *Helicobacter pylori* was associated with reduced risk of oesophageal adenocarcinoma odds ratio 0.56 (95% confidence interval 0.45–0.69). The effect was also seen in the single study undertaken in a non-Western country (Iran), (but the result of this small study just fell short of statistical significance). The odds ratio of development of oesophageal adenocarcinoma with the Cag A strain was 0.56 (95% confidence interval 0.46–0.68) and no difference was seen between *Helicobacter* negative subjects and Cag A negative *Helicobacter pylori* positive subjects. No significant effect was seen with squamous cell carcinoma (Derakhshan et al. 2008).

A further large case-control study from Taiwan (where squamous cell carcinoma accounts for 95% of oesophageal cancers) has demonstrated that the odds ratio of *Helicobacter pylori* infection with squamous cell carcinoma of the oesophagus was 0.470 (95% confidence interval 0.340–0.648) and 0.375 (0.277–0.508) when compared to two hospital control groups and 0.802 (95% confidence interval 0.591–1.089) compared to a community control group (Wu et al. 2009b).

Within patients with established columnar-lined oesophagus there does not appear to be a difference in the risk of cancer development between those who had evidence of *Helicobacter pylori* infection and those who had not been infected (Ramus et al. 2007).

Overall it is possible that the protective effects are secondary to *Helicobacter pylori* induced gastric atrophy and hypochlorhydria, both of which reduce acid exposure of the lower oesophagus (Blaser 2008) and the overall results demonstrate that infection with *Helicobacter pylori* and particularly the Cag A strain are associated with reduced risk of oesophageal adenocarcinoma development, but no clear effect is seen on the risk of squamous cell carcinoma development.

Eradication of *Helicobacter pylori* would subsequently be likely to increase the risk of oesophageal adenocarcinoma, but eradication also reduces the risk of gastric cancers. Using an algorithm based on data from a systematic review, Nakajima and Hattori (2004) estimated that in patients with atrophic gastritis (the macroscopic state most closely associated with development of gastric cancer), eradication of *Helicobacter pylori* would reduce the annual incidence of gastric adenocarcinoma by 5.9 times. The annual incidence of oesophageal cancer was modelled at 1% per annum with 16.5% of patients who had undergone eradication developing gastro-oesophageal reflux disease and 12% of these patients developing columnar metaplasia of the oesophagus. The overall risk of development of oesophageal adenocarcinoma was 0.18% per annum in patients who had undergone eradication. In the presence of atrophic gastritis and columnar metaplasia of the oesophagus, there was still an overall benefit seen in eradication with the combined incidence of gastric and oesophageal cancers being reduced from 1.4% to 1% per annum (Nakajima and Hattori 2004). Anand and Graham (1999) estimated that the risk of development of oesophageal adenocarcinoma following *Helicobacter pylori* eradication was 10–60-fold lower than the risk of development of gastric adenocarcinoma if eradication was not undertaken.

1.2.1 Viral, Parasitic and Fungal Infection

Expression of JC viral protein has been shown in a small study of oesophageal cancer cells, but not in normal oesophageal cells (where viral DNA was also found). The authors suggest that JC virus may have a role in oesophageal cancer development (Del Valle et al. 2005). Studies have not shown a relationship between Epstein Barr Virus infection and risk of oesophageal cancer (Eslick 2010).

No studies have examined the role of human herpes simplex virus in oesophageal cancer development (Eslick 2010). Human papilloma virus has been linked with squamous cell cancer of the oesophagus, with HPV 16 being the type most strongly associated and frequently studied (Eslick 2010).

Chaga's disease (protozoal infection with *Trypanosoma cruzi*) has been associated with both higher and lower rates of oesophageal cancer (Garcia et al. 2003; de Rezende et al. 1985) This occurs several decades after the initial infection with dysfunction of the nervous control of the gastrointestinal tract with development of a dilated mega-oesophagus with poor peristaltic function and oesophageal emptying (Matsuda et al. 2009). However, there is a common finding of coinfection with *Helicobacter pylori* (Barbosa et al. 1993; de Rezende et al. 1985; Eslick et al. 1999; El-Omar et al. 2000) and the overall number of cancers caused by this protozoan is likely to be small compared to the effects of *Helicobacter pylori* infection on oesophageal cancer development.

The data on fungal causes of oesophageal cancer are largely circumstantial, with linkage of several mycotoxins to oesophageal cancer, but no good epidemiological studies (Eslick 2010).

1.3 Gastric Cancer

A hypothesis for the sequence of changes that lead from normal gastric mucosa to gastric cancer was first proposed by Correa et al. (1975). Although this sequence has since been added to and changed, the essential hypothesis (shown in Fig. 1.1) remains the same. Bacterial colonisation/infection would appear to play a role by two different pathways. One pathway is normal mucosa progressing to gastric atrophy, at which stage the stomach would become hypochlorhydric resulting in chronic bacterial colonisation, and the production of N-nitroso compounds. The other pathway is as a result of *Helicobacter pylori* infection.

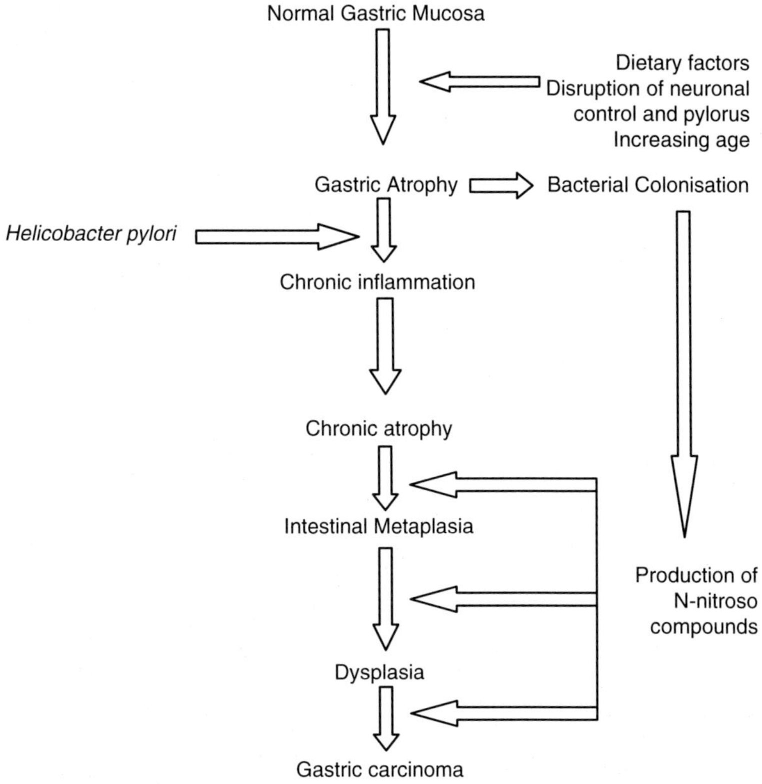

Fig. 1.1 The pathogenesis of gastric cancer

1.3.1 Helicobacter pylori Infection

Helicobacter pylori is a gram-negative bacterium which colonises gastric epithelium. It has evolved the ability to overcome the highly acidic environment of the stomach by metabolising urea to ammonia, thus generating a neutral environment (Wroblewski et al. 2010). *Helicobacter pylori* infection is associated with low socioeconomic status and crowded living conditions, especially in childhood (Malaty and Graham 1994). Approximately half the world's population is infected (with most children in developing countries being infected by the age of 10) (Smith and Parsonnet 1998) with the majority of these developing coexisting chronic inflammation (Wroblewski et al. 2010). In contrast, in developed countries, infection in children is uncommon and only 40–50% of adults are affected. There is a clear age-related increase in prevalence which is probably due to a cohort effect in that, H. *pylori* infection in childhood was more common in the past than it is today (Parsonnet et al. 1992; Banatvala et al. 1993). The route of transmission of *Helicobacter pylori* remains controversial with circumstantial evidence suggesting it probably occurs through person to person transmission.

Studies comparing rates of *Helicobacter pylori* infection in different populations with rates of gastric cancer in the same populations have mostly correlated well (Forman et al. 1993). In addition, as *Helicobacter pylori* infection has declined over time so has the rate of gastric cancer incidence (Parsonnet et al. 1992; Banatvala et al. 1993). It is considered that the gastric inflammatory response due to colonisation by *Helicobacter pylori* is the single strongest risk factor for peptic ulceration and gastric cancer. However only a fraction of those colonised go on to develop cancer (Peek et al. 2010).

Retrospective studies should be viewed with caution in view of the hypothesis that the cancerous stomach may lose its ability to harbour *Helicobacter pylori* (Osawa et al. 1996) but evidence from 2 meta-analyses of all case-control studies (Huang et al. 1998; Eslick and Talley 1998) indicate a 2-fold increase in the risk of gastric cancer in instances of *Helicobacter pylori* infection.

Prospective case-control using stored serum from populations, and thus knowing that infection by *Helicobacter pylori* preceded gastric cancer, has provided more concrete evidence of a link (Parsonnet et al. 1991; Forman et al. 1991; Lin et al. 1995; Siman et al. 1997). It has also been shown that in those infected with *Helicobacter pylori*, and followed up for a period of 10 years or more, risk of gastric cancer was increased 8-fold (Forman et al. 1994).

In a recent review of *Helicobacter pylori* infection and gastric cancer in the Middle East, Hussein (2010) reported that although *Helicobacter* infection rates in childhood were high, gastric cancer rates differ markedly from very high in Iran (26.1/100,000) to low in Israel (12.5/100,000) and very low in Egypt (3.4/100,000). Atherton (2006) concluded that *H. pylori* infection, distribution of virulence factors, diet and smoking could not explain the differences in gastric cancer rate even taking into account the accuracy of the data due to differences in diagnostic methods, limitations in medical services etc.

Whether eradication of *Helicobacter pylori* is an effective strategy for prevention of gastric cancer is still controversial (Selgrad et al. 2010). Some studies show this to be the case (Malfertheiner et al. 2005; Fry et al. 2007) but others do not (De Vries and Kuipers 2007). The effectiveness of *Helicobacter pylori* eradication as a means of protection against gastric cancer is dependant on the extent of preneoplastic changes (gastric atrophy, intestinal metaplasia etc.) at the time (Selgrad et al. 2010). Wu et al. (2009a) reported that the earlier *Helicobacter pylori* is eradicated after peptic ulcer disease, the smaller the risk of gastric cancer. Development of a vaccine to be used as primary prevention, especially with infant vaccination was discussed (Selgrad et al. 2010).

1.3.2 Chronic Bacterial Overgrowth of the Stomach

The normal stomach is acidic with a pH of 2. However in certain pathological conditions such as pernicious anaemia (caused by a lack of intrinsic factor and thus a failure to secrete gastric acid) and surgery for peptic ulcer, or as part of the ageing process, the gastric pH may rise to 4.5 or above on a permanent basis. This would result in chronic bacterial overgrowth of the stomach.

In the case of peptic ulcer, the aim of surgical treatment, either by gastrectomy or by vagotomy, was to decrease acid secretion in order to allow the ulcer to heal. In the case of gastrectomy the lower, acid secreting, part of the stomach was removed by a variety of procedures, and in vagotomy the vagal nerves, which control acid secretion, were severed. Both these procedures resulted in loss of gastric acidity within a year, and in both there was an increased risk of gastric cancer (Caygill et al. 1984, 1986).

1.3.3 Pernicious Anaemia

An increased risk of gastric cancer had been reported in several series of pernicious anaemia patients (Blackburn et al. 1968; Brinton et al. 1989). A study by Caygill et al. (1990) showed an overall 5-fold excess risk of gastric cancer in pernicious anaemia patients. It was not possible to ascertain the onset of pernicious anaemia accurately from patients records as it may be present for some years before diagnosis, therefore the period after diagnosis was divided into 0–19 years and 20+ years and it was found that the excess risk of gastric cancer was 4-fold in the first time period and 11-fold in the second time period.

1.3.4 Surgery for Peptic Ulcer

Table 1.1 is a summary of cohort studies which have shown an increased risk of gastric cancer in peptic ulcer patients who have undergone surgery to remove the ulcer, or in the case of vagotomy to stop secretion of stomach acid. In the study by Caygill et al. (1986), cancer risk for those undergoing a gastrectomy for gastric ulcer was analysed separately from those who had the operation for duodenal ulcer. The risk was analysed by time interval. They found that in the case of duodenal ulcer there was a decrease in risk in the first 19 years followed by an increase in risk thereafter. In contrast in the gastric ulcer patients there was a 3-fold increase in risk immediately after, and presumably prior to surgery, and this rose to over 5-fold 20 or more years after surgery. The pattern of an initial decrease in risk in those operated for duodenal ulcer has been confirmed by Arnthorsson et al. (1988), Moller and Toftgaard (1991), Lundegardh et al. (1988) and Eide et al. (1991). This difference in behaviour between duodenal ulcer and gastric ulcer patients needs to be rationalised. It was hypothesised that prior to surgery duodenal ulcer patients would have good acid secretion and the effect of surgery would be to induce hypochlohydric within a year of surgery. On the other hand many gastric ulcer patients would be hypochlorhydric for varying number of years prior to the operation.

Table 1.1 Cohort studies examining gastric cancer risk following surgery for peptic ulcer

References	Study population (*n*)	Excess risk	Latency (years)
Ross et al. (1982)	779	None	19
Watt et al. (1984)	735	3-fold	15
Tokudome et al. (1984)	3,827	None	–
Caygill et al. (1986)	4,466	4-fold	20
Viste et al. (1986)	3,479	3-fold	20
Arnthorsson et al. (1988)	1,795	2-fold	15
Lundegardh et al. (1988)	6,459	3-fold	30
Offerhaus et al. (1988)	2,633	5-fold	15 females 3 males
Toftgaard (1989)	4,131	2-fold	25
(vagotomy cohort) Caygill et al. (1991)	1,643	1.6	20

1.3.5 Possible Mechanism for Gastric Carcinogenesis in Instances of Hypochlorhydria

The histopathological seqence from the normal to the neoplastic stomach proposed by Correa et al. (1975) and reviewed by Correa (1988) has been generally accepted. They postulated that the first stage, gastric atrophy, progresses to chronic atrophic gastritis. Atrophic gastritis is at increased risk of developing intestinal metaplasia which in turn carries an elevated risk of progressing through increasingly severe dysplasia to cancer. It was suggested that this progression was a result of the action of carcinogenic N-nitroso compounds. Gastric atrophy results in the loss of gastric acid secretion allowing bacterial colonisation of the stomach. The bacteria react with nitrate, present in many foods and in drinking water, and convert it to nitrite. The nitrite further reacts with nitrosatable amines to form a variety of N-nitroso compounds. If this hypothesis is correct then the loss of gastric acidity, with consequent chronic bacterial overgrowth, from any cause (surgical, metabolic, clinical, genetic or environmental) should, after a latency period of 20 years or more, lead to an increased risk of gastric cancer as has been shown in patients with pernicious anaemia and those undergoing gastrectomy or vagotomy (see Fig 1.1). This also offers an explanation for the difference in cancer risk in those operated on for a gastric ulcer or for duodenal ulcer. Gastric ulcer patients, as a result of their hypoacidity prior to operation, will have bacterial overgrowth for variable lengths of time which would contribute to the latency period, whereas those with duodenal ulcer will only become hypochlorhydric after their operation, thus their increase in risk would only start to manifest itself 20 years later.

1.4 Colorectal Cancer

As in many cancers the progression from the normal epithelium to malignancy is a multi-stage process. There are at least three distinct histological stages prior to malignancy and metastatic disease. These are adenoma formation, adenoma growth and increasingly severe dysplasia (Hill et al. 1978, 2001; Hill 1991). The evidence for this was reviewed by Morson (1974) and Morson et al. (1983). Benign adenomas are very common in both men and women in western populations and their prevalence has been found to be approximately 50% in males and 30% in females by the age of 70 in post mortem studies. Most are very small (around 3–5 mm) but some can be greater than 20 mm in diameter. The risk of finding malignant cells in a small adenoma is very small (less than 1 per 1,000 for those with a diameter less than 3 mm) but high in those with a diameter greater than 20 mm (Morson et al. 1983). Thus one of the most important steps in the adenoma-carcinoma sequence is adenoma growth.

1.4.1 Faecal Bacteria Present in the Colon

There are differences within the colon in subsite distribution of small adenomas, large adenomas and colorectal cancers. A very large number of postmortem studies have shown that small adenomas are evenly distributed around the colon and rectum whereas large adenomas and cancers are concentrated in the distal colon and rectum, (Hill 1986). The implication being that the causal agents are delivered via the vascular system; and indeed the colon lumen is a rich source of potential carcinogens, produced *in situ* by bacterial action on benign substrates (Caygill and Hill 2005). Although not proved, this is consistent with the hypothesis that the factors causing adenomas to increase in size and in severity of epithelial dysplasia are luminal products of bacterial metabolism. There is further support for this by the fact that adenomas regress after diversion of the faecal stream.

1.4.2 Streptococcus bovis

Several *Streptococci* have been linked to chronic infections of the colon and subsequent increased risk of colorectal cancer (Kim et al. 2002; Siegert and Overbosch 1995). An association between *Streptococcus bovis* and colorectal cancer was first reported by Roses et al. (1974) and has been validated by more recent studies (Biarc et al. 2004; Gold et al. 2004). The incidence of *Streptococcus bovis* associated with colorectal cancer has been determined as being between 18% and 62% (Zarkin et al. 1990).

1.4.3 E. coli and Inflammatory Bowel Disease

The intestinal flora in patients with inflammatory bowel disease (Crohn's disease and ulcerative colitis) differs from control subjects with increased *E. coli* (Martin et al. 2004). These patients have a marked increase in rate of colorectal cancer development which is highest in those with chronic severe inflammation (Munkholm 2003). Small studies have demonstrated increased mucosa-associated and intramucosal bacteria in Crohn's disease (79% and 71% respectively) and colon cancer (71% and 57% respectively) compared to non inflamed controls (42% and 29% respectively), but no difference between controls and ulcerative colitis. These *E. coli* commonly expressed haemagglutinins (39% Crohn's, 38% cancers, and 4% controls) and the resulting pro-inflammatory cytokines may be implicated in carcinogenesis (Martin et al. 2004).

1.5 Gallbladder Cancer

Cancer of the gallbladder has a very poor prognosis. The highest incidence is in the Andean countries of South and Central America and in American Indian groups (Misra et al. 2003) but is rare in Western countries. The etiology is not well understood, but the major risk factor is the presence of gallstones which are involved in 70–80% of cases (Lazcano-Ponce et al. 2001). Risk factors include obesity, reproductive factors and environmental exposure to certain chemicals (Lazcano-Ponce et al. 2001; Wistuba and Gazdar 2004).

However, the major risk factors are those which involve chronic bacterial infection such are previous polya partial gastrectomy for peptic ulcer, gallstone carriage, chronic infection with *Salmonella typhi/paratyphi* and with *Helicobacter species*.

1.5.1 Gallstones

Although it has been known for some years that gallstones are the most important risk factor for gallbladder cancer (Devor 1982; Zatonski et al. 1997; Randi et al. 2006), the nature of this association is not clear. Gallstones are, however associated with bacterial infection of the gallbladder (England and Rosenblatt 1977).

1.5.2 Polya Partial Gastrectomy

The routine treatment for persistent peptic ulcer, gastric or duodenal, was surgery using a variety of partial gastrectomy operations. These remove much of the lower part (including most of the acid secreting section) of the stomach. As a result, the stomach became hypochlorhydric attaining a pH of around 4.5.

This is a perfect milieu for bacterial overgrowth and formation of N-Nitroso compounds (Hill 1996) which have been shown to be carcinogenic in all species in which they have been studied. Polya partial gastrectomy is associated with a 10-fold excess risk of gallbladder cancer with a 20 year latency period (Caygill et al. 1988).

1.5.3 Infection with Salmonella typhi/paratyphi

There is a growing body of evidence that typhoid carriers are at an increased risk of biliary tract cancer. The New York City Health Department conducted a very large case–control study of 471 registered carriers and 942 age- and sex-matched controls which showed that chronic carriers were six times as liable to die of hepatobiliary cancer as controls (Welton et al. 1979). This finding has been confirmed by others (Mellemgaard and Gaarslev 1988; Caygill et al. 1994; Nath et al. 1997, 2008; Shukla et al. 2000).

Caygill and co-workers studied long-term cancer risk in two Scottish cohorts – one a cohort of 386 acute typhoid cases from a single outbreak which occurred in Aberdeen in 1964 and the other 83 typhoid carriers from a number of different outbreaks (Caygill et al. 1994, 1995). In case of acute infection in Aberdeen, there was neither excess risk for cancer of the gallbladder nor indeed for any other cancer.

In the cohort of patients with chronic infection there was an almost 200-fold excess risk of cancer of the gallbladder and an excess risk of cancer of the pancreas (Table 1.2).

1.5.4 Infection with Helicobacter species

Helicobacter species colonising the biliary tract have been associated with gallbladder cancer (Leong and Sung 2002; Kobayashi et al. 2005).

There is no doubt that gallbladder cancer has a multi-factorial aetiology. Although a proportion of any risk may well be an individuals environmental and life style exposure, the most important risk factor appears to be exposure to chronic, but not acute bacterial infection.

1.6 Pancreatic Cancer

Cancer of the pancreas has a relative low incidence but a very poor prognosis even if diagnosed early and ranks eighth in a world listing of cancer mortality. International incidence rates vary in different countries, implying that environmental factors are important. Smoking is the best documented etiologic agent and accounts for approximately about 25% of all cases. Little is known about dietary factors. The incidence

Table 1.2 Deaths from "cancer" in patients with chronic infection with typhoid/paratyphoid (Caygill et al. 1994)

Site of cancer	ICD no	Observed (O)	Expected (E)	O/E	95% CI
Gallbladder	1,560	5	0.03	167*	(54–391)
Pancreas	157	3	0.37	8.1*	(1.7–23.7)
Colorectum	152–4	3	1.00	3.8	(0.6–8.8)
Lung	162	5	1.98	2.5	(0.8–5.9)
All neoplasms	140–208	20	7.80	2.6*	(1.6–4.0)

ICD International Classification of Disease, *O* observed, *E* expected
$^*P<0.001$

is strongly age dependent thus as the population of western countries ages we can anticipate an increasing number of cases (Lowenfels and Maisonneuve 2006). Cancer of the pancreas is also linked with bacterial infection.

1.6.1 Surgery for Peptic Ulcer

Surgery for peptic ulcer with the resultant hypochlorhydria results in bacterial overgrowth of the stomach. The bacteria thus formed react with ingested nitrates in food and converts them to nitrites. This is the perfect milieu for the formation of N-nitroso compounds which are formed when nitrite and nitrosatable amines are present together. These highly reactive compounds which can combine with nitrosatable amines, also present in food, and form a range of nitrosamines (Caygill et al. 1984; Preussmann 1984). Nitrosamines have been found to be carcinogenic in a number of animals (Pour and Lawson 1984) and are both species and target organ specific. This could well be the explanation for the finding of an excess risk for cancer of the pancreas after surgery for peptic ulcer (Caygill et al. 1987; Mack et al. 1986; Eide et al. 1991; Tersmette et al. 1990; Ross et al. 1982; Luo et al. 2007), the excess risk being greater in gastric ulcer than in duodenal ulcer patients (Caygill et al. 1987).It must be noted however that, Inokuchi et al. (1984), Watt et al. (1984) and Moller and Toftgaard (1991) did not find an excess risk of cancer of the pancreas in patients who had undergone operations for peptic ulcer.

1.6.2 Helicobacter species Infection

In recent years there have been a number of studies investigating a possible association between *Helicobacter species* infection and cancer of the pancreas. *Helicobacter species* ribosomal DNA was detected in the pancreas of 75% of pancreatic cancer patients (Nilsson et al. 2006) and *Helicobacter pylori* was found to be associated with an increased risk of pancreatic cancer in studies by Raderer et al. (1998) and

Stolzenberg-Solomon et al. (2001). A study by Risch et al. (2010) also found an association but only in individuals with non-O blood types. However studies by de Martel et al. (2008) and Lindkvist et al. (2008) could find no such association. Luo et al. (2007) found a modest increased risk of pancreatic cancer in patients with gastric ulcer or gastric resection and hypothesised that colonisation of the corpus by *H. pylori*, together with atrophic gastritis resulting in bacterial overgrowth and nitrosamine formation may contribute to pancreatic carcinogenesis.

1.6.3 Typhoid Carriage

In a study examining cancer risk in those infected with typhoid and in typhoid carriers, Caygill et al. (1994) found a large excess (23-fold) in cancer of the pancreas in a cohort of 83 typhoid carriers, not in 386 acute cases of typhoid who did not become carriers. The mechanism is uncertain, but pancreatic cancer has been associated with bile reflux from the common bile duct (Wynder 1975).

1.7 Bladder Cancer

Industrial exposure to naphthylamines, benzidine and a range of aromatic amines, contained in chemical dyes, has long been known to be associated with cancer of the bladder and explained the reason why men in industrialised countries were most at risk. However, a proportion of bladder cancer cases do not have an industrial origin. Early anecdotal evidence suggested an excess risk of bladder cancer following chronic bladder infection and this was confirmed by Radomski et al. (1978). Bladder infections are very common, and often asymptomatic (Sinclair and Tuxford 1971); the data on cancer risk reported by Radomski et al. (1978) concerns chronic symptomatic infection resistant to therapy, but many of his controls might have had asymptomatic bladder infections and so the magnitude of the excess risk would have been underestimated.

There is copious evidence that carcinogenic N-nitroso compounds are produced *in situ* in the bladder by infecting organisms, which is to be expected since the urine is the route of excretion of the substrates for N-nitroso compounds production – nitrate and nitrosatable amines. Thus Radomski et al. (1978) suggested that N-nitroso compounds, produced by bacterial action on these substrates were the cause of the cancer.

1.7.1 Schistosoma haematobium

Bilharzial (*Schistosoma haematobium)* infection is a major risk factor for bladder cancer, and such infections are accompanied by a profuse secondary bacterial infection of the bladder. Hicks et al. (1977) showed strong evidence that the bladder cancer associated with bilharzial infection was in fact due to the N-nitroso compounds produced by the secondary bacterial infection. This has been supported by others who have shown a similar association (El-Mawla et al. 2001; Bedwani et al. 1998; Saad et al. 2006). Hicks et al. (1977) also produced evidence that the excess risk of bladder cancer in paraplegia was due to the same mechanism – N-nitroso compounds produced by a chronic bacterial infection of the bladder.

1.7.2 Tuberculosis

Increased bladder cancer risk has also been found amongst tuberculosis sufferers in Korea, a country where the prevalence of tuberculosis is particularly high (Kim et al. 2000).

1.8 Lung Cancer

The major risk factor for lung cancer is smoking, however infection by a number of bacteria also has a role.

1.8.1 Pulmonary Tuberculosis

Before 1950 most TB patients died when relatively young, thus any risk of lung cancer would not have became manifest. It was not till TB treatment was sufficiently successful to give the patient a reasonable life-expectancy that the association was noted. Indeed, as a result of early studies there was a theory (Rokitansky 1854) that the two diseases were antagonistic. Since then there have been numerous reviews of the association between tuberculosis and subsequent lung cancer. Aoki (1993) reviewed the epidemiological studies between 1960 and 1990 and confirmed that patients with active pulmonary tuberculosis have an excess risk of dying of lung cancer even though they already had a high mortality from tuberculosis. The excess was 5–10-fold depending on age, and was greater in women than in men. Patients with active disease were the most likely to develop lung cancer and he also found that they also had an excess risk of other cancers such as colon, lymphoma, myeloma etc.

The mechanism for the association is not clear, and there are no good hypotheses to explain it. Attempts to stimulate the immune system in animal with BCG resulted in an increased, rather than decreased, cancer risk (Martin et al. 1977). This may explain the reason for the increased risk of cancer at distant sites seen by Aoki (1993).

1.8.2 *Chlamydia pneumonia*

There have been a number of reports of a connection between lung cancer and infection with *Chlamydia pneumonia* (Laurila et al. 1997; Kocazeybek 2003; Littman et al. 2004). However accurate assessment of past *Chlamydia pneumonia* infection is difficult as there is no serological test to specifically identify persons with chronic infection (Littman et al. 2005).

In 2010, Chaturvedi et al. evaluated the relationship of *Chlamydia pneumoniae* infection with prospective lung cancer risk using serologic markers for both chronic and acute Chlamydial infection and concluded that chronic infection 2–5 years before was associated with an increased risk of lung cancer. They highlight the potential for lung cancer reduction through treatments targeted towards *Chlamydia pneumoniae* infections.

1.8.3 *Helicobacter pylori Infection*

Lung cancer has been associated with *Helicobacter pylori* infection in a number of studies (Gocyk et al. 2000; Ece et al. 2005; Zhou et al. 1992), however Philippou et al. (2004) found no such association. The mechanism is unknown but *Helicobacter pylori* may contribute by upregulating gastrin and COX-2 thus stimulating tumour growth. Also increased plasma gastrin concentrations may increase the risk of lung cancer by inducing proliferation of mucosal cells in the bronchial epithelium (Kanbay et al. 2007).

1.9 Conclusion

Bacteria play a significant role in the aetiology of cancer development, but less than viral infection. The strongest association has been seen in gastric cancer with *Helicobacter pylori* and this bacterium has been associated with the development of other tumours as well as having an inverse association with the development of oesophageal cancer. Chronic infection with ongoing insult from the infecting bacterium has been most strongly demonstrated with typhoid infection and also *Helicobacter pylori.* Eradication of bacterial agents which are causes of cancer may result in a reduction in one quarter of cancers in developing countries and a smaller proportion in developed countries.

References

Allum WH, Griffin SM, Watson A et al (2002) Guidelines for the management of oesophageal and gastric cancer. Gut 50(Suppl V):v1–v23

Anand BS, Graham DY (1999) Ulcer and gastritis. Endoscopy 31(2):215–225

Anderson LA, Murphy SJ, Johnston BT et al (2008) Relationship between *Helicobacter pylori* infection and gastric atrophy and the stages of oesophageal inflammation, metaplasia, adenocarcinoma sequence: results from the FINBAR case-control study. Gut 57(6):734–739

Aoki K (1993) Excess incidence of lung cancer among primary tuberculosis cases. Jpn J Clin Oncol 23(4):205–220

Arnthorsson G, Tulinuis H, Egilsson V et al (1988) Gastric cancer after gastrectomy. Int J Cancer 42:365–367

Atherton JC (2006) The pathogenesis of *Helicobacter pylori*-induced gastro-duodenal diseases. Annu Rev Pathol 1:63–96

Banatvala N, Mayo K, Megraud F et al (1993) The cohort effect and *Helicobacter pylori*. J Infect Dis 168(1):219–221

Barbosa AJ, Queiroz DM, Nogueira AM et al (1993) Chronic gastritis and *Helicobacter pylori* in digestive form of Chaga's disease. Rev Inst Med Trop Sao Paulo 35(2):117–121

Bedwani R, Renganathan E, El-Kwhsky F et al (1998) Schistosomiasis and the risk of bladder cancer in Alexandria, Egypt. Br J Cancer 77(7):1186–1189

Biarc J, Nguyen IS, Pini A et al (2004) Carcinogenic properties of proteins with pro-inflammatory activity from *Streptococcus infantarius* (formally *S. bovis*). Carcinogenesis 25(8):1477–1484

Blackburn E, Callender S, Dacie JV et al (1968) Possible association between pernicious anaemia and leukaemia: a prospective study of 1,625 patients with a note on very high incidence of stomach cancer. Int J Cancer 3(1):163–170

Blaser MJ (2008) Disappearing microbiota: *Helicobacter pylori* protection against esophageal adenocarcinoma. Cancer Prev Res (Phila) 1(5):308–311

Bosch A, Frias Z, Caldwell WL (1979) Adenocarcinoma of the esophagus. Cancer 43(4):1557–1561

Brinton LA, Gridley G, Hrubec Z et al (1989) Cancer risk following pernicious anaemia. Br J Cancer 59(5):810–813

Caygill C, Hill M (2005) Bacteria and cancer. In: Lax A (ed) Advances in molecular and cellular pathology. Cambridge University Press, Cambridge, p 211

Caygill C, Hill M, Craven J et al (1984) Relevance of achlorhydria to human carcinogenesis. In: O'Neill I, Van Borstel R, Miller C et al (eds) N-nitroso compounds: occurrence, biological effects and relevance to human cancer. IARC Sci Pub, Lyon, pp 895–900

Caygill CP, Kirkham JS, Hill M et al (1986) Mortality from gastric cancer following gastric surgery for peptic ulcer. Lancet 327(8487):929–931

Caygill CP, Hill MJ, Hall CN et al (1987) Increased risk of cancer at multiple sites after gastric surgery for peptic ulcer. Gut 28(8):924–928

Caygill C, Hill M, Kirkham J et al (1988) Increased risk of biliary tract cancer following gastric surgery. Br J Cancer 57(4):434–436

Caygill CP, Knowles RL, Hill MJ (1990) The relationship between pernicious anaemia and gastric cancer. Dtsch Zeit Fur Onkologie 22:120–122

Caygill CP, Knowles RL, Hall R (1991) Increased risk of cancer mortality after vagotomy for peptic ulcer: a preliminary analysis. Eur J Cancer Prev 1(1):35–37

Caygill CP, Hill MJ, Braddick M et al (1994) Cancer mortality in chronic typhoid and paratyphoid carriers. Lancet 343(8889):83–84

Caygill CP, Braddick M, Hill MJ et al (1995) The association between typhoid carriage, typhoid infection and subsequent cancer at a number of sites. Eur J Cancer Prev 4(2):187–193

Chalasani N, Wo JM, Waring JP (1998) Racial differences in the histology, location and risk factors of esophageal cancer. J Clin Gastroenterol 26(1):11–13

Chaturvedi AK, Gaydos CA, Agreda P et al (2010) *Chlamydia pneumoniae* infection and risk for lung cancer. Cancer Epidemiol Biomarkers Prev 19(6):1498–1505

Corley DA, Kubo A, Levin TR et al (2008) *Helicobacter pylori* infection and the risk of Barrett's oesophagus: a community-based study. Gut 57(6):727–733

Correa P (1988) Precancerous lesions of the stomach phenotypic changes and their determinants. In: Reed P, Hill M (eds) Gastric carcinogenesis. Exerpta Medica, Amsterdam/New York/Oxford, pp 127–136

Correa P, Haenszel W, Cuello C et al (1975) Model for gastric cancer epidemiology. Lancet 2(7924):58–60

Curado MP, Edwards B, Shin HR et al (2007) Cancer incidence in five continents, vol IX. IARC Sci Pub, Lyon

Daly JM, Karnell LH, Menck HR (1996) National cancer database report on esophageal carcinoma. Cancer 78(8):1820–1828

de Martel C, Llosa AE, Friedman GD et al (2008) *Helicobacter pylori* infection and development of pancreatic cancer. Cancer Epidemiol Biomarkers Prev 17(5):1188–1194

de Rezende JM, Rosa H, Vaz Mda G et al (1985) Endoscopy in megaesophagus. Prospective study of 600 cases. Arq Gastroenterol 22(2):53–62

De Vries AC, Kuipers EJ (2007) Review article: *Helicobacter pylori* eradication for the prevention of gastric cancer. Aliment Pharmacol Ther 26(Suppl 2):25–35

Del Valle L, White MK, Ename S et al (2005) Detection of JC virus DNA sequences and expression of viral T antigen and agnoprotein in esophageal carcinoma. Cancer 103(3):516–527

Derakhshan MH, Malekzadeh R, Watabe H et al (2008) Combination of gastric atrophy, reflux symptoms and histological subtype indicates two distinct aetiologies of gastric cardia cancer. Gut 57(3):298–305

Devesa SS, Blot WJ, Fraumeni JF (1998) Changing patterns in the incidence of esophageal and gastric carcinoma in the United States. Cancer 83(10):2049–2053

Devor E (1982) Ethnogeographic patterns in gallbladder cancer. In: Correa P, Haenszel W (eds) Epidemiology of cancer of the digestive tract. Martinus Nijhof, The Hague, pp 197–225

Ece F, Hatabay NF, Erdal N et al (2005) Does *Helicobacter pylori* infection play a role in lung cancer? Respir Med 99(10):1258–1262

Eide TJ, Viste A, Andersen A et al (1991) The risk of cancer at all sites following gastric operation for benign disease: a cohort of 4,224 patients. Int J Cancer 48(3):333–339

El-Mawla NG, El-Bolkainy MN, Khaled HM (2001) Bladder cancer in Africa: update. Semin Oncol 28(2):174–178

El-Omar EM, Oien K, Murray LS et al (2000) Increased prevalence of precancerous changes in relatives of gastric cancer patients: critical role of *H. pylori*. Gastroenterology 118(1):22–30

England DM, Rosenblatt JE (1977) Anaerobes in human biliary tracts. J Clin Microbiol 6(5):494–498

Eslick GD (2010) Infectious causes of esophageal cancer. Infect Dis Clin North Am 24(4):845–852

Eslick GD, Talley NJ (1998) *Helicobacter pylori* infection and gastric carcinoma: a meta-analysis. Gastroenterology 114:2871

Eslick GD, Lim LL, Byles JE et al (1999) Association of *Helicobacter pylori* infection with gastric carcinoma: a meta-analysis. Am J Gastroenterol 94(9):2373–2379

Forman D, Newell DG, Fullerton F et al (1991) Association between infection with *Helicobacter pylori* and risk of gastric cancer: evidence from a prospective investigation. Br Med J 302(6788):1302–1305

Forman D, Debacker G, Elder J et al (1993) Epidemiology of and risk factors for *Helicobacter pylori* infection among 3194 asymptomatic subjects in 17 populations. The EUROGAST study group. Gut 34(12):1672–1676

Forman D, Webb P, Parsonnet J (1994) *H. pylori* and gastric cancer. Lancet 343(8891):243–244

Fry L, Monkemuller K, Malfertheiner P (2007) Prevention of gastric cancer: a challenging but feasible task. Acta Gastroenterol Latinoam 37(2):110–117

Garcia SB, Aranha AL, Garcia FR et al (2003) A retrospective study of histopathological findings in 894 cases of megacolon: what is the relationship between megacolon and colonic cancer? Rev Inst Med Trop Sao Paulo 45(2):91–93

Gatenby PA, Bann SD (2009) Antireflux surgery. Minerva Chir 64(2):169–181

Gatenby PA, Caygill CP, Ramus JR et al (2008) Barrett's columnar-lined oesophagus: demographic and lifestyle associations and adenocarcinoma risk. Dig Dis Sci 53(5):1175–1185

Gatenby PAC, Hainsworth AJ, Caygill CPJ et al (2011) Projections for oesophageal cancer incidence in England to 2033. Eur J Cancer Prev 20(4):283–286

Gelfand GA, Finley RJ, Nelems B (1992) Transhiatal esophagectomy for carcinoma of the esophagus and cardia. Arch Surg 127(10):1164–1167

Gocyk W, Nikliński T, Olechnowicz H et al (2000) *Helicobacter pylori*, gastrin and cyclooxygenase-2 in lung cancer. Med Sci Monit 6(6):1085–1092

Gold JS, Bayer S, Salem RR (2004) Association of *Streptococcus bovis* bacteremia with colonic neoplasia and extracolonic malignancy. Arch Surg 139(7):760–765

Hansen S, Wiig J, Giercksky KE et al (1997) Esophageal and gastric carcinoma in Norway 1958–1992: incidence time trend variability according to morphological subtypes and organ subsites. Int J Cancer 71(3):340–344

Hicks RM, Walters CL, Elsebai I et al (1977) Demonstration of nitrosamines in human urine: preliminary observations on a possible etiology for bladder cancer in association with chronic urinary tract infection. Proc R Soc Med 70(6):413–417

Higginson J (1968) The theoretical possibilities of cancer prevention in man. Proc R Soc Med 61(7):723–726

Hill MJ (1986) Microbes and human carcinogenesis. Hodder Arnold, London

Hill MJ (1991) Bile acids and colorectal cancer: hypothesis. Eur J Cancer Prev 1(Suppl 2):69–74

Hill MJ (1996) Endogenous N-nitrosation. Eur J Cancer Prev 5(Suppl 1):47–50

Hill MJ, Morson BC, Bussey HJ (1978) Etiology of the adenoma-carcinoma sequence in the large bowel. Lancet 1(8058):245–247

Hill MJ, Davies GJ, Giacosa A (2001) Should we change our dietary advice on cancer prevention? Eur J Cancer Prev 10(1):1–6

Huang JQ, Sridhar S, Chen Y et al (1998) Meta-analysis of the relationship between *Helicobacter pylori* seropositivity and gastric cancer. Gastroenterology 114(6):1169–1179

Hussein NR (2010) *Helicobacter pylori* and gastric cancer in the Middle East: a new enigma? World J Gastroenterol 16(26):3226–3234

IARC Working Group on the Evaluation of Carcinogenic Risks to Humans (1994) Schistosomes, liver flukes and *Helicobacter pylori*. IARC Monogr Eval Carcinog Risks Hum 61:1–241

Inokuchi K, Tokudome S, Ikeda M et al (1984) Mortality from carcinoma after partial gastrectomy. Gann 75(7):588–594

Islami F, Kamangar F (2008) *Helicobacter pylori* and esophageal cancer risk: a meta-analysis. Cancer Prev Res (Phila) 1(5):329–338

Johnston BJ, Reed PI (1991) Changing pattern of oesophageal cancer in a general hospital in the UK. Eur J Cancer Prev 1(1):23–25

Kanbay M, Kanbay A, Boyacioglu S (2007) *Helicobacter pylori* infection as a possible risk factor for respiratory system disease: a review of the literature. Respir Med 101(2):203–209

Kim WJ, Lee HL, Lee SC et al (2000) Polymorphisms of N-acetyltransferase 2, glutathione S-transferase mu and theta genes as risk factors of bladder cancer in relation to asthma and tuberculosis. J Urol 164(1):209–213

Kim NH, Park JP, Jeon SH et al (2002) Purulent pericarditis caused by group G streptococcus as an initial presentation of colon cancer. J Korean Med Sci 17(4):571–573

Kobayashi T, Harada K, Miwa K et al (2005) *Helicobacter* genus DNA fragments are commonly detectable in bile from patients with extrahepatic biliary diseases and associated with their pathogenesis. Dig Dis Sci 50(5):862–867

Kocazeybek B (2003) Chronic *Chlamydia pneumoniae* infection in lung cancer, a risk factor: case-control study. J Med Microbiol 52(Pt 8):721–726

Kocher HM, Linklater K, Patel S et al (2001) Epidemiological study of oesophageal and gastric cancer in south-east England. Br J Surg 88(9):1249–1257

Labenz J, Blum AL, Bayerdorffer E et al (1997) Curing *Helicobacter pylori* infection in patients with duodenal ulcer may provoke reflux oesophagitis. Gastroenterology 112(5):1442–1447

Lagergren J, Bergstrom R, Lindgren A et al (1999) Symptomatic gastroesophageal reflux as a risk factor for esophageal adenocarcinoma. N Engl J Med 340(11):825–831

Laurila AL, Anttila T, Läärä E et al (1997) Serological evidence of an association between *Chlamydia pneumonia* infection and lung disease. Int J Cancer 74(1):31–34

Lazcano-Ponce EC, Miquel JF, Munoz N et al (2001) Epidemiology and molecular pathology of gallbladder cancer. CA Cancer J Clin 51(6):349–364

Leong RW, Sung JJ (2002) Review article: *Helicobacter* species and hepatobiliary diseases. Aliment Pharmacol Ther 16(6):1037–1045

Liabeuf A, Faivre J (1997) Time trends in oesophageal cancer incidence in Cote d'Or (France), 1976–93. Eur J Cancer Prev 6(1):24–30

Lin JT, Wang LY, Wang JT et al (1995) A nested case-control study on the association between *Helicobacter pylori* infection and gastric cancer risk in a cohort of 9,775 men in Taiwan. Anticancer Res 15(2):603–606

Lindkvist B, Johansen D, Borgstrom A et al (2008) A prospective study of *Helicobacter pylori* in relation to the risk for pancreatic cancer. BMC Cancer 8:321

Littman AJ, White E, Jackson LA et al (2004) *Chlamydia pneumoniae* infection and risk of lung cancer. Cancer Epidemiol Biomarkers Prev 13(10):1624–1630

Littman AJ, Jackson LA, Vaughan TL (2005) *Chlamydia pneumoniae* and lung cancer: epidemiological evidence. Cancer Epidemiol Biomarkers Prev 14:773–778

Lowenfels AB, Maisonneuve P (2006) Epidemiology and risk factors for pancreatic cancer. Best Pract Res Clin Gastroenterol 20(2):179–209

Lundegardh G, Adami HO, Helmick C et al (1988) Stomach cancer after partial gastrectomy for benign ulcer disease. N Engl J Med 319(4):195–200

Luo J, Nordenvall C, Nyren O et al (2007) The risk of pancreatic cancer in patients with gastric or duodenal ulcer disease. Int J Cancer 120(2):368–372

MacFarlane S, Furrie E, Macfarlane GT et al (2007) Microbial colonization of the upper gastrointestinal tract in patients with Barrett's esophagus. Clin Infect Dis 45(1):29–38

Mack TM, Yu MC, Hanisch R et al (1986) Pancreas cancer and smoking, beverage consumption, and past medical history. J Natl Cancer Inst 76(1):49–60

Malaty HM, Graham DY (1994) Importance of childhood socioeconomic status on the current prevalence of *Helicobacter pylori* infection. Gut 35(6):742–745

Malfertheiner P, Sipponen P, Naumann M et al (2005) *Helicobacter pylori* eradication has the potential to prevent gastric cancer: a state-of-the-art critique. Am J Gastroenterol 100(9):2100–2115

Martin MS, Martin F, Justrabo E et al (1977) Immunoprophylaxis and therapy of grafted rat colonic carcinoma. Gut 18(3):232–235

Martin HM, Campbell BJ, Hart CA et al (2004) Enhanced *Escherichia coli* adherence and invasion in Crohn's disease and colon cancer. Gastroenterology 127(1):80–93

Matsuda NM, Miller SM, Evora PR (2009) The chronic gastrointestinal manifestations of Chagas disease. Clinics (Sao Paulo) 64(12):1219–1224

McKinney P, Sharp L, Macfarlane GJ et al (1995) Oesophageal and gastric cancer in Scotland 1960–90. Br J Cancer 71(2):411–415

Mellemgaard A, Gaarslev K (1988) Risk of hepatobiliary cancers in carriers of *Salmonella typhi*. J Natl Cancer Inst 80(4):288

Misra S, Chaturvedi A, Misra NC et al (2003) Carcinoma of the gallbladder. Lancet Onco 4(3):167–176

Moller H (1992) Incidence of cancer of oesophagus, cardia and stomach in Denmark. Eur J Cancer Prev 1(2):159–164

Moller H, Toftgaard C (1991) Cancer occurrence in a cohort of patients surgically treated for peptic ulcer. Gut 32(7):740–744

Moller H, Fairley L, Coupland V et al (2007) The future burden of cancer in England: incidence and numbers of new patients in 2020. Br J Cancer 96(9):1484–1488

Morita E, Narikiyo M, Yano A et al (2003) Different frequencies of *Streptococcus anginosus* infection in oral cancer and esophageal cancer. Cancer Sci 94(6):492–496

Morson BC (1974) Evolution of cancer of the colon and rectum. Cancer 34(3):845–849

Morson BC, Bussey HJR, Day DW et al (1983) Adenomas of the large bowel. Cancer Surv 2:451–478

Munkholm P (2003) Review article: the incidence and prevalence of colorectal cancer in inflammatory bowel disease. Aliment Pharmacol Ther 18(Suppl 2):1–5

Nakajima S, Hattori T (2004) Oesophageal adenocarcinoma or gastric cancer with or without eradication of *Helicobacter pylori* infection in chronic atrophic gastritis patients: a hypothetical opinion from a systematic review. Aliment Pharmacol Ther 20(Suppl 1):54–61

Narikiyo M, Tanabe C, Yamada Y et al (2004) Frequent and preferential infection of *Treponema denticola, Streptococcus mitis*, and *Streptococcus anginosus* in esophageal cancers. Cancer Sci 95(7):569–574

Nath G, Singh H, Shukla VK (1997) Chronic typhoid carriage and carcinoma of the gallbladder. Eur J Cancer Prev 6(6):557–559

Nath G, Singh YK, Kumar K et al (2008) Association of carcinoma of the gallbladder with typhoid carriage in a typhoid endemic area using nested PCR. J Infect Dev Ctries 2(4):302–307

Newnham A, Quinn MJ, Babb P et al (2003) Trends in oesophageal and gastric cancer incidence, mortality and survival in England and Wales 1971–1998/1999. Aliment Pharmacol Ther 17(5):655–664

Nilsson HO, Stenram U, Ihse I et al (2006) *Helicobacter* species ribosomal DNA in the pancreas, stomach and duodenum of pancreatic cancer patients. World J Gastroenterol 12(19):3038–3043

Offerhaus GJ, Tersmette AC, Huibregtse K et al (1988) Mortality caused by stomach cancer after remote partial gastrectomy for benign conditions: 40 years of follow-up of an Amsterdam cohort of 2,633 postgastrectomy patients. Gut 29(11):1588–1590

Osawa H, Inoue F, Yoshida Y (1996) Inverse relation of serum *Helicobacter pylori* antibody titres and extent of intestinal metaplasia. J Clin Pathol 49(2):112–115

Ouwehand A, Vaughan EE (2006) Gastrointestinal microbiology. Taylor & Francis, New York

Parkin DM (2001) Global cancer statistics in the year 2000. Lancet Oncol 2(9):533–543

Parkin DM (2006) The global health burden of infection-associated cancers in the year 2002. Int J Cancer 118(12):3030–3044

Parsonnet J, Friedman GD, Vandersteen DP et al (1991) *Helicobacter pylori* infection and the risk of gastric carcinoma. N Engl J Med 325(16):1127–1131

Parsonnet J, Blaser MJ, Perez-Perez GI et al (1992) Symptoms and risk factors of *Helicobacter pylori* infection in a cohort of epidemiologists. Gastroenterology 102(1):41–46

Patyar S, Joshi R, Byrav DS et al (2010) Bacteria in cancer therapy: a novel experimental strategy. J Biomed Sci 17(1):21

Peek RM, Fiske C, Wilson KT (2010) Role of innate immunity in *Helicobacter pylori*-induced gastric malignancy. Physiol Rev 90(3):831–858

Philippou N, Koursarakos P, Anastasakou E et al (2004) *Helicobacter pylori* seroprevalence in patients with lung cancer. World J Gastroenterol 10(22):3342–3344

Pour PM, Lawson T (1984) Pancreatic carcinogenic nitrosamines in Syrian hamsters. In: O'Neill I, Van Borstel R, Miller C et al (eds) N-nitroso compounds: occurrence, biological effects and relevance to human cancer, 57. IARC, Lyon, pp 683–688

Powell JJ, McConkey CC (1992) The rising trend in oesophageal adenocarcinoma and gastric cardia. Eur J Cancer Prev 1(3):265–269

Preussmann R (1984) Occurence and exposure to N-nitroso compounds and precursors. In: O'Neill I, Van Borstel R, Miller C et al (eds) N-nitroso compounds: occurence, biological effects and relevance to human cancer, 57. IARC, Lyon, pp 3–15

Puestow CB, Gillesby WJ, Guynn VL (1955) Cancer of the esophagus. AMA Arch Surg 70(5):662–671

Putnam JB, Suell DM, McMurtrey MJ (1994) Comparison of three techniques of oesophagectomy within a residency training program. Ann Thorac Surg 57(2):319–325

Raderer M, Wrba F, Kornek G et al (1998) Association between *Helicobacter pylori* infection and pancreatic cancer. Oncology 55(1):16–19

Radomski J, Greenwald D, Hearn WL et al (1978) Nitrosamine formation in bladder infections and its role in the etiology of bladder cancer. J Urol 120(1):48–50

Rahamim J, Cham CW (1993) Oesophagogastrectomy for carcinoma of the oesophagus and cardia. Br J Surg 80(10):1305–1309

Ramus JR, Gatenby PA, Caygill CP et al (2007) *Helicobacter pylori* infection and severity of reflux-induced oesophageal disease in a cohort of patients with columnar-lined oesophagus. Dig Dis Sci 52(10):2821–2825

Randi G, Franceschi S, La Vecchia C (2006) Gallbladder cancer worldwide: geographical distribution and risk factors. Int J Cancer 118(7):1591–1602

Ries LAG, Eisner MP, Kosary CL et al (2004) SEER cancer statistics review, 1975–2001. National Cancer Institute, Bethesda

Risch HA, Yu H, Lu L et al (2010) ABO blood group, *Helicobacter pylori* seropositivity, and risk of pancreatic cancer: a case-control study. J Natl Cancer Inst 102(7):502–505

Robins G, Crabtree JE, Bailey A et al: EUROGAST Study Group (2008) International variation in *Helicobacter pylori* infection and rates of oesophageal cancer. Eur J Cancer 44(5):726–732

Rokitansky C (1854) A manual of pathological anatomy. Sydenham Soc, London

Rokkas T, Pistiolas D, Sechopoulos P, Robotis I, Margantinis G (2007) Relationship between *Helicobacter pylori* infection and esophageal neoplasia: a meta-analysis. Clin Gastroenterol Hepatol 5(12):1413–1417

Roses DF, Richman H, Localio SA (1974) Bacterial endocarditis associated with colorectal carcinoma. Ann Surg 179(2):190–191

Ross A, Smith M, Andersen J et al (1982) Late mortality after surgery for peptic ulcer. N Engl J Med 307(9):519–522

Ruprecht K, Ferreira H, Flockerzi A et al (2008) Human endogenous retrovirus family HERV-K (HML-2) RNA transcripts are selectively packaged into retroviral particles produced by the human germ cell tumor line Tera-1 and originate mainly from a provirus on chromosome 22q11.21. J Virol 82(20):10008–10016

Saad AA, O'Connor PJ, Mostafa MH et al (2006) Bladder tumor contains higher N-7 methylguanine levels in DNA than adjacent normal bladder epithelium. Cancer Epidemiol Biomarkers Prev 15(4):740–743

Selgrad M, Bornschein J, Rokkas T et al (2010) Clinical aspects of gastric cancer and *Helicobacter pylori* - screening, prevention and treatment. Helicobacter 15(Suppl 1):40–45

Shukla VK, Singh H, Pandey MK et al (2000) Carcinoma of the gallbladder: is it a sequel of typhoid? Dig Dis Sci 45(5):900–903

Siegert C, Overbosch D (1995) Carcinoma of the colon presenting as *Streptococcus sanguis* bacteremia. Am J Gastroenterol 90(9):1528–1529

Siman JH, Fosgren A, Berglund G et al (1997) Association between *Helicobacter pylori* and gastric carcinoma in the city of Malmo, Sweden. A prospective study. Scand J Gastroenterol 32(12):1215–1221

Sinclair T, Tuxford AF (1971) The incidence of urinary tract infection and asymptomatic bacteriuria in a semi-rural practice. Practitioner 207(273):81–84

Sjostedt S (1989) The upper gastrointestinal microbiota in relation to gastric diseases and gastric surgery. Acta Chir Scand Suppl 551:1–57

Smith KL, Parsonnet J (1998) *Helicobacter pylori*. In: Evans AS, Brachman PS (eds) Bacterial infections of humans: epidemiology and control, 3rd edn. Plenum Publishing, New York, pp 337–353

Stolzenberg-Solomon RZ, Blaser MJ, Limburg PJ et al (2001) *Helicobacter pylori* seropositivity as a risk factor for pancreatic cancer. J Natl Cancer Inst 93(12):937–941

Tersmette AC, Offerhaus JG, Giardiello FM et al (1990) Occurrence of non-gastric cancer in the digestive tract after remote partial gastrectomy - analysis of an Amsterdam cohort. Int J Cancer 46(5):792–795

Toftgaard C (1989) Gastric cancer after peptic ulcer surgery - A historic prospective cohort investigation. Ann Surg 210(2):159–164

Tokudome S, Kono S, Ikeda M et al (1984) A prospective study on primary gastric stump cancer following partial gastrectomy for benign gastroduodenal diseases. Cancer Res 44(5):2208–2212

Turnbull ADM, Goodner JT (1968) Primary adenocarcinoma of the esophagus. Cancer 22(5):915–918

Tuyns AJ (1992) Oesophageal cancer in France and Switzerland: recent time trends. Eur J Cancer Prev 1(3):275–278

Viste P, Opheim P, Thunold J et al (1986) Risk of carcinoma following gastric operations for benign disease. Lancet 328(8505):502–505

Watt PC, Patterson CC, Kennedy TL (1984) Late mortality after vagotomy and drainage for duodenal ulcer. Br Med J 288(6427):1335–1338

Webb J, Busuttil A (1978) Adenocarcinoma of the oesophagus and of the oesophagogastric junction. Br J Surg 65(7):475–479

Welton J, Marr J, Friedman S (1979) Association between hepatobiliary cancer and typhoid carrier state. Lancet 313(8120):791–794

Wistuba II, Gazdar AF (2004) Gallbladder cancer: lessons from a rare tumour. Nat Rev Cancer 4(9):695–706

Wroblewski LE, Peek RM, Wison KT (2010) *Helicobacter pylori* and gastric cancer: factors that modulate disease risk. Clin Microbiol Rev 23(4):713–739

Wu CY, Kuo KN, Wu MS et al (2009a) Early *Helicobacter pylori* eradication decreases risk of gastric cancer in patients with peptic ulcer disease. Gastroenterology 137(5):1641–1648

Wu IC, Wu DC, Yu FJ et al (2009b) Association between *Helicobacter pylori* seropositivity and digestive tract cancers. World J Gastroenterol 15(43):5465–5471

Wynder EL (1975) An epidemiological evaluation of the causes of cancer of the pancreas. Cancer Res 35(8):2228–2233

Zarkin BA, Lilliemoe KD, Cameron JL et al (1990) The triad of *Streptococcus bovis* bacteremia, colonic pathology, and liver disease. Ann Surg 211(6):786–791

Zatonski WA, Lowenfels AB, Boyle P et al (1997) Epidemiologic aspects of gallbladder cancer: a case-control study of the SEARCH program of the International Agency for Research on Cancer. J Natl Cancer Inst 89(15):1132–1138

Zhou Q, Zhang H, Pang X (1992) Pre- and postoperative sequential study on the serum gastrin level in patients with lung cancer. J Surg Oncol 51(1):22–25

Zhuo X, Zhang Y, Wang Y et al (2008) *Helicobacter pylori* infection and oesophageal cancer risk: association studies via evidence-based meta-analyses. Clin Oncol 20(10):757–762

zur Hausen H (2009) The search for infectious causes of human cancers: where and why (Nobel lecture). Angew Chem Int Ed Engl 48(32):5798–5808

Chapter 2
Gastric Cancer and *Helicobacter pylori*

Amedeo Amedei and Mario M. D'Elios

Abstract *Helicobacter pylori* (*H. pylori*) is a gram-negative bacterium that chronically infects the stomach of more than 50% of the human population and represents a major cause of gastric cancer, gastric lymphoma, gastric autoimmunity and peptic ulcer diseases. The International Agency for Research on Cancer classifies *H. pylori* as a human carcinogen for gastric cancer. Eradicating the bacterium in high-risk populations reduces the incidence of gastric cancer. Likewise, antibiotic treatment leads to regression of gastric mucosa-associated lymphoid tissue lymphoma. Gastric inflammation induced by *H. pylori* is the main singular risk factor for gastric malignancies. This chapter outlines the bacterial and host factors involved in the genesis of gastric cancer and gastric lymphoma. Treatment options for patients with an advanced gastric malignancy are still limited, but the introduction of an effective vaccine will be the best tool for preventing both *H. pylori* infection, gastric cancer and gastric lymphoma.

Keywords *Helicobacter pylori* • Gastric cancer • Gastric lymphoma • MALT • Mucosal immunity • T cells • T helper 1 • T helper 2 • T helper 17 • Cancerogenesis • Lymphoma genesis • Cytotoxicity • B-cell help • Perforin • Fas ligand • Antibiotic therapy • Vaccines

---

A. Amedei • M.M. D'Elios (✉)
Department of Internal Medicine, University of Florence,
viale Morgagni 85, 50134 Florence, Italy

Department of Biomedicine, Policlinico AOU Careggi,
Florence, Italy
e-mail: delios@unifi.it

A.A. Khan (ed.), *Bacteria and Cancer*, DOI 10.1007/978-94-007-2585-0_2,

Abbreviations

AIG	Autoimmune gastritis
BER	Base excision repair
BMDC	Normal bone marrow stem cells
EGFR	Epithelial Growth Factor Receptor
GC	Gastric Cancer
HbEGF	Heparin-binding EGF-like growth factor
HP	*Helicobacter pylori*
IFNGR	Interferon gamma receptor
IL-1β	Interleukin-1 beta
KO	Knockout
LPS	Lipopolysaccharide
MALT	Mucosa-associated lymphoid tissue
MALToma	MALT lymphoma
MBL2	Mannose binding lectin-2 gene
mmP1	matrix metalloproteinase 1
MMR	Mismatch repair pathway
mtDNA	mitochondrial DNA
NFAT	Nuclear factor of activated T cells
NoD1	Nucleotide-binding oligomerization domain-containing protein 1
omP	outer membrane proteins
Pi3K	Phosphotidyl inositol 3 kinase
PTPRZ1	Protein receptor-type tyrosine protein phosphatase-ζ
ROS	Reactive oxygen species
SHP-2	Src homology 2 domain–containing tyrosine phosphatase 2
T4SS	Type 4 Secretion system
TLR	Toll-like receptor
TNF	Tumour Necrosis Factor
TPM	Tyrosine phosphorylation motifs
TRAIL	TNF-related apoptosis-inducing ligand
VNTR	Variable number of tandem repeats

2.1 Introduction

Gastric cancer (GC) is currently the second leading cause of death due to cancer worldwide, with high incidence in China, South America, Eastern Europe, Korea, and Japan (Yao et al. 2003). Surgical tumour resection remains the primary curative treatment for gastric cancer. Nevertheless, the overall 5-years survival rate remains poor, ranging between 15 and 35%. Among patients who relapse after curative surgery, the 87% have locoregional recurrences (Roukos and Kappas 2005). These facts have prompted many studies addressing surgical issues, as well as exploring the role of adjuvant and neoadjuvant treatments, such as perioperative chemotherapy, which is associated with a 5 years overall survival benefit in 13% cases for stages II or III operable GC patients (Cunningham et al. 2006). Despite the improvements in GC management, estimated cure rates for patients with advanced stages remain poor throughout the world (Liakakos and Roukos 2008). These data highlight the urgent need of new therapeutic strategies to combat GC, such as Epithelial Growth Factor Receptor (EGFR) inhibitors (Vanhoefer et al. 2004), anti-angiogenetic agents (Shah et al. 2006), apoptosis promoters (Ocean et al. 2006) and specific immunotherapy (Elkord et al. 2008; Amedei et al. 2009). The mechanisms that account for the observed geographic and temporal incidence patterns have not yet been established and a number of factors are known to suppress or promote gastric cancer (Saikawa and Kitajima 2009). The intake of fruits and vegetables have a protective role, as well as modern food processing and storage, that had reduced both spoilage and the use of salt-curing and pickling also found to be protective for gastric cancer. Several gastric cancer promoting factors have been defined, such as natural carcinogens or precursors (nitrates in food), carcinogens produced during the grilling of meats, and other carcinogens are synthesized from dietary precursors in the stomach. *Helicobacter pylori* (*H. Pylori*) infection is the major GC-promoting factor. In 1994, the International Agency for Research on Cancer declared *H. pylori* to be a type I carcinogen, or a definite cause of cancer in humans (IARC Working Group 1994) and epidemiological studies have determined that the attributable risk for gastric cancer conferred by *H. pylori* is approximately 75% (Parkin 2006). If this is accurate, *H. pylori* would be responsible for as many as 5.5% of all cancers, making it the leading infectious cause of cancer worldwide and second only to smoking as a defined cause of malignancy. *H. pylori* is a gram negative, spiral-shaped, microaerophilic, urease-positive bacillus that is acquired during childhood, probably via the fecal/oral or gastric/oral routes. Once acquired, the infection persists throughout life unless treated with antibiotics. Common wisdom until 1980 suggested that the stomach, with its low pH, was a sterile environment. Then, endoscopy of the stomach became common and, in 1984, the Nobel Laureates Robin Warren and Barry Marshall found and cultured *H. pylori* from the gastric epithelium of patients with gastritis (inflammation of the stomach) and ulcer disease (Marshall and Warren 1984). Soon, the medical community understood that *H. pylori* is the major cause of stomach inflammation, which, in some infected individuals, precedes peptic ulcer disease (10–20%), distal gastric adenocarcinoma (1–2%), and gastric mucosal-associated lymphoid tissue (MALT) lymphoma (<1%) (Kusters et al. 2006). Although, the bacteria mainly reside on the surface mucus gel layer with little invasion of the gastric glands, the host responds with an impressive humoral and cell-mediated immune response. Despite this sophisticated immune response, most infections become chronically established with little evidence that spontaneous clearance occurs (Amieva and El-Omar 2008). Two histologically distinct variants of gastric adenocarcinoma have been described, each with different pathophysiological features. *Diffuse-type gastric adenocarcinoma* more commonly affects younger people and consists of individually infiltrating neoplastic cells that do not form glandular structures. The more prevalent form of gastric adenocarcinoma, *intestinal-type adenocarcinoma* which progresses through a series of histological steps that are initiated by the transition from normal mucosa to chronic superficial gastritis, which then leads to atrophic gastritis and intestinal metaplasia, and finally to dysplasia and adenocarcinoma (Correa 1992).

Although, *H. pylori* significantly increases the risk of developing both diffuse-type and intestinal-type gastric adenocarcinoma, chronic inflammation is not required for the development of diffuse-type cancers, suggesting that mechanisms underpinning the ability of *H. pylori* to induce malignancy are different for these cancer subtypes. However, only a small proportion of HP-infected people ever develop neoplasia, and disease risk involves well-choreographed interactions between pathogen and host, which are in turn dependent on strain-specific bacterial factors and/or host genotypic traits. These observations underscore the importance and timelines of reviewing mechanisms that regulate the biological interactions of *H. pylori* with its hosts and that promote carcinogenesis.

In this chapter we analyzed the role of *H. pylori* and host factors leading to gastric adenocarcinoma and gastric MALT lymphoma and discussed the epidemiological and pathophysiological data that led to the consensus that *H. pylori* is one of the world's most important causes of cancer.

2.2 Pathophysiology of *H. pylori* Infection

Helicobacter species inhabit the gastrointestinal tract of mammals and birds. These species are mostly host specific, implying coevolution of the bacteria with their hosts. By comparing nucleotide sequences of different strains, it is possible to determine the minimal time that *H. pylori* and its host have shared a common ancestor. Genetic diversity among *H. pylori* strains diminish with distance from East Africa, just like genetic diversity decreases among humans (Atherton and Blaser 2009). Taken together these data show that *H. pylori* has coevolved with humans, at least since their joint exodus from Africa 60,000 years ago and likely throughout their evolution.

In physiological terms, the stomach may be divided into two main compartments: an acidic proximal corpus that contains the acid-producing parietal cells and a less acidic distal antrum that does not have parietal cells but contains the endocrine cells that control acid secretion. Both animal and human ingestion studies suggest that successful colonization of the gastric mucosa is best achieved with the aid of acid suppression (Danon et al. 1995). Furthermore, the pharmacological inhibition of acid secretion in infected patients leads to the redistribution of the infection and its associated gastritis from an antral to a corpus-predominant pattern (Kuipers et al. 1996). Thus, lack of gastric acid extends the colonization area and maximizes the tissue damage of colonization. The key pathophysiological event in *H. pylori* infection is the start of an inflammatory response that is triggered by different bacterial products, such as CagA, HP-NAP, VacA, lipopolysaccharide (LPS), urease with the ensuing inflammatory response mediated by cytokines (D'Elios et al. 1997b; Israel and Peek 2001; Amedei et al. 2006). The cytokine repertoire comprises a multitude of pro- and anti-inflammatory mediators whose function is to co-ordinate an effective immune response against invading pathogens without damage to the host (D'Elios et al. 2005). In addition to their inflammatory properties, some *H. pylori*-induced cytokines, such as interleukin-1 beta (IL-1β), IFN-γ, TNF-α, and IL-4 may exert a variety of effects on the gastric epithelial cells that might lead to gastric pH alterations or activation of oncogenes (El-Omar et al. 2000; El-Omar 2001; Robert et al. 1991). Normal bone marrow stem cells (BMDCs) are frequently recruited to sites of tissue injury and inflammation and a recent study (Houghton et al. 2004) demonstrated that BMDCs might also represent a potential source of malignant cells in gastric cancer, while epithelial cancers are believed to originate from the transformation of tissue stem cells. They showed that chronic infection of C57BL/6 mice with *Helicobacter* induced repopulation of the stomach with BMDCs, and subsequently these cells progressed through metaplasia and dysplasia to intraepithelial cancer. These findings suggest that epithelial cancers can originate from marrow-derived sources. Subsequently, others studies demonstrated the possible relationship of BMDCs to carcinogenesis (Takaishi et al. 2008; Dittmar et al. 2006). Houghton et al. (2004) speculated that, the plasticity of BMDCs might contribute to epithelial cancers, particularly those associated with chronic inflammation. The model of gastric cancer in *Helicobacter felis* infected C57BL/6 mice represents an ideal system for evaluating the effects of chronic inflammation on BMDC recruitment and engraftment in the stomach. Inflammation (maximal 2–3 months after infection) subsequently continues at a moderate level for the remainder of the life of the animal. In this model, chronic inflammation and tissue injury may be associated with BMDC engraftment within the gastric epithelium, and the resulting microenvironment is strongly linked with the progression of inflammation-associated cancer. BMDCs may also contribute to established cancers through cell mimicry or cell fusion as suggested for hepatocytes or intestinal cells (Wang et al. 2003; Rizvi et al. 2006), or they may initiate cancer directly. Interestingly, acute gastric infection with *Helicobacter* species, acute ulceration, or drug-induced parietal cell loss does not lead to the recruitment of BMDCs, while severe chronic inflammation may lead to BMDC-related carcinogenesis. The latter process probably upregulates proinflammatory cytokines such as IL-1β, IL-6, and TNF (Tumour Necrosis Factor)-α, and chemokines such as CXCL12, contributing to the recruitment of progenitors.

Therefore as result of chronic *H. pylori* infection there are three main gastric phenotypes: (a) the commonest by far is a mild pan-gastritis that is not associated with significant human disease; (b) a corpus-predominant gastritis associated with progressive gastric atrophy and multifocal intestinal metaplasia and increased risk of gastric cancer (El-Omar et al. 1997); and (c) an antral-predominant gastritis associated with high gastric acid secretion and increased risk of duodenal ulcer disease (El-Omar et al. 1995). It is fascinating to note that the determinants of these outcomes are the severity and extent of the gastritis, which in turn are determined by bacterial, host and environmental factors.

2.3 *H. pylori* Factors That Interact with the Host and Mediate Carcinogenesis

2.3.1 *CagA and the cag Pathogenicity Island*

The association of *H. pylori* infection with gastric cancer raises the interesting question of whether *H. pylori* encodes one or more oncogenes? Oncogenic viruses initiate and promote cellular transformation by integrating virally encoded oncogenes into the host genome (Maeda et al. 2008; Howley and Livingston 2009). By contrast, *H. pylori* remains primarily extracellular and does not integrate its genome into the host DNA. However, the bacterium can still affect the function of host cells by translocating a bacterial protein, CagA, which is a component of the *cag* pathogenicity island. *cag*+strains significantly augment the risk for distal gastric cancer compared with *cag* – strains (Peek and Blaser 2002). In last few years, the research in this field has advanced, particularly that concerning how effects are localized within the epithelial cell? (Kwok et al. 2007; Higashi et al. 2002a; Amieva et al. 2003); how the CagA effector protein varies between strains? (Argent et al. 2004; Basso et al. 2008), and how CagA can directly induce carcinogenesis?

In *H. pylori* strains, the *cag* PaI may be present, absent, or disrupted and thus nonfunctional. But, *cag* PaI is usually present and functional in *H. pylori*, it contains approximately 30 genes, including those that collectively encode a type IV secretion system (T4SS) – a conduit that connects the cytoplasm of the bacterial and host epithelial cells (Odenbreit et al. 2000). This is brilliantly designed for the human stomach: an antigenically variable, acid-stable structural protein (CagY) coats the "syringe," conferring stability and allowing evasion from the host immune response (Algood et al. 2007). Subsequent contact of the tip protein (CagL) with the epithelial cell and delivery and activation of the effector protein (CagA) stimulate local signaling and effects at the site of attachment (Backert and Selbach 2008) (Fig. 2.1).

CagA may also have more well-known cellular effects, including activation of NF-kB (Brandt et al. 2005) (Fig. 2.1). CagA is polymorphic and from different *H. pylori* strains has different numbers and types of activating tyrosine phosphorylation motifs (TPMs), leading to different effects on cellular signaling and differing risks of disease; for example the D-type TPM found in East Asian strains binds to Src homology 2 domain–containing tyrosine phosphatase 2 (SHP-2) strongly and stimulates very marked cellular effects (Higashi et al. 2002a, b; Argent et al. 2008). This may, in part explain the high rate of *H. pylori* associated disease in Japan and parts of China.

In host cells, CagA interacts with a number of cellular complexes implicated in carcinogenesis (Bourzac and Guillemin 2005; Hatakeyama 2006) and appears to be very important in *H. pylori* induced gastric carcinogenesis, regardless of inflammation status. Among transgenic mice engineered to express CagA constitutively, some developed B cell lymphomas and some developed gastric adenocarcinomas (Ohnishi et al. 2008) with no gastritis. This implies that the underlying mechanism

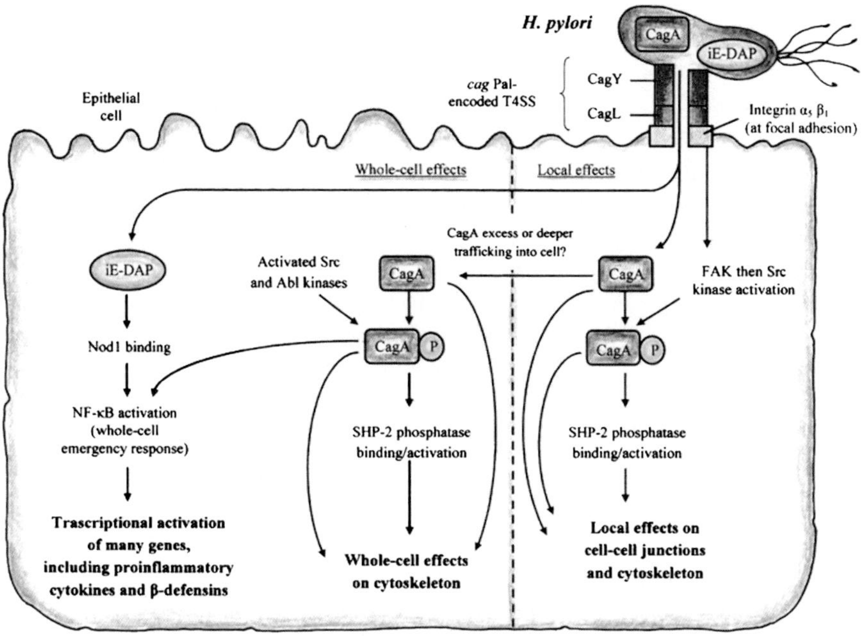

Fig. 2.1 Local and whole-cell effects of the *H. pylori* protein CagA. *H. pylori* with an intact *cag* PaI, forms a T4SS, which injects CagA into epithelial cells. The CagL protein of T4SS complex binds to and activates integrin α5β1, resulting in local activation of focal adhesion kinase (*FAK*) and then Src kinase. Activated kinases phosphorylate CagA, in turn activating local Src homology 2 domain–containing tyrosine phosphatase 2 (*SHP-2*) and therefore local signaling. A soluble component of bacterial peptidoglycan, γ-D-glutamyl-*meso*-diaminopimelic acid (*iE-DAP*) also enters the cell and is recognized by the intracellular innate immune pattern-recognition receptor Nod1, leading to stimulation of NF-κB. In addition, phosphorylated CagA itself also may activate NF-κB and have other whole-cell effects. The **bold** text indicates cellular effects

did not involve chronic inflammatory damage, although early time points were not examined. The implications for human gastric intestinal-type carcinogenesis, which appears to arise on a background of inflammation through a stepwise progression of gastritis-atrophy-metaplasia-carcinoma, remain to be fully elucidated. As the cag T4SS also induces proinflammatory cytokines via the intracellular bacterial peptidoglycan recognition molecule nucleotide-binding oligomerization domain-containing protein 1(NoD1).

NoD1 activation by *H. pylori* peptidoglycan stimulates NF-κB, p38 and ERK, culminating in the expression of the chemokines CXCL2 and CXCL8 (Viala et al. 2004; Allison et al. 2009). The delivery of peptidoglycan components into host cells induces additional epithelial responses with carcinogenic potential, such as the activation of Pi3K and cell migration. The *H. pylori* gene *slt* encodes a soluble lytic transglycosylase that is required for peptidoglycan turnover and release (Viala et al. 2004), thereby regulating the amount of peptidoglycan translocated into host cells. Inactivation of *slt* has been shown to inhibit *H. pylori* induced Pi3K signalling and cell migration (Nagy et al. 2009). The protein encoded by the *H. pylori* gene *HP0310* deacetylates *N*-acetylglucosamine peptidoglycan residues and is required for normal peptidoglycan synthesis (Wang et al. 2009). Loss of *HP0310*, which leads to decreased peptidoglycan production, reciprocally augments the delivery of the other major *cag* secretion system substrate, CagA, into host cells. This suggests that functional interactions occur between *H. pylori* translocated effectors (Franco et al. 2009). These findings indicate that contact between *cag*+strains and host cells activates multiple signalling pathways that regulate oncogenic cellular responses, which may heighten the risk for transformation.

Thus, while CagA may not promote cancer itself, exposure to CagA and inflammatory insults may select for heritable host cell changes (genetic or epigenetic) that together contribute to cancer progression. Eradication of *H. pylori* in the human atrophic stomach does not greatly reduce the proportion of people who develop cancer over a 5-year time frame (Wong et al. 2004), implying that CagA effects must be mediated relatively early in the carcinogenic process.

2.3.2 The H. pylori Vacuolating Cytotoxin

The *H. pylori* gene *vacA* encodes a secreted protein (vacA) that was initially identified on the basis of its ability to induce vacuolation in cultured epithelial cells (Telford et al. 1994). However, vacA also exerts other effects on host cells (Boncristiano et al. 2003), and *vacA* is a specific locus linked with gastric malignancy.

All strains contain *vacA*, but there is marked variation in *vacA* sequences among strains with the regions of greatest diversity localized to the 5′ signal terminus, the mid-region and the intermediate region (Rhead et al. 2007); and *vacA* sequence diversity corresponds to variations in vacuolating activity. VacA is secreted and undergoes proteolysis to yield two fragments, p33 and p55 (Fig. 2.2a): the p33 domain contains a hydrophobic sequence that is involved in pore formation, whereas the p55 fragment contains cell-binding domains. VacA binds multiple epithelial cell-surface components, including the transmembrane protein receptor-type tyrosine protein phosphatase-ζ (PTPRZ1), fibronectin, epidermal growth factor receptor (EGFR), various lipids and sphingomyelin, as well as CD18 (integrin β2) on T cells (Polk and Peek 2010).

The protein vacA not only induces vacuolation but also can stimulate apoptosis in gastric epithelial cells (Fig. 2.2b). Transient expression of p33 or full-length vacA induces cytochrome-*c* release from mitochondria, leading to the activation of caspase 3, and vacA proteins that contain a s1 signal allele induce higher levels of apoptosis than vacA proteins that contain a s2 allele or vacA mutants lacking the hydrophobic amino terminus region (Cover and Blanke 2005).

In addition, vacA exerts effects on the host immune response that permit long-term colonization with an inherent increased risk of transformation. vacA binding to integrin β2 blocks antigen-dependent proliferation of transformed T cells by

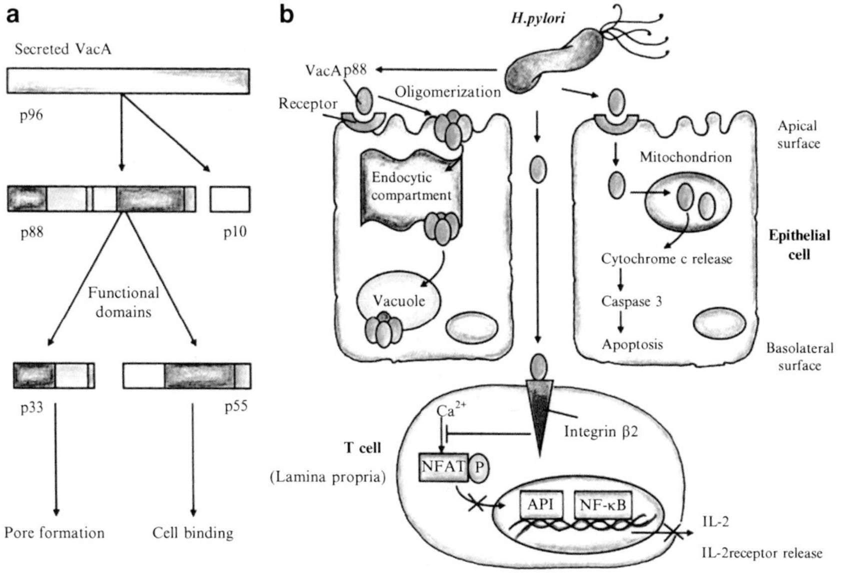

Fig. 2.2 Constitution and functional effects of *H. pylori* Vaca protein. (**a**) VacA is secreted as a 96 kDa protein, which is rapidly cleaved into a 10 kDa passenger domain (*p10*) and an 88 kDa mature protein (*p88*). The p88 fragment contains two domains, designated p33 and p55, which are VacA functional domains. (**b**) The p88 monomeric form of VacA binds to epithelial cells nonspecifically and through specific receptor binding. Following binding, VacA monomers form oligomers, which are then internalized and form anion-selective channels in endosomal membranes; vacuoles arise owing to the swelling of endosomal compartments. The biological consequences of vacuolation are currently undefined, but VacA also induces other effects, such as apoptosis, partly by forming pores in mitochondrial membranes, allowing cytochrome *c* release. VacA has also been identified in the lamina propria, and probably enters by traversing epithelial paracellular spaces, where it can interact with integrin β2 on T cells and inhibit the transcription factor nuclear factor of activated T cells (*NFAT*), leading to the inhibition of IL-2 secretion and blockade of T cell activation and proliferation

interfering with IL-2 (interleukin-2)-mediated signaling through the inhibition of Ca^{2+} mobilization and down regulation of the Ca^{2+} dependent phosphatase calcineurin (Boncristiano et al. 2003; Gebert et al. 2003) (Fig. 2.2). This in turn inhibits the activation of the transcription factor nuclear factor of activated T cells (NFAT) and its target genes *IL2* and the high-affinity IL-2 receptor-α (*IL2RA*). vacA exerts effects on primary human CD4⁺ T cells that are different from its effects on transformed T cell lines by suppressing IL-2-induced cell cycle progression and proliferation in an NFAT-independent manner (Sundrud et al. 2004) Collectively, these results suggest that vacA inhibits the expansion of T cells that are activated by bacterial antigens, thereby allowing *H. pylori* to evade the specific immune response.

Several epidemiological studies have evidenced linking vacA production to gastric cancer. *H. pylori* strains that express forms of vacA, that are active *in vitro* are associated with a higher risk of GC than the strains that express inactive forms of vacA (Gerhard et al. 1999; Miehlke et al. 2000;). This relationship is consistent with studies that have examined the distribution of *vacA* genotypes throughout the world. In the regions that have high rate of distal gastric cancer, such as Colombia and Japan, most *H. pylori* strains contain *vacA* s1 and m1 alleles (Van Doorn et al. 1999).

2.3.3 Outer Membrane Proteins

Generally, most *H. pylori* reside within the semi-permeable mucous gel layer of the stomach blanketing the apical surface of the gastric epithelium, but about 20% bind to gastric epithelial cells and genome analysis of *H. pylori* strains has revealed that an unusually high proportion of identified open reading frames encode proteins that reside in the outer, as well as the inner, membrane of the bacterium (known as outer membrane proteins (omPs) (Oh et al. 2006; McClain et al. 2009)). Consistent with genomic studies, *H. pylori* strains express multiple paralogous omPs, several of which bind to defined receptors on gastric epithelial cells, and strains differ in both expression and binding properties of certain omPs (Fig. 2.3).

A member of highly conserved omPs, the protein babA (encoded by the strain-specific gene *babA2*), is an adhesin that binds the Lewis histo-blood-group antigen leb on gastric epithelial cells (Solnick et al. 2004). Gerhard et al. (1999) have demonstrated that *H. pylori babA2+* strains are associated with an increased risk of GC. SabA is an *H. pylori* adhesin that binds the sialyl-lewisx (lex) antigen, which is an established tumour antigen and a marker of gastric dysplasia that is up regulated by chronic gastric inflammation (Mahdavi et al. 2002). Exploitation of host lewis antigens is further evidenced by data demonstrating that the antigen of *H. pylori* lipopolysaccharide (LPS) contains various human lewis antigens (lex, ley, lea and leb); and the inactivation of lex and ley encoding genes prevents *H. pylori* from colonizing mice (Monteiro et al. 2000). Approximately 85% of *H. pylori* clinical isolates express lex and ley, and although both can be detected on individual strains but one antigen usually predominates (Wirth et al. 1999). *In vivo* studies using mice have demonstrated that the lewis antigen expression pattern of colonizing bacteria is directly altered in response to the expression pattern of their cognate host. In leb expressing transgenic or wild-type control mice challenged with an *H. pylori* strain that expressed lex and ley, only bacterial populations recovered from leb positive mice expressed leb, and this was mediated by a putative galactosyltransferase gene (β-*(1,3)galT*) (Pohl et al. 2009). This suggests that lewis antigens facilitate molecular mimicry and allow *H. pylori* to escape host immune defenses by preventing the formation of antibodies against shared bacterial and host epitopes.

Another differentially expressed omP of *H. pylori* is oipA that has been linked to disease outcome (Yamaoka et al. 2002). Expression of oipA is regulated by slipped strand mispairing within a CT-rich dinucleotide repeat region located in the 5′ terminus of the gene. Several reports have demonstrated that the expression of proinflammatory cytokines/chemokines (such as CXCL8, IL-6, CCL5) is co-regulated by oipA, as well as other effector proteins that may have a role in pathogenesis, such as matrix metalloproteinase 1 (mmP1) (Yamaoka et al. 2002; Wu et al. 2006).

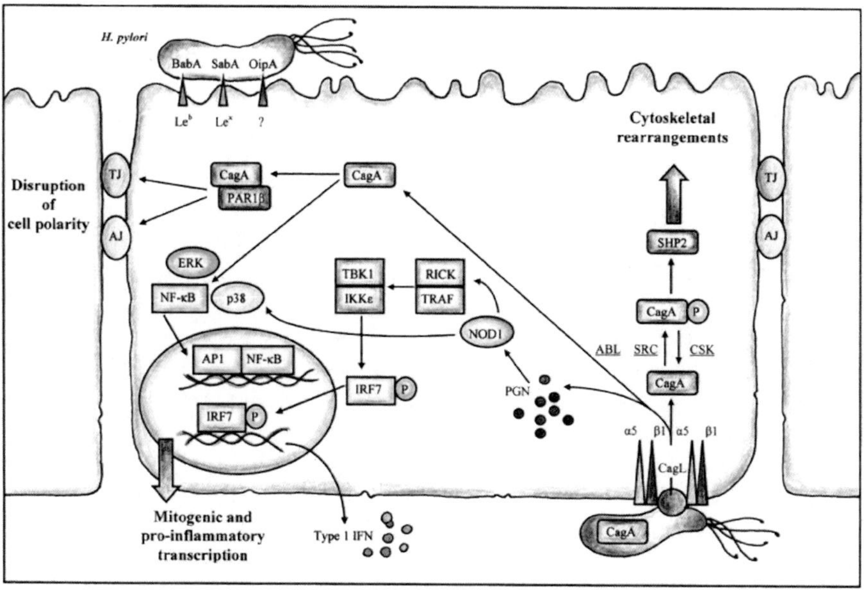

Fig. 2.3 Relations between gastric epithelial cells and *H. pylori*. Several adhesions (*BabA, SabA and OipA*) mediate binding of *H.pylori* to gastric epithelial cells, probably through the apical surface. After adherence, *H. pylori* can translocate effector molecules such as CagA and peptidoglycan (*PGN*) into the host cell. PGN is sensed by the intracellular receptor nucleotide-binding oligomerization domain-containing protein 1 (*NOD1*), which activates nuclear factor-κB (*NF-κB*), p38, ERK and IRF7 to induce the release of pro-inflammatory cytokines. After translocation, CagA is quickly phosphorylated (*P*) by SRC and ABL kinases, leading to cytoskeletal rearrangements. Unphosphorylated CagA can trigger different signalling cascades such as the activation of NF-κB and the disruption of cell–cell junctions, which may contribute to the loss of epithelial barrier function. Injection of CagA seems to be dependent on basolateral integrin α5β1. *AJ* adherens junction, *TJ* tight junction, *CSK* c-src tyrosine kinase, *IKKε* IκB kinase-ε, *IRF7* interferon regulatory factor 7, *RICK* receptor-interacting serine-threonine kinase 2, *TBK1* TANK-binding kinase 1

2.4 Influence of *Helicobacter pylori* Infection on DNA Damage and Repair

2.4.1 DNA Repair

Genetic instability is a hallmark of cancer. Therefore, one would expect that if *H. pylori* infection causes damage to DNA or decreases the activity of DNA repair pathways, it will allow accumulation of mutations that can cause inactivation of tumour suppressor genes or activation of oncogenes, which with time will increase the risk for growth of GC (Touati et al. 2003).

MMR (Mismatch repair pathway) is one of the most studied DNA repair mechanism. Human MMR is initiated by the binding of a heterodimeric protein complex, MutSα or MutSβ to a mismatch. MutSα (formed by MSH2 and MSH6 proteins) binds preferentially to base-base mismatches and small insertion-deletion loops. MutSβ combines MSH2 an MSH3 proteins and binds to larger insertion-deletion loops.

After the binding of one of these complexes to the mismatch, MutLα or MutLβ is recruited. MutLα is formed by MLH1 and PMS2 proteins, and MutLβ consists of the MLH1 and PMS1 proteins. These complexes signal downstream MMR components that proceed with the excision of the DNA containing the mismatch and resynthesis of the newly synthesized strand (Li 2003). Concerning the effect of *H. pylori* infection on the MMR pathway in GC, a recent study showed that, *in vitro* expression of MLH1, PMS1, PMS2, MSH2 and MSH6 proteins decreased (dose-dependent) after *H. pylori* infection. Interestingly, the decrease in protein expression correlated with mRNA down-regulation for MSH2 and MSH6. The levels of MLH1 in cells that had undergone *H. pylori* eradication, returned to values similar to the non-infected cells suggesting a reversible inhibition of MMR gene expression (Kim et al. 2002).

To explore an association between *H. pylori* infection and decrease in MMR expression *in vivo*, Park et al. (2005) studied *H. pylori*-infected gastritis and peptic ulcer disease tissue samples before and after eradication treatment. The results obtained elucidated that *H. pylori* eradication treatment increased MLH1 and MSH2 levels, suggesting that *H. pylori* gastritis might lead to a deficiency of MMR in gastric epithelium that may increase the risk of mutation accumulation in the gastric mucosa cells during chronic *H. pylori* infection.

In order to characterize the biological role of the bacteria-induced decrease in MMR gene/protein expression, Machado et al. (2009) investigated the effect of *H. pylori* infection of gastric cells on major repair pathways. In AGS gastric epithelial cells as well as in C57BL/6 mice infected with *H. pylori*, MMR gene/protein expression decreased. This decrease regarding overall MMR is not dependent on the virulence factors of the bacteria and occurs during the early stages of infection. It was observed that MMR down-regulation in mice occurred after 3 months of infection with *H. pylori* but not after 12 months.

Another major repair pathway critical for the maintenance of genome stability is, Base excision repair (BER) that repairs a number of endogenously generated DNA lesions. BER removes various forms of base damage such as oxidation, methylation and deamination and is initiated by DNA glycosylases that recognize and cleave the damaged bases, creating abasic (AP) sites. The AP sites are cytotoxic and mutagenic, and therefore further processed by DNA glycosylases with AP-lyase activity or by APE1 (Guillet and Boiteux 2002; Fortini et al. 2003). The single nucleotide gap is filled and the nick sealed to complete the repair reaction. APE1 expression was down-regulated in gastric cells infected with *H. pylori*, while OGG1 (a DNA glycosylase repairing oxidative DNA damage) remained constant. These results suggest that *H. pylori* infection causes an imbalance between generation and repair of AP sites, which is highly mutagenic (Machado et al. 2009; Glassner et al. 1998).

All together the published literature suggests that increased levels of cellular damage and death due to for example reactive oxygen species (ROS) would lead to increased inflammation and consequently to the production of more ROS and tumour-promoting cytokines. It also strongly indicates that one mechanism underlying genetic instability caused by *H. pylori* infection is deregulation of central DNA repair pathways.

2.4.2 DNA Damage

H. pylori infection causes oxidative DNA damage of the host cells, which results in mutagenesis and disease development. Chronic infections that induce an inflammatory response are a great source of ROS. Various studies demonstrated an association between increased levels of oxidized bases and cancer or inflammatory diseases including hepatitis, cirrhosis and *H. pylori* infection (Marnett 2000). The cellular consequences of DNA oxidation by ROS are several since it can lead to a number of different types of damage, such as oxidized bases, single and double-strand breaks (De Bont and van Larabeke 2004). Polyunsaturated fatty acid residues of phospholipids are also very sensitive to oxidation and the first products derived from fatty acid oxidation can either be reduced into harmless fatty acid alcohols or react with metal, generating substances that damage DNA by forming exocyclic adducts that block the DNA base pairing region. The levels of DNA adducts are increased by chronic infections and inflammation. Various studies have revealed a connection between *H. pylori* infection and oxidative DNA damage (De Bont and van Larabeke 2004). It has been shown that the bacterial infection affects the antioxidant defenses of gastric cells, suggesting that this response is one of the cellular mechanisms to survive attack of ROS (Obst et al. 2000; Khanzode et al. 2003).

In vitro it was shown that *H. pylori* infection induced microsatellite instability (MSI), which correlated with a decrease in expression of the MMR proteins MLH1 and MSH2 (Yao et al. 2006). The decrease in MLH1 expression could be explained by the CpG methylation detected in the MLH1 promoter region from infected cells. It has been suggested that *H. pylori* is responsible for methylation of promoters in early steps of gastric transformation and eradication of infection may result in reversal of methylation (Perri et al. 2007; Leung et al. 2006). In cancer, DNA methylation is often aberrant and methylation can lead to mutations due to the formation of mutagenic adducts (Stern et al. 2000). Alkylated DNA adducts can generate, for example, mutagenic AP sites and imidazole ring opening, which can be responsible for the blocking of DNA replication (Tudek et al. 1992). With the purpose of elucidate the nature of mutations introduced by *H. pylori* infection, Touati et al. (2003) used mice carrying the λ phage shuttle vector containing a cII reporter to show that mutation frequencies increased after 6 months of infection. The mutation spectrum in infected mice was dominated by mutations that could be explained to originate from oxidative damage, supporting the idea that the bacterial infection induces oxidative DNA damage, which is associated with the host-inflammatory response.

These results add novelty to the field of *H. pylori* pathogenesis by showing that *H. pylori* infection induces a decrease in repair activity and a transient mutator phenotype, contributing to epithelial gastric genomic instability and to neoplastic transformation.

2.4.3 mtDNA Instability and *H. pylori* Infection

There is evidence that *H. pylori* plays a role in the appearance of mutations in mitochondrial DNA (mtDNA) (Hiyama et al. 2003; Lee et al. 2007). In gastric cancer, mtDNA mutations occur both in the non coding D-loop region and the coding genes (Wu et al. 2005; Han et al. 2003). Recently it has been demonstrated, that *H. pylori* infection resulted in increased mutations in the non-coding D-loop as well as the coding genes ND1 and COI of mtDNA of gastric cells (Machado et al. 2009). The increase in the number of mutations was mainly attributed to a rise of transitions, possibly a consequence of oxidative damage. The increase in mtDNA mutations was dependent on the bacterial virulence factors. *H. pylori* positive chronic gastritis patients also showed that transitions were the main mutational event and patients harboring mtDNA mutations were frequently infected by *H. pylori* with cagA+ and vacA s1/m1 genotype (Machado et al. 2009).

In view of the studies cited, *H. pylori* infection is able to induce mtDNA mutations both *in vitro* and *in vivo*, suggesting that the mitochondrial genome is highly susceptible to bacterial infections.

2.5 Pro-tumorigenesis Host Factors

2.5.1 Human Gene Polymorphisms

One of the paradoxes of *H. pylori* infection is its association with mutually exclusive clinical outcomes such as GC and duodenal ulcer disease. As we have already discussed previously, various bacterial virulence factors (e.g., *cagA*, *vacA*, *BabA* and *OipA*) have an undoubted role in the pathogenesis of these diseases, but they do not readily distinguish between the two key outcomes of GC and duodenal ulcer. Also, hereditary factors clearly increase the risk of gastric cancer and this malignancy is part of a number of familial cancer syndromes. The most celebrated familial case of gastric cancer is that of Napoleone Bonaparte (Marchall and Windsor 2005). All these considerations prompted some researchers to consider the host genetic factors that may be relevant to this process.

But what are the genes considered to be important and where to focus resources? Since *H. pylori* achieve most damage through induction of chronic inflammation, it is reasonable to consider genes that control this process as appropriate candidates.

2.5.1.1 IL-1 Gene

The *IL-1* gene cluster on chromosome 2q contains three related genes within a 430 kb region: *IL-1A*, *IL-1B* and *IL-1RN*, which encode for the pro-inflammatory cytokines IL-1α and IL-1β, as well as their endogenous receptor antagonist IL-1ra, respectively (Dinarello 1996). IL-1β is upregulated in the presence of *H. pylori* and plays a central role in initiating and amplifying the inflammatory response to this infection (El-Omar 2001).

Three diallelic polymorphisms in *IL-1B* have been reported, all representing C–T or T–C transitions, at positions–511, −31, and +3954 bp from the transcriptional start site (Bidwell et al. 2001). The *IL-1RN* gene has a penta-allelic 86 bp tandem repeat (VNTR) in intron 2, of which the less common allele 2 (*IL-1RN**2) is associated with a wide range of chronic inflammatory and autoimmune conditions (Bidwell et al. 2001). There are several epidemiologic studies that have tested the role of these candidate loci. El-Omar et al. (2000) first studied the correlation of high IL-1β genotypes (two polymorphisms in the *IL-1B* and *IL-1RN* genes) with hypochlorhydria and gastric atrophy in a Caucasian population of gastric cancer. These relatives were known to be at increased risk of developing the same cancer and had a high prevalence of precancerous abnormalities (hypochlorhydria and gastric atrophy) but only in the presence of *H. pylori* infection (El-Omar et al. 2000). The association between the same IL-1β genetic polymorphisms and gastric cancer itself was subsequently examined using two independent Caucasian case control studies from Poland and the USA (El-Omar et al. 2000, 2003). In the above studies the pro-inflammatory *IL-1* genotypes were associated with an increased risk of both intestinal and diffuse types of gastric cancer; however, the risk was restricted to the non cardia sub-site.

Indeed, the *IL-1* markers had no effect on risk of cardia gastric adenocarcinomas, esophageal adenocarcinomas or esophageal squamous cell carcinomas (El-Omar et al. 2003). The latter findings are entirely in keeping with the proposed mechanism for the effect of these polymorphisms in gastric cancer, namely the reduction of gastric acid secretion. Thus, a high IL-1β genotype appears to increase the risk of non-cardia gastric cancer, a disease characterized by hypochlorhydria, while it has no effect on cancers associated with high acid exposure such as esophageal adenocarcinomas and some cardia cancers.

The association between *IL-1* gene cluster polymorphisms and gastric cancer and its precursors has been confirmed independently by other groups covering Caucasian, Asian and Hispanic populations (Rad et al. 2003: Palli et al. 2005; Furuta et al. 2002; Zeng et al. 2003). Machado et al. (2001) were the first to confirm the association between *IL-1* markers and gastric cancer in Caucasians and reported similar modest odds ratios to those reported by El-Omar (Machado et al. 2001) Furthermore, the same group subsequently reported on the combined effects of pro-inflammatory *IL-1* genotypes and *H. pylori* bacterial virulence factors (*cagA* positive, *VacA s1* and *VacA m1*). This study on combined effects reported, that for each combination of bacterial and host genotype the odds of having a gastric carcinoma were greatest in those with both bacterial and host high-risk genotypes (Figueiredo et al. 2002). This highlights a potentially important interaction between host and bacterium in the pathogenesis of gastric cancer.

A decisive piece of evidence that confirmed the apparent role of IL-1β in *H. pylori*-induced gastric carcinogenesis came from a transgenic mouse model in which IL-1β over production was targeted to the stomach by the H^+/K^+-ATPase beta promoter (Tu et al. 2008). With the over expression of IL-1β confined to the stomach, these transgenic mice had a thickened gastric mucosa, produced lower amounts of gastric acid and developed severe gastritis followed by gastric atrophy, intestinal metaplasia, dysplasia and adenocarcinomas. Importantly, these IL-1β transgenic mice proceeded through a multistage process that mimicked human gastric neoplasia. These changes occurred even in the absence of *H. pylori* infection, which, when introduced, led to an acceleration of these abnormalities. Most interestingly, the pathological changes, including the progression to gastric cancer were prevented by infusion of IL-1 receptor antagonist, proving beyond doubt that IL-1β is responsible for the pathological effects.

Another crucial piece of evidence came from well-designed recent study (Stoicov et al. 2009): T-bet is a central regulator of the cytokine environment during *Helicobacter* infection and T-bet knockout (KO) mice maintain infection for 15 months at levels similar to wild type mice and develop significant inflammation with a blunted Th1 (T helper 1). Furthermore, this blunted response is associated with the preservation of parietal and chief cells and protection from the development of gastric cancer. Crucially however, T-bet KO mice respond to *Helicobacter* infection with a markedly blunted IL-1β and TNF-α and elevated IL-10 levels. This mirrors the situation in humans who are protected against gastric cancer, and merits further research.

2.5.1.2 TNF-a, IL-10 and CXCL8 Genes

Soon after the identification of *IL-1* gene cluster polymorphisms as risk factors for gastric cancer, the proinflammatory genotypes of *TNF- a* and *IL-10* were reported as independent additional risk factors for non cardia gastric cancer (El-Omar et al. 2000).

TNF- *a* is another powerful pro-inflammatory cytokine that is produced in the gastric mucosa in response to *H. pylori* infection (D'Elios et al. 1997a). The *TNF-a*-308 G>A polymorphism is known to be involved in a number of inflammatory conditions. Carrying the pro-inflammatory A allele increased the risk of non cardia gastric cancer. This association was independently confirmed by a study from Machado et al. (2003). The same group also showed that the association between the *TNF-a*-308 G>A polymorphism and increased risk of gastric carcinoma is dependent on a linkage disequilibrium with an yet unidentified locus (Canedo et al. 2008a, b). Another interleukin, IL-10, is an anti-inflammatory cytokine that down regulates IL-1β, TNF-α, interferon-γ and other pro-inflammatory cytokines. A relative deficiency of IL-10 may result in a Th1-driven hyper-inflammatory response to *H. pylori* with greater damage to the gastric mucosa. A recent study reported that homozygosity for the low-IL-10 *ATA* haplotype (based on three promoter polymorphisms at positions −592, −819 and −1,082) increased the risk of non cardia gastric cancer with an odds ratio of 2.5 (95% *CI*: 1.1–5.7) (El-Omar et al. 2003).

Interestingly, there seems to be a cumulative effect in carrying more than one pro-inflammatory cytokine gene polymorphism. The same authors studied the effect of having multiple pro-inflammatory genotypes (*IL-1*β-511*T, *IL-1RN**2*2, *TNF*-α-308*A and *IL-10* ATA/ATA) on the risk of non-gastric cancer. With the addition of each relevant polymorphism the risk of non-gastric cancer increased progressively such that when three to four of these polymorphisms were present, the risk for gastric cancer was increased 27-fold (El-Omar et al. 2003). The fact that *H. pylori* is a prerequisite for the association of these polymorphisms with malignancy demonstrates that in this situation, infection–induced inflammation may indeed be contributing to carcinogenesis.

A chemokine that has an important role in the pathogenesis of *H. pylori* induced diseases is CXCL8. This chemokine is a potent chemoattractant for neutrophils and lymphocytes. It also has effects on cell proliferation, migration and tumour angiogenesis. The gene has a well established promoter polymorphism at position −251 (*IL-8*–251 T>A). The A allele is associated with increased production of CXCL8 in *H. pylori* infected gastric mucosa (Smith et al. 2004). It has been reported to increase the risk of severe inflammation and precancerous gastric abnormalities in Caucasian and Asian populations (Smith et al. 2004; Taguchi et al. 2005).

2.5.1.3 Immune Response Genes

In the last few years, association of other genes (particularly gene correlate with immune response) polymorphisms with the increased risk of GC have been studied vigorously. The seminal study by D'Elios et al. (1997a) demonstrated that *H. pylori* typically induces a Th1 response in infected individuals. A preferential activation of Th1 responses has been reported in different animal models, such as mice, beagle dogs, monkeys, and gerbils experimentally infected with *H. pylori* or *H. felis* (Mohammadi et al. 1997; Rossi et al. 2000; Mattapallil et al. 2000; Wiedemann et al. 2009). A large number of studies agree that *H. pylori* elicits Th1 response and is associated with more severe disease, including precancerous and cancerous lesions (D'Elios et al. 1997b; Bamford et al. 1998; Sommer et al. 1998; Fox et al. 2000; Lehmann et al. 2002; Del Giudice et al. 2001; Tomita et al. 2001; de Jonge et al. 2004; Wen et al. 2004). The major Th1 promoting factor of *H. pylori* is HP-NAP that stimulates IL-12 production via TLR2 (Amedei et al. 2006). However, other Th1 driving factors also exist, such as products of *cag,* as well as VacA, hsp90, outer membrane protein 18, cysteine-rich protein A, and lipopolysaccharide (D'Elios et al. 1997a; Guiney et al. 2003; Deml et al. 2005; Voland et al. 2006; Taylor et al. 2006; Takeshima et al. 2011). Hou et al. (2007) studied the association between gastric cancer and several variants in genes responsible for Th1 cell-mediated response, that is typical in *H. pylori* infection, (D'Elios et al. 1997a; Amedei et al. 2006), and in particular the results suggest that a polarized Th1 response may play a role in the genesis of severe clinical forms of disease (D'Elios et al. 1997a,b), Hou et al. (2007) subsequently reported that carrying the C allele of the interferon gamma receptor 2 (*IFNGR2*) (Ex7–128) rs4986958 polymorphism was associated with an increased risk of gastric cancer compared with the TT genotype. Furthermore, there was an additive effect when the *IFNGR2* polymorphism was combined with the *TNF-a*-308 TT genotype (*OR* 5.5, 95% *CI* 1.5–19.4). The interferon gamma receptor 1 (*IFNGR1*) –56C/T gene polymorphism was studied by Canedo et al. (2008a, b) in patients with early onset GC (less than 40 years of age at the time of diagnosis) there was a significant over-representation of the *IFNGR1* –56*T/*T homozygous genotype with an *OR* of 4.1 (95% *CI* 1.6–10.6).

A recent study evaluated polymorphisms in the Th1 *IL7R* gene and one polymorphism in the Th2 *IL5* gene (Mahajan et al. 2008). The *OR* for *IL7R* rs1494555 were 1.4 (95% *CI* 1.0–1.9) for A/G and 1.5 (95% *CI* 1.0–2.4) for G/G carriers compared with A/A carriers ($P=0.04$). The *OR* for *IL5* rs2069812 were 0.9 (95% *CI* 0.7–1.3) for C/T and 0.6 (95% *CI* 0.3–1.0) T/T carriers compared with C/C carriers ($P=0.03$). These results suggest that *IL5* rs2069812 and *IL7R* rs1389832, rs1494556 and rs1494555 polymorphisms may contribute to the risk of GC.

Genetic polymorphisms of the cytokines and chemokines discussed above clearly play an important role in the risk of *H. pylori*-induced gastric adenocarcinomas. However, *H. pylori* is initially handled by the innate immune response and it is conceivable that functionally relevant polymorphisms in the genes of this arm of the immune system may affect the magnitude and subsequent direction of the host's response against the infection. Most *H. pylori* cells do not invade the gastric mucosa but the inflammatory response against it is triggered through the attachment of *H. pylori* to the gastric epithelia, mainly via TLR2 or TLR4 (Amedei et al. 2006; Segal et al. 1997). Toll-like receptor 4 (TLR4) was initially identified as the potential signaling receptor for *H. pylori* on gastric epithelial cells (Su et al. 2003) and recently, Hold et al. (2007) have shown that the *TLR4*+896A>G polymorphism was associated with an exaggerated and destructive chronic inflammatory phenotype in *H. pylori* infected patients. This phenotype was characterized by gastric atrophy and hypochlorhydria, the hallmarks of subsequent increased risk of gastric cancer.

The association of *TLR4*+896 A>G polymorphism with both GC and its precursor lesions implies that it is relevant to the entire multistage process of gastric carcinogenesis that starts with the *H. pylori* colonization of the gastric mucosa. Patients with this polymorphism have an increased risk of severe inflammation and subsequent development of hypochlorhydria and gastric atrophy, which are regarded as the most important precancerous abnormalities. This severe inflammation is initiated by *H. pylori* infection but it is entirely feasible that subsequent co-colonization of an achlorhydric stomach by a variety of other bacteria may sustain and enhance the microbial inflammatory stimulus and continue to drive the carcinogenic process.

Recently another *TLR4* polymorphism is associated to increase of the risk of intestinal type gastric cancer. The *TLR4* Thr399Ile was linked with an increased hazard ratio of 5.38, (95% *CI* 1.652–8.145), $P=0.006$ (Santini et al. 2008).

There are other reports of innate immune response gene polymorphisms being associated with increased risk of gastric cancer: mannose binding lectin is an antigen recognition molecule involved in systemic and mucosal innate immunity. It is able to bind to a range of microbes and subsequently kill them by activating the complement system and promoting complement independent opsonophagocytosis. A latest work showed that polymorphisms in the mannose binding lectin-2 gene

(*MBL2*) were associated with increased risk of GC. In haplotype analysis, the HYD haplotype was associated with an increased risk of stomach cancer when compared with HYA, the most common haplotype ($OR = 1.9$, 95% CI 1.1–3.2; $P = 0.02$) (Baccarelli et al. 2006).

2.5.2 β-Catenin in H. pylori Carcinogenesis

A specific host molecule that may influence carcinogenic responses in conjunction with *H. pylori* is β-catenin, a ubiquitously expressed protein that has distinct functions within host cells. Membrane bound β-catenin is a component of adherens junctions that link cadherin receptors to the actin cytoskeleton. Cytoplasmic β-catenin is a downstream component of the Wnt signal transduction pathway (Fig. 2.4a). In the absence of Wnt ligand, the inhibitory complex induces the degradation of β-catenin and maintains low steady state levels of free β-catenin either in the cytosol or the nucleus. After binding of wnt to its receptor Frizzled, β-catenin translocate to the nucleus and activate the transcription of target genes that are involved in carcinogenesis (Fig. 2.4b).

Increased β-catenin expression or *APC* mutations are present in up to 50% of GC specimens when compared with non-transformed gastric mucosa (Tsukashita et al. 2003), and the nuclear accumulation of β-catenin is increased in gastric adenomas and foci of dysplasia (Cheng et al. 2004), These studies suggest that, aberrant activation of β-catenin precedes the development of GC. *H. pylori* increases the expression of β-catenin target genes in colonized mucosa and during co-culture with gastric epithelial cells *in vitro*. Therefore, it is likely that the activation of β-catenin signalling is a central component in the regulation of pre-malignant epithelial responses to *H. pylori*.

H. pylori isogenic mutant studies have revealed that the translocation of CagA into gastric epithelial cells induces the nuclear accumulation and functional activation of β-catenin (Cheng et al. 2004; Franco et al. 2005). Murata-Kamiya et al. (2007) demonstrated that intracellular CagA interacts with E-cadherin, disrupts the formation of E-cadherin–β-catenin complexes and induces nuclear accumulation of β-catenin, all of which are independent of CagA phosphorylation (Fig. 2.4a). Consequences of CagA-dependent β-catenin activation include the upregulation of target genes that influence GC, such as caudal type homeobox 1 (*CDX1*), which encodes a transcription factor that is required for the development of intestinal metaplasia (Murata-Kamiya et al. 2007).

Recently, additional pathways have been demonstrated to regulate β-catenin activation in response to *H. pylori*. Activation of Pi3K and AKT leads to the phosphorylation and inactivation of GSK3β, permitting β-catenin to accumulate in the cytosol and the nucleus. A recent study have shown that CagA Cm motifs interact with mET, leading to the sustained induction of Pi3K–AKT signalling in response to *H. pylori* and the subsequent activation of β-catenin *in vitro* and *in vivo* (Fig. 2.4b) (Suzuki et al. 2009)

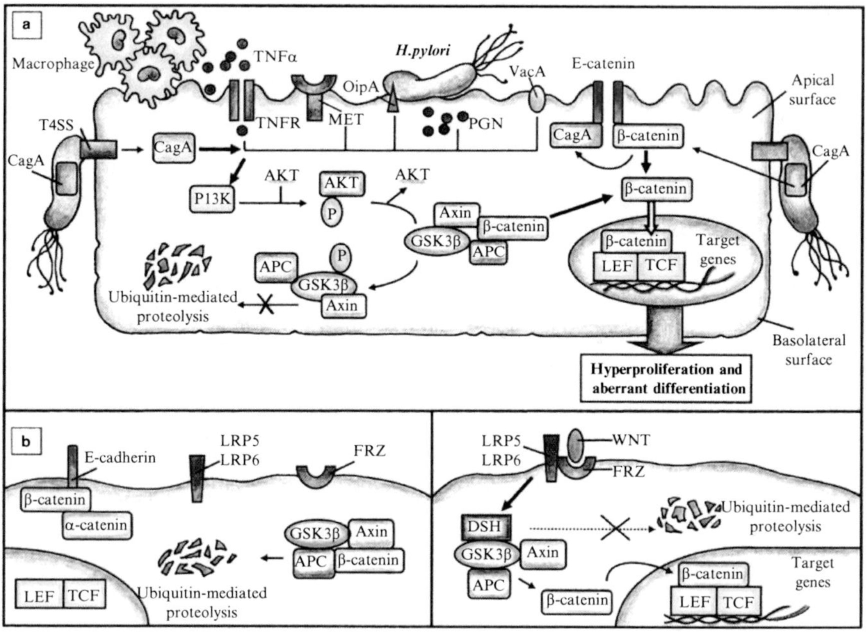

Fig. 2.4 Aberrant β-catenin activation by *Helicobacter pylori.* (**a**) Membrane-bound β-catenin links cadherin receptors to the actin cytoskeleton, and in non-transformed epithelial cells β-catenin is primarily localized to E-cadherin complexes. Cytoplasmic β-catenin is a downstream component of the Wnt pathway; in the absence of Wnt, cytosolic β-catenin remains bound within a multi-protein inhibitory complex comprised of glycogen synthase kinase-3β (*GSK3β*), the adenomatous polyposis coli (*APC*) tumour suppressor protein and axin. Under unstimulated conditions, β-catenin is phosphorylated (*P*) by GSK3β, ubiquitylated and degraded. (**b**) Binding of Wnt to its receptor, Frizzled (*FRZ*; *lower panel*), activates dishevelled (*DSH*) and Wnt co-receptors, low density lipoprotein receptor-related protein 5 (*LRP5*) and LRP6, which then interact with axin and other members of the inhibitory complex, leading to the inhibition of the kinase activity of GSK3β141. These events inhibit the degradation of β-catenin, leading to its nuclear accumulation and formation of heterodimers with the transcription factor lymphocyte enhancer factor/T cell factor (*LEF/TCF*), resulting in the transcriptional activation of target genes that influence carcinogenesis. Injection of CagA results in the dispersal of β-catenin from β-catenin–E-cadherin complexes at the cell membrane, allowing β-catenin to accumulate in the cytosol and nucleus. CagA, potentially by binding MET or other *H. pylori* constituents such as OipA, VacA and peptidoglycan (*PGN*) as well as TNFα, which is produced by infiltrating macrophages, can activate PI3K, leading to the phosphorylation and inactivation of GSK3β. This liberates β-catenin to translocate to the nucleus and upregulate genes, leading to increased proliferation and aberrant differentiation; *TNFR* TNF receptor

Studies focused on Pi3K and AKT have revealed that other *H. pylori* constituents may also influence β-catenin activation. Nakayama et al. (2009) reported that vacA can activate Pi3K-dependent β-catenin activation, and oipA has also been implicated in aberrant nuclear localization of β-catenin, although the specific mechanism underpinning this observation has not yet been delineated.

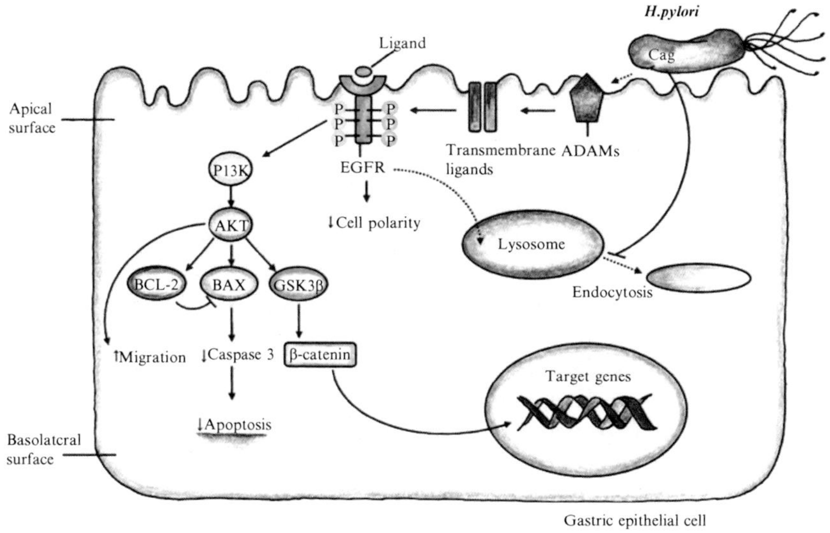

Fig. 2.5 eGFr transactivation by H. pylori and induced cellular effects with carcinogenic potential. *H. pylori* transactivates epidermal growth factor receptor (*EGFR*) through cleavage, which is dependent on a disintegrin and metalloproteinase (*ADAM*) family proteinases, of EGFR ligands, such as heparin-binding EGF-like growth factor (*HBEGF*) in gastric epithelial cells. One down-stream target of EGFR transactivation is PI3K–AKT, which leads to AKT-dependent cell migration, inhibition of apoptosis and β-catenin activation. *BAX* BCL-2-associated X protein, *GSK3β* glycogen synthase kinase-3β

2.5.3 Transactivation of EGFR by H. pylori

EGFR is an important target for the treatment of several malignancies other than gastrointestinal cancers. Phosphorylation and activation of EGFR increases the transcriptional activity of β-catenin by the inactivation of GSK3β. *H. pylori* infection, gastric epithelial hyperplasia and gastric atrophy are strongly linked to the dysregulation of EGFR and/or cognate ligands, such as heparin-binding EGF-like growth factor (HbEGF) in human animal and cell culture models (Romano et al. 1998; Wong et al. 2001). The *in vitro* transactivation of EGFR by *H. pylori* is dependent on genes in the *cag* pathogenicity island and secreted proteins as well as host factors such as TLR4 and NoD1 (Keates et al. 2005; Basu et al. 2008).

EGFR can be activated by direct interaction with ligands, which initiate dimerization and increased kinase activity (Fig. 2.5). Cytokines, such as TNFα, and other stimuli are present in the gastric mucosa following *H. pylori* infection (D'Elios et al. 1997a) and transactivate EGFR in gastric epithelial cells (Pece and Gutkind 2000). EGFR transactivation by these elements is mediated through met alloproteinase dependent cleavage of EGFR (Erbb family) ligands in a manner similar to *H. pylori* induced EGFR transactivation (Pece and Gutkind 2000). The required metalloproteinases are likely to be members of disintegrin and metalloproteinase (ADAm) family.

Given a requirement for metalloproteinase activity in *H. pylori* initiated HbEGF release, ADAm17, a multidomain type I transmembrane protein that contains an extracellular zinc-dependent protease domain is an ideal candidate enzyme for the regulation of this pathway (Pece and Gutkind 2000). ADAm17 was the first ADAm to have a defined physiological substrate, the precursor transmembrane form of TNFα. In fact inhibitors of ADAm17 block the release of soluble TNFα.

Although ADAm17 is ubiquitously expressed in the gastrointestinal tract and is a target of drug development for inflammatory conditions, the disorganized and inflamed nature of the gastrointestinal tract that develops in ADAm17-deficient mice suggests that this metalloproteinase may also have an important role in gut epithelial homeostasis, perhaps through the regulation of EGFR ligands (Sunnarborg et al. 2002). Therefore, a better understanding of the function of ADAm17 during *H. pylori*-induced gastric epithelial injury could provide insights into its potential role in gastric carcinogenesis.

H. pylori specifically amplifies EGFR signalling by both activating EGFR and decreasing EGFR degradation by blocking endocytosis (Bauer et al. 2009). The transactivation of EGFR by this pathogen mediates several cellular responses with pre-malignant potential (Fig. 2.5). Alterations in apoptosis have been implicated in the pathogenesis of *H. pylori* induced injury before the development of GC and the ability of *H. pylori* to induce apoptosis in gastric epithelial cells has been well demonstrated *in vitro* (Cover et al. 2003) However, chronically infected mongolian gerbils harbouring *cag*+strains exhibit increased gastric epithelial cell proliferation without a concordant increase in apoptosis (Peek et al. 2000) which may contribute to the augmented risk for gastric cancer that is associated with *cag*+strains.

H. pylori has been shown to induce anti-apoptotic pathways in gastric epithelial cells through *cag* mediated EGFR transactivation (Maeda et al. 2002) (Fig. 2.5). Altered cell polarity and migration are phenotypic responses to *H. pylori* infection. Although they may acutely promote gastric mucosal repair, long term stimulation of these responses has been linked to transformation and tumorigenesis (Nagy et al. 2009).

As the biological responses to EGFR activation include increased proliferation, reduced apoptosis, the disruption of cell polarity and enhanced migration, transactivation of EGFR by *H. pylori* is an attractive target for studying early events that may precede transformation.

2.5.4 H. pylori, Gastric Autoimmunity and Gastric Atrophy

A strong association between *H. pylori* infection and gastric autoimmunity has been highlighted by a number of clinical and epidemiological studies indicating that most of patients with autoimmune gastritis (AIG) have or had *H. pylori* infection

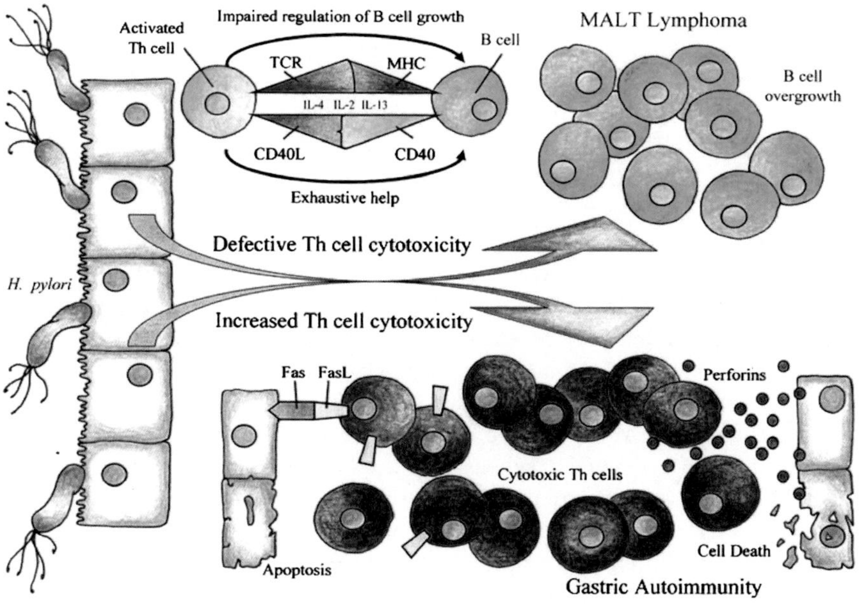

Fig. 2.6 Different effector functions of T cells in *H. pylori*-related gastric lymphoma and gastric autoimmunity. T cells are essential for defence against infection, but inappropriate Th responses can be harmful for the host. In *H. pylori*-infected patients with gastric lymphoma, gastric *H. pylori*-specific Th cells display deficient cytotoxic control (both perforin and Fas-Fas-ligand mediated) of B-cell growth. Such cytolytic defects, associated with the chronic delivery of costimulatory signals by Th cells and by the production of cytokines with B-cell growth factor activity, together with chronic exposure to *H. pylori* antigens, would result in overgrowth of B cells of gastric low-grade MALT B-cell lymphoma. Conversely, *H. pylori* induces gastric autoimmunity via molecular mimicry by the expansion of *H. pylori*-specific T cells that cross-react with H^+,K^+-ATPase epitopes. Cross-reactive T cells would result in destruction of gastric mucosa, by the long-lasting activation of both Fas-ligand (*FasL*)-induced apoptosis and perforin-mediated cytotoxicity

(D'Elios et al. 2004). *H. pylori* associated autoimmune gastritis is characterized by an inflammatory infiltrate of the gastric mucosa, including T cells, macrophages and B cells. It mainly affects the corpus and the fundus, and it is accompanied by loss of gastric parietal and zymogenic cells. We have characterized at molecular level the gastric T-cell mediated responses to *H. pylori* and to the H^+, K^+-ATPase autoantigen in a series of *H. pylori*-infected patients with gastric autoimmunity (Amedei et al. 2003) (Fig. 2.6).

Among gastric Th clones, a number proliferated to *H. pylori*, but not to the *H. pylori* proteins CagA, VacA, hsp, urease nor to H^+, K^+-ATPase. Some other Th clones proliferated to H^+, K^+-ATPase and not to *H. pylori* (autoreactive), and a third group of clones was found that proliferated to both *H. pylori* and H^+, K^+-ATPase (cross-reactive) (Amedei et al. 2003). All the Th clones able to proliferate to H^+, K^+-ATPase were studied for their ability to respond to the 261 overlapping 15-mer peptides covering the amino acid sequence of α and β chain of the human H^+, K^+-ATPase. In the series of cross-reactive Th clones 11 recognized their epitope in the α chain and two clones in the β chain. In the subgroup of autoreactive Th clones 6 recognized their epitope in the α chain and 9 in the β chain of the proton pump. Therefore, some "shared" H^+, K^+-ATPase epitopes, mainly in the α chain, are cross-reactive with epitopes of *H. pylori* antigens, whereas others can be considered as "private" epitopes of H^+, K^+-ATPase. A cross-reactive *H. pylori* peptide could be found for each of the 10 H^+, K^+-ATPase/*H. pylori* cross-reactive gastric Th clones. Overall, that study led to the identification of nine different *H. pylori* proteins (such as lipopolysaccharide biosynthesis protein, histidine kinase, porphobilinogen deaminase, dimethyl-adenosine transferase, glucose-inhibited division protein A, VirB4 homolog, phosphoglucosamine mutase, acetate kinase, penicillin-binding protein-2) each harboring a T-cell peptide suitable for cross-reaction with T-cell epitopes of gastric H^+, K^+-ATPase α chain. Interestingly, none of the bacterial epitopes recognized by cross-reactive Th clones belong to the known *H. pylori* immunodominant antigens, such as CagA, VacA and urease, which are major targets of gastric T-cell responses in *H. pylori* infected patients with peptic ulcer (D'Elios et al. 1997a). Two possibilities can be considered: these peptides are implicated in cross-reactivity because of their structural properties or alternatively a physiological relevance implicating these particular nine proteins can be postulated. All the cross-reactive and autoreactive H^+, K^+-ATPase-specific Th clones after activation were able to induce cell death via either Fas–Fas ligand-mediated apoptosis or perforin-mediated cytotoxicity against target cells (Amedei et al. 2003). This ability to induce apoptosis in T cells might give a selective advantage that can promote survival and persistence of bacteria, allowing *H. pylori* to escape the host immune response. On the other hand, the relevance of cross-reactive and autoreactive cytolytic Th effector cells in the genesis of AIG is consistent with data in the mouse model that Fas-related death is required for the development of full-blown destructive autoimmune gastritis (Marshall et al. 2002). Based on these results, it is tempting to speculate that in the "gastric autoimmune inflammatory scenario" in which cross-reactive and autoreactive Th clones are activated, parietal cells might become target of the pro-apoptotic and cytotoxic activity of cross-reactive and autoreactive gastric Th cells. The end point of this process would be gastric atrophy, which might lead to gastric cancer.

2.6 *H. pylori* and Immunity in MALT Lymphoma

Extranodal marginal zone B cell lymphoma of mucosa-associated lymphoid tissue (MALT lymphoma) represents the third commonest form of non-Hodgkin lymphoma (Armitage 1997; Du 2007). The most frequent site of MALT lymphomas is the stomach, where they were first recognized as a distinct entity (Isaacson and Wright 1983). A link of *H. pylori* infection with gastric MALT lymphoma was provided by the identification of *H. pylori* in the majority of the lymphoma specimens (Wotherspoon et al. 1991). *H. pylori* related low-grade gastric MALT lymphoma represents a model for studying the interplay between chronic infection, immune response, and lymphoma genesis (Fig. 2.6). This lymphoma represents the first described neoplasia susceptible of regression following antibiotic therapy resulting in *H. pylori* eradication (Wotherspoon et al. 1991). A prerequisite for lymphoma genesis is the development of secondary inflammatory MALT induced by *H. pylori*. Tumour cells of low-grade gastric MALT lymphoma (MALToma) are memory B cells still responsive to differentiation signals, such as CD40 costimulation and cytokines produced by antigen-stimulated Th cells, and dependent for their growth on the stimulation by *H. pylori* specific T cells (Hussell et al. 1993, 1996; Greiner et al. 1997; D'Elios et al. 1999). In early phases, this tumour is sensitive to withdrawal of *H. pylori* induced T-cell help, providing an explanation for both the tumour's tendency to remain localized to its primary site and its regression after *H. pylori* eradication. The analysis of the antigen induced B-cell help exerted by *H. pylori* reactive gastric T-cell clones provided detailed information on the molecular and cellular mechanisms associated with the onset of low-grade gastric MALToma. In the stomach of MALToma patients, a high percentage of Th cells were specific for *H. pylori* (between 3% and 20% in each case). In particular, 25% were specific for urease, 4% for VacA, and none for CagA or HSP; 71% of Th clones proliferated in response to *H. pylori* antigens, different from Urease, CagA, VacA, or HSP (D'Elios et al. 1999). Each *H. pylori*-specific Th clone derived from gastric MALToma produced IL-2 and a variety of B-cell-stimulating cytokines, such as IL-4 and IL-13 (D'Elios et al. 1999). *In vitro* stimulation with the appropriate *H. pylori* antigens induced *H. pylori* specific Th clones derived from gastric MALToma to express powerful help for B-cell activation and proliferation (D'Elios et al. 1999). B cells from MALToma patients proliferate in response to *H. pylori*, but the B-cell proliferation induced by *H. pylori* antigens was strictly T-cell-dependent because it could not take place with *H. pylori* and without T helper cells (D'Elios et al. 1999, 2003; Bergman and D'Elios 2010). In chronic gastritis patients, either with or without ulcer, the helper function towards B cells exerted by *H. pylori* antigen-stimulated gastric T-cell clones was negatively regulated by the concomitant cytolytic killing of B cells (D'Elios et al. 1997a). In contrast, gastric T-cell clones from MALToma were unable to down modulate their antigen-induced help for B-cell proliferation. Indeed, none of these clones was able to express perforin-mediated cytotoxicity against autologous B cells. Moreover, the majority of Th clones from uncomplicated chronic gastritis induced Fas-Fas ligand mediated apoptosis in target cells, whereas only a small fraction of *H. pylori* specific gastric clones from MALToma were able to induce apoptosis in target cells, including autologous B cells (D'Elios et al. 1999). Both defective perforin mediated cytotoxicity and poor ability to induce Fas-Fas ligand mediated apoptosis were restricted to MALToma-infiltrating T cells, since *H. pylori*-specific Th cells derived from the peripheral blood of the same patients expressed the same cytolytic potential and proapoptotic activity as that shown by Th cells from chronic gastritis patients (D'Elios et al. 1999). Accordingly, mice lacking T cell and NK cell cytotoxic effector pathways have also been shown to develop spontaneous tumours (Swann and Smyth 2007; Trapani and Smyth 2002; Smyth et al. 2000; Street et al. 2001, 2002, 2004; Zerafa et al. 2005; Davidson et al. 1998; Liu et al. 2004; Mitra-Kaushik et al. 2004).

For example, mice that lack perforin, a cytotoxic molecule used by cytotoxic cells such as $CD8^+$ T cells and NK cells to form membrane pores in target cells, develop lymphomas with age. These spontaneous lymphomas are of B cell origin, develop in older mice (>1 year of age) regardless of the mouse strain (Smyth et al. 2000; Street et al. 2002), and when transplanted into WT mice, are rejected by $CD8^+$ T cells (Smyth et al. 2000). B cell lymphomas also arise in mice lacking both perforin and β_2m, and tumour onset is earlier and occurs with increased prevalence compared with mice lacking only perforin. In addition, B cell lymphomas derived from mice lacking both perforin and β_2m are rejected by either NK cells or $\gamma\delta$ T cells following transplantation to WT mice, rather than by $CD8^+$ T cells (as in tumours derived from mice lacking only perforin), demonstrating that cell surface expression of MHC class I molecules by tumour cells can be an important factor in determining which effector cells mediate immune protective effects (Street et al. 2004). Intriguingly, mutations in the gene encoding perforin have also been identified in subsets of lymphoma patients (Clementi et al. 2005), although it is not clear whether this contributes to disease. Mice lacking the death-inducing molecule TNF-related apoptosis inducing ligand (TRAIL) or expressing a defective mutant form of the death-inducing molecule FasL have also been shown to be susceptible to spontaneous lymphomas that develop with late onset (Zerafa et al. 2005; Davidson et al. 1998). These aging studies have clearly demonstrated a critical role for cytotoxic pathways in immunoregulation and/or immunosuppression of spontaneous tumour development in mice. The reason why gastric T cells of MALToma, while delivering powerful help to B cells, are deficient in mechanisms involved in the control of B-cell growth still remains unclear. It has been shown that VacA toxin inhibits antigen processing in APCs and T cells, but not the exocytosis of perforin-containing granules of NK cells (Molinari et al. 1998; Boncristiano et al. 2003). It is possible that, in some *H. pylori*-infected individuals, some bacterial components affect the development or the expression in gastric T cells of regulatory cytotoxic mechanisms on B-cell proliferation, allowing exhaustive and unbalanced B-cell help and lymphoma genesis to occur (D'Elios et al. 1999; Lehours et al. 2004, 2009).

2.7 Conclusion

H. pylori is an important organism and, while it may not cause any clinical problems in most infected patients, it has the potential to leave the host with devastating consequences. *Helicobacter pylori* is able to induce a huge variety of responses in the stomach, due to host genetics, age, sex, different bacterial and environmental factors, or other concomitant infections. Sporadic gastric cancer is a common cancer with a grave prognosis, particularly in China, South America, Eastern Europe, Korea, and Japan. A major advance in the fight against this global killer came with the recognition of the role of *H. pylori* infection in its pathogenesis and the acquisition of new techniques and biological markers to identify high risk subpopulations.

The cancer represents a classic example of an inflammation induced malignancy. Host genetic factors interacting with bacterial virulence and environmental factors play an important role in the pathogenesis of cancer.

In this chapter we have analyzed the different ways in which *H. pylori* can contribute to both gastric cancer, as the consequences of *H. pylori* infection on the integrity of DNA in the host cells, and gastric MALT lymphoma.

By down-regulating major DNA repair pathways, *H. pylori* infection has the potential to generate mutations. In addition, *H. pylori* infection can induce direct changes on the DNA of the host, such as oxidative damage, methylation, chromosomal instability, microsatellite instability, and mutations. It's very interesting that *H. pylori* infection can generate genetic instability not only in nuclear but also in mitochondrial DNA.

Based on the analyzed literature we can declare that *H. pylori* infection promotes gastric carcinogenesis by at least three different mechanisms: (I) a combination of increased endogenous DNA damage and decreased repair activities, (II) induction of mutations in the mitochondrial DNA, and (III) generation of a transient mutator phenotype that induces mutations in the nuclear genome.

In addition, *H. pylori* can directly help the gastric carcinogenesis by some component such as CagA, that as mentioned previously in detail, may not promote cancer itself, but the exposure to them and inflammatory insults may select for heritable host cell changes (genetic or epigenetic) that together contribute to cancer progression. Several epidemiological studies have evidenced that another *H. pylori* protein, vacA, is linking to gastric cancer. *H. pylori* strains that express forms of vacA that are active *in vitro* are associated with a higher risk of gastric cancer than the strains that express inactive forms of vacA and this relationship is consistent with studies that have examined the distribution of *vacA* genotypes throughout the world; in the regions that have high rate of distal gastric cancer, such as Colombia and Japan, most *H. pylori* strains contain *vacA* s1 and m1 alleles.

Host factors in particular, genetic polymorphisms in the adaptive and innate immune response genes seem to increase the risk of gastric cancer, largely through induction of severe gastritis, which progresses to atrophy and hypochlorhydria. As noted previously, the most relevant and consistent genetic factors, among those considered and analyzed in the literature, uncovered thus far are in the IL-1 and TNF-A gene clusters. These cytokines appear to play a key role in the pathophysiology of gastric cancer and their roles have been confirmed in animal models that mimic human gastric neoplasia.

Furthermore, the cytolytic and helper effector functions of gastric *H. pylori* specific T cells are extremely different between patients with autoimmune gastritis or MALT lymphoma. In some patients, due to genetic and environmental factors not yet fully elucidated, *H. pylori* infection triggers an abnormal activation at gastric level of cytotoxic, and proapoptotic cross-reactive T cells leading to gastric autoimmunity via molecular mimicry. Conversely in a minority of infected patients, *H. pylori* is able to induce the development of specific T cells defective of both perforin- and Fas ligand-mediated cytotoxicity, which consequently promotes both B-cell overgrowth and exhaustive B-cell proliferation, finally leading to the onset of low-grade gastric MALT lymphoma.

In conclusion, future research must focus on defining a more comprehensive genetic profile (human and bacterial) that better predicts the clinical outcome of *H. pylori* infection, including gastric cancer and gastric lymphoma. Besides, the delineation of pathways activated by *H. pylori*-human interactions will not only improve our understanding of gastric carcinogenesis and lymphoma genesis, but will also facilitate identification of potential therapeutic targets for prevention and more effective treatment of these malignant diseases.

Acknowledgment The authors thank Ente Cassa di Risparmio di Firenze and the Italian Ministry of University and Research, for support of their studies, and Dr. Chiara Della Bella for the artwork.

Financial & Competing Interests Disclosure

Mario M. D'Elios and Amedeo Amedei are applicants of the EU Patent 05425666.4 for HP-NAP as a potential therapeutic agent in cancer, asthma, allergic and infectious diseases. The authors have no other relevant affiliations or financial involvement with any organization or entity with a financial interest in or financial conflict with the subject matter.

References

Algood HM, Torres VJ, Unutmaz D et al (2007) Resistance of primary murine CD4+ T cells to *Helicobacter pylori* vacuolating cytotoxin. Infect Immun 75(1):334–341

Allison CC, Kufer TA, Kremmer E et al (2009) *Helicobacter pylori* induces MAPK phosphorylation and AP-1 activation via a NOD1- dependent mechanism. J Immunol 183(12):8099–8109

Amedei A, Bergman MP, Appelmelk BJ et al (2003) Molecular mimicry between Helicobacter pylori antigens and H+, K+-adenosine triphosphatase in human gastric autoimmunity. J Exp Med 198(8):1147–1156

Amedei A, Cappon A, Codolo G et al (2006) The neutrophil-activating protein of *Helicobacter pylori* promotes Th1 immune responses. J Clin Invest 116(4):1092–1101

Amedei A, Niccolai E, Della Bella C et al (2009) Characterization of tumor antigen peptide-specific T cells isolated from the neoplastic tissue of patients with gastric adenocarcinoma. Cancer Immunol Immunother 58(11):1819–1830

Amieva MR, El-Omar EM (2008) Host–bacterial interactions in *Helicobacter pylori* infection. Gastroenterology 134(1):306–323

Amieva MR, Volgemann R, Covacci A et al (2003) Disruption of the epithelial apical-junctional complex by *Helicobacter pylori* CagA. Science 300(5624):1430–1434

Argent RH, Kidd M, Owen RJ et al (2004) Determinants and consequences of different levels of CagA phosphorylation for clinical isolates of *Helicobacter pylori*. Gastroenterology 127(2):514–523

Argent RH, Hale JL, El-Omar EM et al (2008) Differences in *Helicobacter pylori* CagA tyrosine phosphorylation motif patterns between western and East Asian strains, and influences on interleukin-8 secretion. J Med Microbiol 57(Pt 9):1062–1067

Armitage JO (1997) A clinical evaluation of the International Lymphoma Study Group classification of non-Hodgkin's lymphoma. The Non-Hodgkin's Lymphoma Classification Project. Blood 89(11):3909–3918

Atherton JC, Blaser MJ (2009) Coadaptation of *Helicobacter pylori* and humans: ancient history, modern implications. J Clin Invest 119(9):2475–2487

Baccarelli A, Hou L, Chen J et al (2006) Mannose-binding lectin-2 genetic variation and stomach cancer risk. Int J Cancer 119(8):1970–1975

Backert S, Selbach M (2008) Role of type IV secretion in *Helicobacter pylori* pathogenesis. Cell Microbiol 10(8):1573–1581

Bamford KB, Fan X, Crowe SE et al (1998) Lymphocytes in the human gastric mucosa during *Helicobacter pylori* have a T helper cell 1 phenotype. Gastroenterology 114(3):482–492

Basso D, Zambon CF, Letley DP et al (2008) Clinical relevance of *Helicobacter pylori* cagA and vacA gene polymorphisms. Gastroenterology 135(1):91–99

Basu S, Pathak SK, Chatterjee G et al (2008) *Helicobacter pylori* protein HP0175 transactivates epidermal growth factor receptor through TLR4 in gastric epithelial cells. J Biol Chem 283(47):32369–32376

Bauer B, Bartfeld S, Meyer TF (2009) *H. pylori* selectively blocks EGFR endocytosis via the non-receptor kinase c-Abl and Cag A. Cell Microbiol 11(1):156–169

Bergman MP, D'Elios MM (2010) Cytotoxic T cells in *H. pylori*-related gastric autoimmunity and gastric lymphoma. J Biomed Biotechnol 2010:104918

Bidwell J, Keen L, Gallagher G et al (2001) Cytokine gene polymorphism in human disease: on-line databases, supplement 1. Genes Immun 2(2):61–70

Boncristiano M, Paccani SR, Barone S et al (2003) The *Helicobacter pylori* vacuolating toxin inhibits T-cell activation by two independent mechanisms. J Exp Med 198(12):1887–1897

Bourzac KM, Guillemin K (2005) *Helicobacter pylori*-host cell interactions mediated by type IV secretion. Cell Microbiol 7(7):911–919

Brandt S, Kwok T, Hartig R et al (2005) NF-kappaB activation and potentiation of proinflammatory responses by the *Helicobacter pylori* CagA protein. Proc Natl Acad Sci USA 102(26):9300–9305

Canedo P, Corso G, Pereira F et al (2008a) The interferon gamma receptor 1 (IFNGR1) –56 C/T gene polymorphism is associated with increased risk of early gastric carcinoma. Gut 57(11):1504–1508

Canedo P, Duraes C, Pereira F et al (2008b) Tumor necrosis factor alpha extended haplotypes and risk of gastric carcinoma. Cancer Epidemiol Biomarkers Prev 17(9):2416–2420

Cheng XX, Sun Y, Chen XY et al (2004) Frequent translocalization of beta-catenin in gastric cancers and its relevance to tumor progression. Oncol Rep 11(6):1201–1207

Clementi R, Locatelli F, Dupre L et al (2005) A proportion of patients with lymphoma may harbor mutations of the perforin gene. Blood 105(11):4424–4428

Correa P (1992) Human gastric carcinogenesis: a multistep and multifactorial process - first American Cancer Society award lecture on cancer epidemiology and prevention. Cancer Res 52(24):6735–6740

Cover TL, Blanke SR (2005) *Helicobacter pylori* VacA, a paradigm for toxin multifunctionality. Nat Rev Microbiol 3(4):320–332

Cover TL, Krishna US, Israel DA et al (2003) Induction of gastric epithelial cell apoptosis by *Helicobacter pylori* vacuolating cytotoxin. Cancer Res 63(5):951–957

Cunningham D, Allum WH, Stenning SP et al (2006) Perioperative chemotherapy versus surgery alone for resectable gastroesophageal cancer. N Engl J Med 355(1):11–20

D'Elios MM, Manghetti M, de Carli M et al (1997a) T helper 1 effector cells specific for *Helicobacter pylori* in the gastric antrum of patients with peptic ulcer disease. J Immunol 158(2):962–967

D'Elios MM, Manghetti M, Almerigogna F et al (1997b) Different cytokine profile and antigen-specificity repertoire in *Helicobacter pylori*-specific T cell clones from the antrum of chronic gastritis patients with or without peptic ulcer. Eur J Immunol 27(7):1751–1755

D'Elios MM, Amedei A, Manghetti M et al (1999) Impaired T-cell regulation of B-cell growth in *Helicobacter pylori* - related gastric low-grade MALT lymphoma. Gastroenterology 117(5):1105–1112

D'Elios MM, Amedei A, Del Prete G (2003) *Helicobacter pylori* antigen-specific T-cell responses at gastric level in chronic gastritis, peptic ulcer, gastric cancer and low-grade mucosa associated lymphoid tissue (MALT) lymphoma. Microbes Infect 5(8):723–730

D'Elios MM, Appelmelk BJ, Amedei A et al (2004) Gastric autoimmunity: the role of *Helicobacter pylori* and molecular mimicry. Trends Mol Med 10(7):316–323

D'Elios MM, Amedei A, Benagiano M et al (2005) *Helicobacter pylori*, T cells and cytokines: the "dangerous liaisons". FEMS Immunol Med Microbiol 44(2):113–119

Danon SJ, O'Rourke JL, Moss ND et al (1995) The importance of local acid production in the distribution of *Helicobacter felis* in the mouse stomach. Gastroenterology 108(5):1386–1395

Davidson WF, Giese T, Fredrickson TN (1998) Spontaneous development of plasmacytoid tumors in mice with defective Fas-Fas ligand interactions. J Exp Med 187(11):1825–1838

De Bont R, van Larabeke N (2004) Endogenous DNA damage in humans: a review of quantitative data. Mutagenesis 19(3):169–185

de Jonge R, Kuipers EJ, Langeveld SC et al (2004) The *Helicobacter pylori* plasticity region locus *jhp0947–jhp0949* is associated with duodenal ulcer disease and interleukin-12 production in monocyte cells. FEMS Immunol Med Microbiol 41(2):161–167

Del Giudice G, Covacci A, Telford JL et al (2001) The design of vaccines against *Helicobacter pylori* and their development. Annu Rev Immunol 19:523–563

Deml L, Aigner M, Decker J et al (2005) Characterization of the *Helicobacter pylori* cysteine-rich protein A as a T-helper cell type 1 polarizing agent. Infect Immun 73(8):4732–4742

Dinarello CA (1996) Biologic basis for interleukin-1 in disease. Blood 87(6):2095–2147

Dittmar T, Seidel J, Zaenker KS et al (2006) Carcinogenesis driven by bone marrow-derived stem cells. Contrib Microbiol 13:156–169

Du MQ (2007) MALT lymphoma: recent advances in aetiology and molecular genetics. J Clin Exp Hematop 47(2):31–42

Elkord E, Hawkins RE, Stern PL (2008) Immunotherapy for gastrointestinal cancer: current status and strategies for improving efficacy. Expert Opin Biol Ther 8(4):385–395

El-Omar EM (2001) The importance of interleukin 1beta in *Helicobacter pylori* associated disease. Gut 48(6):743–747

El-Omar EM, Penman ID, Ardill JE et al (1995) *Helicobacter pylori* infection and abnormalities of acid secretion in patients with duodenal ulcer disease. Gastroenterology 109(3):681–691

El-Omar EM, Oien K, El Nujumi A et al (1997) *Helicobacter pylori* infection and chronic gastric acid hyposecretion. Gastroenterology 113(1):15–24

El-Omar EM, Carrington M, Chow WH et al (2000) Interleukin-1 polymorphisms associated with increased risk of gastric cancer. Nature 404(6776):398–402

El-Omar EM, Rabkin CS, Gammon MD et al (2003) Increased risk of noncardia gastric cancer associated with proinflammatory cytokine gene polymorphisms. Gastroenterology 124(5):1193–1201

Figueiredo C, Machado JC, Pharoah P et al (2002) *Helicobacter pylori* and interleukin 1 genotyping: an opportunity to identify high-risk individuals for gastric carcinoma. J Natl Cancer Inst 94(22):1680–1687

Fortini P, Pascucci B, Parlanti E et al (2003) The base excision repair: mechanisms and its relevance for cancer susceptibility. Biochimie 85(11):1053–1071

Fox JG, Beck P, Dangler CA et al (2000) Concurrent enteric helminth infection modulates inflammation and gastric immune responses and reduces Helicobacter-induced gastric atrophy. Nat Med 6(5):536–542

Franco AT, Israel DA, Washingthon MK et al (2005) Activation of β-catenin by carcinogenic *Helicobacter pylori*. Proc Natl Acad Sci USA 102(30):10646–10651

Franco AT, Friedman DB, Nagy TA et al (2009) Delineation of a carcinogenic *Helicobacter pylori* proteome. Mol Cell Proteomics 8(8):1947–1958

Furuta T, El-Omar EM, Xiao F et al (2002) Interleukin 1beta polymorphisms increase risk of hypochlorhydria and atrophic gastritis and reduce risk of duodenal ulcer recurrence in Japan. Gastroenterology 123(1):92–105

Gebert B, Fischer W, Weiss E et al (2003) *Helicobacter pylori* vacuolating cytotoxin inhibits T lymphocyte activation. Science 301(5636):1099–1102

Gerhard M, Lehn N, Neumayer N et al (1999) Clinical relevance of the *Helicobacter pylori* gene for blood-group antigen binding adhesin. Proc Natl Acad Sci USA 96(22):12778–12783

Glassner BJ, Rasmussen LJ, Najarian MT et al (1998) Generation of a strong mutator phenotype in yeast by imbalanced base excision repair. Proc Natl Acad Sci USA 95(17):9997–10002

Greiner A, Knörr C, Qin Y et al (1997) Low-grade B cell lymphomas of mucosa-associated lymphoid tissue (MALT-type) require CD40-mediated signaling and Th2-type cytokines for in vitro growth and differentiation. Am J Pathol 150(5):1583–1593

Guillet M, Boiteux S (2002) Endogenous DNA abasic sites cause cell death in the absence of Apn1, Apn2 and Rad1/Rad10 in *Saccharomyces cerevisiae*. EMBO J 21(11):2833–2841

Guiney DG, Hasegawa P, Cole SP (2003) *Helicobacter pylori* preferentially induces interleukin 12 (IL-12) rather than IL-6 or IL-10 in human dendritic cells. Infect Immun 71(7):4163–4166

Han CB, Li F, Zhao YJ et al (2003) Variations of mitochondrial D-loop region plus downstream gene 12 S rRNA–tRNA(phe) and gastric carcinomas. World J Gastroenterol 9(9):1925–1929

Hatakeyama M (2006) Helicobacter pylori CagA – A bacterial intruder conspiring gastric carcinogenesis. Int J Cancer 119(6):1217–1223

Higashi H, Tsutsumi R, Fujita A et al (2002a) Biological activity of the *Helicobacter pylori* virulence factor CagA is determined by variation in the tyrosine phosphorylation sites. Proc Natl Acad Sci USA 99(22):14428–14433

Higashi H, Tsutsumi R, Muto S et al (2002b) SHP-2 tyrosine phosphatase as an intracellular target of *Helicobacter pylori* CagA protein. Science 295(5555):683–686

Hiyama T, Tanaka S, Shima H et al (2003) Somatic mutation of mitochondrial DNA in *Helicobacter pylori*-associated chronic gastritis in patients with and without gastric cancer. Int J Mol Med 12(2):169–174

Hold GL, Rabkin CS, Chow WH et al (2007) A functional polymorphism of toll-like receptor 4 gene increases risk of gastric carcinoma and its precursors. Gastroenterology 132(3): 905–912

Hou L, El-Omar EM, Chen J et al (2007) Polymorphisms in Th1-type cell-mediated response genes and risk of gastric cancer. Carcinogenesis 28(1):118–123

Houghton J, Stoicov C, Nomura S et al (2004) Gastric cancer originating from bone marrow-derived cells. Science 306(5701):1568–1571

Howley PM, Livingston DM (2009) Small DNA tumor viruses: large contributors to biomedical sciences. Virology 384(2):256–259

Hussell T, Isaacson PG, Crabtree JE et al (1993) The response of cells from low-grade B-cell gastric lymphomas of mucosa-associated lymphoid tissue to *Helicobacter pylori*. Lancet 342(8871):571–574

Hussell T, Isaacson PG, Crabtree JE et al (1996) *Helicobacter pylori*-specific tumour-infiltrating T cells provide contact dependent help for the growth of malignant B cells in low-grade gastric lymphoma of mucosa-associated lymphoid tissue. J Pathol 178(2):122–127

IARC Working Group (1994) IARC working group on the evaluation of carcinogenic risks to humans: some industrial chemicals Lyon, 15–22 February 1994. IARC Monogr Eval Carcinog Risks Hum 60:1–560

Isaacson P, Wright DH (1983) Malignant lymphoma of mucosa-associated lymphoid tissue. A distinctive type of B-cell lymphoma. Cancer 52(8):1410–1416

Israel DA, Peek RM (2001) Pathogenesis of *Helicobacter pylori*-induced gastric inflammation. Aliment Pharmacol Ther 15(9):1271–1290

Keates S, Keates AC, Nath S et al (2005) Transactivation of the epidermal growth factor receptor by *cag+ Helicobacter pylori* induces up-regulation of the early growth response gene Egr-1 in gastric epithelial cells. Gut 54(10):1363–1369

Khanzode SS, Khanzode SD, Dakhale GN (2003) Serum and plasma concentration of oxidant and antioxidants in patients of *Helicobacter pylori* gastritis and its correlation with gastric cancer. Cancer Lett 195(1):27–31

Kim JJ, Tao H, Carloni E et al (2002) *Helicobacter pylori* impairs DNA mismatch repair in gastric epithelial cells. Gastroenterology 123(2):542–553

Kuipers EJ, Lundell L, Klinkenberg-Knol EC et al (1996) Atrophic gastritis and *Helicobacter pylori* infection in patients with reflux esophagitis treated with omeprazole or fundoplication. N Engl J Med 334(16):1018–1022

Kusters JG, van Vliet AH, Kuipers EJ (2006) Pathogenesis of *Helicobacter pylori* infection. Clin Microbiol Rev 19(3):449–490

Kwok T, Zabler D, Urman S et al (2007) *Helicobacter* exploits integrin for type IV secretion and kinase activation. Nature 449(7164):862–866

Lee S, Shin MG, Jo WH et al (2007) Association between *Helicobacter pylori*-related peptic ulcer tissue and somatic mitochondrial DNA mutations. Clin Chem 53(7):1390–1392

Lehmann FS, Terracciano L, Carena I et al (2002) *In situ* correlation of cytokine secretion and apoptosis in *Helicobacter pylori* associated gastritis. Am J Physiol Gastrointest Liver Physiol 283(2):481–488

Lehours P, Dupouy S, Bergey B et al (2004) Identification of a genetic marker of *Helicobacter pylori* strains involved in gastric extranodal marginal zone B cell lymphoma of the MALT-type. Gut 53(7):931–937

Lehours P, Zheng Z, Skoglund A et al (2009) Is there a link between the lipopolysaccharide of *Helicobacter pylori* gastric MALT lymphoma associated strains and lymphoma pathogenesis? PLoS One 4(10):e7297

Leung WK, Man EP, Yu J et al (2006) Effects of *Helicobacter pylori* eradication on methylation status of E-cadherin gene in noncancerous stomach. Clin Cancer Res 12(10):3216–3221

Li GM (2003) DNA mismatch repair and cancer. Front Biosci 8:997–1017

Liakakos T, Roukos DH (2008) More controversy than ever - challenges and promises towards personalized treatment of gastric cancer. Ann Surg Oncol 15(4):956–960

Liu J, Xiang Z, Ma X (2004) Role of IFN regulatory factor-1 and IL-12 in immunological resistance to pathogenesis of N-methyl-N-nitrosourea-induced T lymphoma. J Immunol 173(2):1184–1193

Machado JC, Pharoah P, Sousa S et al (2001) Interleukin 1B and interleukin 1RN polymorphisms are associated with increased risk of gastric carcinoma. Gastroenterology 121(4):823–829

Machado JC, Figueiredo C, Canedo P et al (2003) A proinflammatory genetic profile increases the risk for chronic atrophic gastritis and gastric carcinoma. Gastroenterology 125(2):364–371

Machado AM, Figueiredo C, Touati E et al (2009) *Helicobacter pylori* infection induces genetic instability of nuclear and mitochondrial DNA in gastric cells. Clin Cancer Res 15(9):2995–3002

Maeda S, Yoshida H, Mitsuno Y et al (2002) Analysis of apoptotic and anti apoptotic signalling pathways induced by *Helicobacter pylori*. Gut 50(6):771–778

Maeda N, Fan H, Yoshikai Y (2008) Oncogenesis by retroviruses: old and new paradigms. Rev Med Virol 18(6):387–405

Mahajan R, El-Omar EM, Lissowska J et al (2008) Genetic variants in T helper cell type 1, 2 and 3 pathways and gastric cancer risk in a Polish population. Jpn J Clin Oncol 38(9):626–633

Mahdavi J, Sondén B, Hurtig M et al (2002) *Helicobacter pylori* SabA adhesin in persistent infection and chronic inflammation. Science 297(5581):573–578

Marchall BJ, Windsor HM (2005) The relation of *Helicobacter pylori* to gastric adenocarcinoma and lymphoma: pathophysiology, epidemiology, screening, clinical presentation, treatment, and prevention. Med Clin North Am 89(2):313–344

Marnett LJ (2000) Oxyradicals and DNA damage. Carcinogenesis 21(3):361–370

Marshall BJ, Warren JR (1984) Unidentified curved bacilli in the stomach of patients with gastritis and peptic ulceration. Lancet 1(8390):1311–1315

Marshall AC, Alderuccio F, Toh BH (2002) Fas/CD95 is required for gastric mucosal damage in autoimmune gastritis. Gastroenterology 123(3):780–789

Mattapallil JJ, Dandekar S, Canfield DR et al (2000) A predominant Th1 type of immune response is induced early during acute *Helicobacter pylori* infection in rhesus macaques. Gastroenterology 118(2):307–315

McClain MS, Shaffer CL, Israel DA et al (2009) Genome sequence analysis of *Helicobacter pylori* strains associated with gastric ulceration and gastric cancer. BMC Genomics 10:3

Miehlke S, Kirsch C, Agha-Amiri K et al (2000) The *Helicobacter pylori vacA* s1, m1 genotype and *cagA* is associated with gastric carcinoma in Germany. Int J Cancer 87(3):322–327

Mitra-Kaushik S, Harding J, Hess J et al (2004) Enhanced tumorigenesis in HTLV-1 tax-transgenic mice deficient in interferon-gamma. Blood 104(10):3305–3311

Mohammadi M, Nedrud J, Redline R et al (1997) Murine CD4 T-cell response to *Helicobacter* infection: TH1 cells enhance gastritis and TH2 cells reduce bacterial load. Gastroenterology 113(6):1848–1857

Molinari M, Salio M, Galli C et al (1998) Selective inhibition of Ii-dependent antigen presentation by *Helicobacter pylori* toxin VacA. J Exp Med 187(1):135–140

Monteiro MA, Zheng P, Ho B et al (2000) Expression of histo-blood group antigens by lipopolysaccharides *of Helicobacter pylori* strains from Asian hosts: the propensity to express type 1 blood-group antigens. Glycobiology 10(7):701–713

Murata-Kamiya N, Kurashima Y, Teishikata Y et al (2007) *Helicobacter pylori* CagA interacts with E-cadherin and deregulates the beta-catenin signal that promotes intestinal transdifferentiation in gastric epithelial cells. Oncogene 26(32):4617–4626

Nagy TA, Frey MR, Yan F et al (2009) *Helicobacter pylori* regulates cellular migration and apoptosis by activation of phosphatidylinositol 3-kinase signaling. J Infect Dis 199(5): 641–651

Nakayama M, Hisatsune J, Yamasaki E et al (2009) *Helicobacter pylori* VacA induced inhibition of GSK3 through the PI3K/Akt signaling pathway. J Biol Chem 284(3):1612–1619

Obst B, Wagner S, Sewing KF et al (2000) *Helicobacter pylori* causes DNA damage in gastric epithelial cells. Carcinogenesis 21(6):1111–1115

Ocean AJ, Schnoll-Sussman F, Keresztes R et al (2006) Phase II study of PS-341 (bortezomib) with or without irinotecan in patients (pts) with advanced gastric adenocarcinomas (AGA). J Clin Oncol 24(18):14040

Odenbreit S, Püls J, Sedlmaier B et al (2000) Translocation of *Helicobacter pylori* CagA into gastric epithelial cells by type IV secretion. Science 287(5457):1497–1500

Oh JD, Kling-Bäckhed H, Giannakis M et al (2006) The complete genome sequence of a chronic atrophic gastritis *Helicobacter pylori* strain: evolution during disease progression. Proc Natl Acad Sci USA 103(26):9999–10004

Ohnishi N, Yuasa H, Tanaka S et al (2008) Transgenic expression of *Helicobacter pylori* CagA induces gastrointestinal and hematopoietic neoplasms in mouse. Proc Natl Acad Sci USA 105(3):1003–1008

Palli D, Saieva C, Luzzi I et al (2005) Interleukin-1 gene polymorphisms and gastric cancer risk in a high-risk Italian population. Am J Gastroenterol 100(9):1941–1948

Park DI, Park SH, Kim SH et al (2005) Effect of *Helicobacter pylori* infection on the expression of DNA mismatch repair protein. Helicobacter 10(3):179–184

Parkin DM (2006) The global health burden of infection-associated cancers in the year 2002. Int J Cancer 118(12):3030–3044

Pece S, Gutkind JS (2000) Signaling from E-cadherins to the MAPK pathway by the recruitment and activation of epidermal growth factor receptors upon cell-cell contact formation. J Biol Chem 275(52):41227–41233

Peek RM Jr, Blaser MJ (2002) *Helicobacter pylori* and gastrointestinal tract adenocarcinomas. Nat Rev Cancer 2(1):28–37

Peek RM Jr, Wirth HP, Moss SF et al (2000) *Helicobacter pylori* alters gastric epithelial cell cycle events and gastrin secretion in Mongolian gerbils. Gastroenterology 118(1):48–59

Perri F, Cotugno R, Piepoli A et al (2007) Aberrant DNA methylation in non-neoplastic gastric mucosa of *H. pylori* infected patients and effect of eradication. Am J Gastroenterol 102(7):1361–1371

Pohl MA, Romero-Gallo J, Guruge JL et al (2009) Host-dependent Lewis (Le) antigen expression in *Helicobacter pylori* cells recovered from Leb-transgenic mice. J Exp Med 206(13): 3061–3072

Polk DB, Peek RM Jr (2010) *Helicobacter pylori*: gastric cancer and beyond. Nat Rev Cancer 10(6):403–414

Rad R, Prinz C, Neu B et al (2003) Synergistic effect of *Helicobacter pylori* virulence factors and interleukin-1 polymorphisms for the development of severe histological changes in the gastric mucosa. J Infect Dis 188(2):272–281

Rhead JL, Letley DP, Mohammadi M et al (2007) A new *Helicobacter pylori* vacuolating cytotoxin determinant, the intermediate region, is associated with gastric cancer. Gastroenterology 133:926–936

Rizvi AZ, Swain JR, Davies PS et al (2006) Bone marrow-derived cells fuse with normal and transformed intestinal stem cells. Proc Natl Acad Sci USA 103(16):6321–6325

Robert A, Olafsson AS, Lancaster C et al (1991) Interleukin-1 is cytoprotective, antisecretory, stimulates PGE2 synthesis by the stomach, and retards gastric emptying. Life Sci 48(2):123–134

Romano M, Ricci V, Di Popolo A et al (1998) *Helicobacter pylori* upregulates expression of epidermal growth factor-related peptides, but inhibits their proliferative effect in MKN 28 gastric mucosal cells. J Clin Invest 101(8):1604–1613

Rossi G, Fortuna D, Pancotto L et al (2000) Immunohistochemical study of the lymphocyte populations infiltrating the gastric mucosa of beagle dogs experimentally infected with *Helicobacter pylori*. Infect Immun 68(8):4769–4772

Roukos DH, Kappas AM (2005) Perspectives in the treatment of gastric cancer. Nat Clin Pract Oncol 2(2):98–107

Saikawa Y, Kitajima M (2009) Gastrointestinal. In: Bland KI, Sarr MG, Buchler MW, Csendes A, Garden OJ, Wong J (eds) General surgery, principles and international practice, 2nd edn. Springer, New York/Berlin, pp 567–576

Santini D, Angeletti S, Ruzzo A et al (2008) Toll-like receptor 4 Asp299Gly and Thr399Ile polymorphisms in gastric cancer of intestinal and diffuse histotypes. Clin Exp Immunol 154(3):360–364

Segal ED, Lange C, Covacci A et al (1997) Induction of host signal transduction pathways by *Helicobacter pylori*. Proc Natl Acad Sci USA 94(14):7595–7599

Shah MA, Ramanathan RK, Ilson DH et al (2006) Multicenter phase II study of irinotecan, cisplatin, and bevacizumab in patients with metastatic gastric or gastroesophageal junction adenocarcinoma. J Clin Oncol 24(33):5201–5206

Smith MG, Hold GL, Rabkin CS et al (2004) The IL-8 – 251 promoter polymorphism is associated with high IL-8 production, severe inflammation and increased risk of pre-malignant changes in *H. pylori* positive subjects. Gastroenterology 126:A23

Smyth MJ, Thia KY, Street SE et al (2000) Perforin-mediated cytotoxicity is critical for surveillance of spontaneous lymphoma. J Exp Med 192(5):755–760

Solnick JV, Hansen LM, Salama NR et al (2004) Modification of *Helicobacter pylori* outer membrane protein expression during experimental infection of rhesus macaques. Proc Natl Acad Sci USA 101(7):2106–2111

Sommer F, Faller G, Konturek P et al (1998) Antrum and corpus mucosa-infiltrating $CD4^+$ lymphocytes in *Helicobacter pylori* gastritis display a Th1 phenotype. Infect Immun 66(11):5543–5546

Stern LL, Mason JB, Selhub J et al (2000) Genomics DNA hypomethylation, a characteristic of most cancers, is present in peripheral leukocytes of individuals who are homozygous for the C677T polymorphism in the methylene tetrahydrofolate reductase gene. Cancer Epidemiol Biomarkers Prev 9(8):849–853

Stoicov C, Fan X, Liu JH et al (2009) T-bet knockout prevents *Helicobacter felis* induced gastric cancer. J Immunol 183(1):642–649

Street SEA, Cretney E, Smyth MJ (2001) Perforin and interferon-γ activities independently control tumor initiation, growth, and metastasis. Blood 97(1):192–197

Street SEA, Trapani JA, MacGregor D et al (2002) Suppression of lymphoma and epithelial malignancies effected by interferon γ. J Exp Med 196(1):129–134

Street SEA, Hayakawa Y, Zhan Y et al (2004) Innate immune surveillance of spontaneous B cell lymphomas by natural killer cells and gamma delta T cells. J Exp Med 199(6):879–884

Su B, Ceponis PJ, Lebel S et al (2003) *Helicobacter pylori* activates Toll-like receptor 4 expression in gastrointestinal epithelial cells. Infect Immun 71(6):3496–3502

Sundrud MS, Torres VJ, Unutmaz D et al (2004) Inhibition of primary human T cell proliferation by *Helicobacter pylori* vacuolating toxin (VacA) is independent of VacA effects on IL-2 secretion. Proc Natl Acad Sci USA 101(20):7727–7732

Sunnarborg SW, Hinkle CL, Stevenson M et al (2002) Tumor necrosis factor-alpha converting enzyme (TACE) regulates epidermal growth factor receptor ligand availability. J Biol Chem 277(15):12838–12845

Suzuki M, Mimuro H, Kiga K et al (2009) *Helicobacter pylori* CagA phosphorylation-independent function in epithelial proliferation and inflammation. Cell Host Microbe 5(1):23–34

Swann JB, Smyth MJ (2007) Immune surveillance of tumors. J Clin Invest 117(5):1137–1146

Taguchi A, Ohmiya N, Shirai K et al (2005) Interleukin-8 promoter polymorphism increases the risk of atrophic gastritis and gastric cancer in Japan. Cancer Epidemiol Biomarkers Prev 14(11):2487–2493

Takaishi S, Okumura T, Wang TC (2008) Gastric cancer stem cells. J Clin Oncol 26(17):2876–2882

Takeshima E, Tomimori K, Teruya H et al (2011) *Helicobacter pylori*-induced interleukin-12 p40 expression. Infect Immun 79(1):546

Taylor JM, Ziman ME, Huff JL et al (2006) *Helicobacter pylori* lipopolysacccharide promotes a Th1 type immune response in immunized mice. Infect Immun 24(23):4987–4994

Telford JL, Ghiara P, Dell'Orco M et al (1994) Gene structure of the *Helicobacter pylori* cytotoxin and evidence of its key role in gastric disease. J Exp Med 179(5):1653–1670

Tomita T, Jackson AM, Hida N et al (2001) Expression of Interleukin-18, a Th1 cytokine, in human gastric mucosa is increased in *Helicobacter pylori* infection. J Infect Dis 183(4):620–627

Touati E, Michel V, Thiberge JM et al (2003) Chronic *Helicobacter pylori* infections induce gastric mutations in mice. Gastroenterology 124(5):1408–1419

Trapani JA, Smyth MJ (2002) Functional significance of the perforin/granzyme cell death pathway. Nat Rev Immunol 2(10):735–747

Tsukashita S, Kushima R, Bamba M et al (2003) Beta-catenin expression in intramucosal neoplastic lesions of the stomach. Comparative analysis of adenoma/dysplasia, adenocarcinoma and signet-ring cell carcinoma. Oncology 64(3):251–258

Tu S, Bhagat G, Cui G et al (2008) Over expression of interleukin-1 [beta] induces gastric inflammation and cancer and mobilizes myeloid-derived suppressor cells in mice. Cancer Cell 14(5):408–419

Tudek B, Boiteux S, Laval J (1992) Biological properties of imidazole ring opened N7- methylguanine in M13mp18 phage DNA. Nucleic Acids Res 20(12):3079–3084

Van Doorn LJ, Figueiredo C, Mégraud F et al (1999) Geographic distribution of *vacA* allelic types of *Helicobacter pylori*. Gastroenterology 116(4):823–830

Vanhoefer U, Tewes M, Rojo F et al (2004) Phase I study of the humanized anti-epidermal growth factor receptor monoclonal antibody EMD72000 in patients with advanced solid tumors that express the epidermal growth factor receptor. J Clin Oncol 22(1):175–184

Viala J, Chaput C, Boneca IG et al (2004) Nod1 responds to peptidoglycan delivered by the *Helicobacter pylori cag* pathogenicity island. Nat Immunol 5(11):1166–1174

Voland P, Zeitner M, Hafsi N et al (2006) Human immune response towards recombinant *Helicobacter pylori* urease and cellular fractions. Vaccine 24(18):3832–3839

Wang X, Willenbring H, Akkari Y et al (2003) Cell fusion is the principal source of bone marrow-derived hepatocytes. Nature 422(6934):897–901

Wang G, Olczak A, Forsberg LS et al (2009) Oxidative stress-induced peptidoglycan deacetylase in *Helicobacter pylori*. J Biol Chem 284(11):6790–6800

Wen S, Felley CP, Bouzourene H et al (2004) Inflammatory gene profiles in gastric mucosa during *Helicobacter pylori* infection. J Immunol 172(4):2595–2606

Wiedemann T, Loell E, Mueller S et al (2009) *Helicobacter pylori* cag-Pathogenicity island-dependent early immunological response triggers later precancerous gastric changes in Mongolian gerbils. PLoS One 4(3):e4754

Wirth HP, Yang M, Peek RM Jr et al (1999) Phenotypic diversity in Lewis expression of *Helicobacter pylori* isolates from the same host. J Lab Clin Med 133(5):488–500

Wong BC, Wang WP, So WH et al (2001) Epidermal growth factor and its receptor in chronic active gastritis and gastroduodenal ulcer before and after Helicobacter pylori eradication. Aliment Pharmacol Ther 15(9):1459–1465

Wong BC, Lam SK, Wong WM et al (2004) *Helicobacter pylori* eradication to prevent gastric cancer in a high-risk region of China: a randomized controlled trial. JAMA 291(2):187–194
Wotherspoon AC, Ortiz-Hidalgo C, Falzon MR et al (1991) *Helicobacter pylori*-associated gastritis and primary B-cell gastric lymphoma. Lancet 338(8776):1175–1176
Wu CW, Yin PH, Hung WY et al (2005) Mitochondrial DNA mutations and mitochondrial DNA depletion in gastric cancer. Genes Chromosomes Cancer 44(1):19–28
Wu JY, Lu H, Sun Y et al (2006) Balance between polyoma enhancing activator 3 and activator protein 1 regulates *Helicobacter pylori*-stimulated matrix metalloproteinase 1 expression. Cancer Res 66(10):5111–5120
Yamaoka Y, Kikuchi S, el-Zimaity HM et al (2002) Importance of *Helicobacter pylori oipA* in clinical presentation, gastric inflammation, and mucosal interleukin 8 production. Gastroenterology 123(2):414–424
Yao JC, Mansfield PF, Pisters PW, Feig BW, Janjan NA, Crane C, Ajani JA (2003) Combined-modality therapy for gastric cancer. Semin Surg Oncol 21(4):223–227
Yao Y, Tao H, Park DI et al (2006) Demonstration and characterization of mutations induced by *Helicobacter pylori* organisms in gastric epithelial cells. Helicobacter 11(4):272–286
Zeng ZR, Hu PJ, Hu S et al (2003) Association of interleukin 1B gene polymorphism and gastric cancers in high and low prevalence regions in China. Gut 52(12):1684–1689
Zerafa N, Westwood JA, Cretney E et al (2005) Cutting edge: TRAIL deficiency accelerates hematological malignancies. J Immunol 175(9):5586–5590

Chapter 3
Streptococcus bovis and Colorectal Cancer

Harold Tjalsma, Annemarie Boleij, and Ikuko Kato

Abstract The most salient feature of *Streptococcus bovis* (SB) is its clinical association with malignancy of the colon and rectum. The relationship between SB and colorectal cancer (CRC) was already recognized in the 1950s and many case reports and retrospective studies on this association have been published since then. SB is an opportunistic pathogen that normally resides asymptomatically in the human intestinal tract. In compromised individuals, however, this bacterium can cause systemic infections most often presenting as bacterial endocarditis. Investigators reported the presence of colorectal tumours in up to 60% of the cases in which a patient was diagnosed with SB endocarditis or bacteremia. Therefore, these infections are nowadays often regarded as indication for full bowel examination in clinical practice. Importantly, recent studies have indicated that the association between *S. gallolyticus* subsp *gallolyticus* (previously called SB biotype I) with CRC seems much more pronounced than that of other known SB biotypes. Nevertheless, the question whether SB has a causal or predominantly incidental involvement with cancer of the colon remains to be answered. Furthermore, still little is known about the precise molecular mechanisms that determine this specific relationship. This chapter aims to summarize the literature on this subject and to illustrate possible mechanisms behind the association of SB with CRC.

H. Tjalsma (✉) • A. Boleij
Department of Laboratory Medicine, Nijmegen Institute for Infection,
Inflammation and Immunity (N4i) & Radboud University Centre for Oncology (RUCO),
Radboud University Nijmegen Medical Centre, P.O. Box 9101,
6500 HB Nijmegen, The Netherlands
e-mail: H.Tjalsma@labgk.umcn.nl

I. Kato
Department of Pathology, Karmanos Cancer Institute, Wayne State University,
Detroit, MI, USA

A.A. Khan (ed.), *Bacteria and Cancer*, DOI 10.1007/978-94-007-2585-0_3,

Keywords Colon • Colorectal cancer • Tumour • Polyp • Carcinoma • Adenoma • Carcinogenesis • *Streptococcus bovis* • *Streptococcus gallolyticus* • Endocarditis • Bacteremia • Inflammation • Intestine • Serology • Diagnosis • Biomarker

Abbreviations

CRC	Colorectal cancer
ELISA	Enzyme-linked immunosorbent assay
HP	*Helicobacter pylori*
IBD	Inflammatory bowel disease
IC-TOF MS	Immunocapture time-of-flight mass spectrometry
MMP	Matrix metalloproteinases
NSAID	Non steroidal anti inflammatory drugs
OR	Odds ratio
PAH	Polycyclic aromatic hydrocarbons
SB	*Streptococcus bovis*
SIC	*S. infantarius* subsp. *coli*
SGG	*S. gallolyticus* subsp. *gallolyticus*
SGM	*S. gallolyticus* subsp. *macedonicus*
SGP	*S. gallolyticus* subsp. *pasteurianus*
SII	*S. infantarius* subsp. *infantarius*

3.1 Colorectal Cancer and Microbial Agents

Colorectal cancer (CRC) is the third most common cancer for men and women in Western society. It is estimated that nearly 150,000 cases were newly diagnosed and 50,000 persons died of this disease in year 2009 in the USA (Horner et al. 2009). The temporal and geographic variations in CRC incidence in US whites and blacks (Horner et al. 2009) and among immigrants (Curado et al. 2007) are best explained by environmental factors rather than genetic predisposition. According to Dr. Parkin's estimate (Parkin 2006), 17.8% of the worldwide cancer incidence is attributable to infectious agents, resulting in approximately 1.9 million cases per year. These include a variety of infectious agents: parasites such as *Schistosoma haematobium* and *Opisthorchis viverrini*, bacteria, such as *Helicobacter pylori* (HP), and, viruses, such as Epstein-Barr virus, Hepatitis virus, and Human herpes, papilloma (HPV), polyma and retro-viruses (IARC 1994; Persing and Prendergast 1999; Del Valle et al. 2002). Several mechanisms have been proposed, including direct effects on host cell proliferation and communication pathways, impairment of host immune system, induction of genomic instability and chronic inflammation (Herrera et al. 2005). Chronic inflammation often accompanies increased host cell turnover, which increases the probability of mutagenic events, and enhanced formation of reactive oxygen and nitrogen species that damage DNA and induce genomic instability (Coussens and Werb 2002; Blaser 2008; Hussain and Harris 2007; Terzić et al. 2010). Thus, inflammatory responses play decisive roles at different stages of tumour development, including initiation, promotion, malignant conversion, invasion, and metastasis (Grivennikov et al. 2007). Genomic instability may arise from inactivation of DNA mismatch repair (MMR) system (MSH1/2), which leads to the development of a specific molecular subtype of CRC termed microsatellite instability high (MSI-H) (Jass 2007). MSI has been observed frequently in long standing ulcerative colitis mucosa (Ishitsuka et al. 2001) as well as in HP-positive gastric cancer (Li et al. 2005) and MSH2 deficient mice are susceptible to inflammation associated colorectal tumours (Kohonen-Corish et al. 2002). In addition, overexpression of a COX-2 receptor protein has been characterized for MSI-H tumours (Baba et al. 2010). The large bowel is indeed the natural habitat for a large, dynamic and highly competitive bacterial community, which is essential for the control of intestinal epithelial homeostasis and human health. Strikingly, the increase in bacterial colonization from the ileum to the colon (six orders of magnitude; Stone and Papas 1997), is paralleled by a marked difference in cancer incidence (by at least a factor of 30) between the small and large intestines. Although bacterial etiologies in sporadic CRC have never been firmly established in humans, studies in germ free mice suggest that intestinal bacteria are indeed required for colorectal carcinogenesis in model systems (Hope et al. 2005; Sinicrope 2007). Finally, there is good evidence that aspirin and non-steroidal anti-inflammatory drugs (NSAIDs) reduce the risk of CRC and its precursor (Rostom et al. 2007; Dubé et al. 2007).

3.2 Microbiological Characteristics of *Streptococcus bovis*

Streptococcus bovis (SB) is a gram-positive bacterium and lower-grade opportunistic pathogen that can cause systemic infections (endocarditis or bacteraemia) in humans. It is a group D streptococcus with the specific ability to grow in 40% bile (Moellering et al. 1974; Roberts 1992). The classification and identification of SB has been problematic for a long time. Based on phenotypic diversity, three SB biotypes (I, II/1, and II/2) have been reported. Recently, based on biochemical traits, DNA homology and divergence in 16S rRNA sequences, Schlegel et al. (2004) suggested to rename SB I into *S. gallolyticus* subsp. *gallolyticus* (SGG), to divide SB biotype II/1 strains into (i) *S. infantarius* subsp. *coli* (SIC) and (ii) *S. infantarius* subsp. *infantarius* (SII) and to rename SB II/2 into *S. gallolyticus* subsp. *pasteurianus* (SGP; see Table 3.1). In addition, the closely related non-pathogenic strain *Streptococcus macedonicus* was reclassified as *S. gallolyticus* subsp. *macedonicus* (SGM). Earlier studies suggest that SGG and SII are the most commonly isolated pathogens from the SB group, with the former being the more virulent in humans and more often associated with endocarditis (Corredoira et al. 2008a). In a recent reexamination of SB bacteremias in a 20-year period in France, the association of colon tumours with SGG was found to be ~50% versus 11% for SII. Strikingly,

Table 3.1 Reappraisal of SB nomenclature

New name	Old name	Association with CRC
S. gallolyticus subsp. *gallolyticus* (SGG)	*S. bovis* biotype I	++++
S. infantarius subsp. *infantarius* (SII)	*S. bovis* biotype II/1	++
S. infantarius subsp. *coli* (SIC)	*S. bovis* biotype II/1	++
S. gallolyticus subsp. *pasteurianus* (SGP)	*S. bovis* biotype II/2	+
S. gallolyticus subsp. *macedonicus* (SGM)	*S. macedonicus*	–

however, for non-colonic cancer the association was 6% for SGG versus 57% for SII. Most of the non-colonic cancers associated with SII were of the pancreas and biliary tract (Corredoira et al. 2008b). Because of the lack of unified terminology information in literature, we refer to both SGG and SII as SB in the rest of this chapter.

3.3 Association of SB with Colorectal Disease

Although SB is a member of normal gastrointestinal flora in ruminants, e.g., cattle, sheep, horses, pigs, camels and deers (Ghali et al. 2004), it can also found in human feces as well as gastric biopsy materials (Schlegel et al. 2000; Ribeiro et al. 2004). Approximately 10% of healthy individuals have been estimated to carry this bacterium asymptomatically in their digestive tract (Schlegel et al. 2000). While fecal-oral or oral-oral is a possible transmission route between humans, it may be acquired through dietary intake of ruminant-derived foods, such as unpasteurized dairy products (Randazzo et al. 2006), red meat and animal organs (Schlegel et al. 2000). In fact SB is a frequently detected contaminant in commercially available meat (Knudtson and Hartman 1993; Thian and Hartman 1981). The correlation between SB and colonic disease has long been recognized. Besides case-reports for the patients who were diagnosed with asymptomatic colorectal neoplasia simultaneously with SB endocarditis or bacteremia (McMahon et al. 1991; Nielsen et al. 2007; Wentling et al. 2006; Gupta et al. 2010; Kahveci et al. 2010; Kim et al. 2010), investigators have reported increased prevalence of colorectal tumours (cancer and polyps) among patients diagnosed with SB endocarditis or bacteremia. As summarized in the Table 3.2, the prevalence of colorectal tumours ranges from 10% to 60% (Corredoira et al. 2005, 2008a; Murray and Roberts 1978; Klein et al. 1979; Reynolds et al. 1983; Pigrau et al. 1988; Ruoff et al. 1989; Clarridge et al. 2001; Gonzlez-Quintela et al. 2001; Gold et al. 2004; Lee et al. 2003; Zarkin et al. 1990; Jean et al. 2004; Alazmi et al. 2006; Giannitsioti et al. 2007; Beck et al. 2008; Vaska and Faoagali 2009), although these are based on diverse study populations in terms of patient demographics and colorectal surveillance methods. These variations may also be due to the heterogenous definition of the cases, as adenomas have been defined as diseased in some studies but not in others (Boleij et al. 2009b). None of these studies, however, have evaluated their results in comparison with expected frequencies in the general population. The second set of evidence is derived from studies comparing

Table 3.2 Summary of studies among SB bacteremia patients

Author (year)	Study location	Patients (n)	Detected colorectal adenomas and carcinomas n	%
Murray and Roberts (1978)	USA	36	4	11%
Klein et al. 1979	USA	29	15	52%
Reynolds et al. (1983)	USA	19	7	37%
Pigrau et al. (1988)	Spain	16	1	6%
Ruoff et al. (1989)	USA	38	15	39%
Zarkin et al. (1990)	USA	43	16	37%
Clarridge et al. (2001)	USA	12	1	8%
Gonzlez-Quintela et al. (2001)	Spain	20	6	30%
Lee et al. (2003)	Hong Kong	37	4	11%
Gold et al. (2004)	USA	45	17	38%
Jean et al. (2004)	Taiwan	19	9	47%
Corredoira et al. (2005)	Spain	124	54	44%
Alazmi et al. (2006)	USA	46	6	13%
Giannitsioti et al. (2007)[a]	France	142	70	49%
Corredoira et al. (2008a)	Spain	107	42	40%
Beck et al. (2008)	Germany	15	7	47%
Vaska and Faoagali (2009)	Australia	20	12	60%

[a]Include other benign lesions e.g., diverticulosis and colitis

SB prevalence among various patient groups with or without colonic diseases (Table 3.3) (Klein et al. 1977; Burns et al. 1985; Darjee and Gibb 1993; Dubrow et al. 1991; Potter et al. 1998; Teitelbaum and Triantafyllopoulou 2006; Tjalsma et al. 2006; Abdulamir et al. 2009). While three small studies including 13–46 controls and corresponding 11 CRC, 47 pediatric inflammatory bowel disease (IBD) and 56 polyp patients failed to show any association (Dubrow et al. 1991; Potter et al. 1998; Teitelbaum and Triantafyllopoulou 2006), five other studies found SB carriage (either in stool or antibodies) rates were significantly higher in cancer patients than in controls. Interestingly, three studies also showed that patients with premalignant lesions (IBD or polyps) had intermediate SB carriage rate between cancer cases and controls (Klein et al. 1977; Teitelbaum and Triantafyllopoulou 2006; Tjalsma et al. 2006). For an update, see our recent literature-based meta-analysis on the association between *S. bovis* and CRC (Boleij et al. 2011).

3.4 Potential Mechanisms in Carcinogenesis

Despite observations discussed above, implications of SB infection on CRC remain largely elusive. There are several possible interpretations that are not necessarily mutually exclusive. First, it has been hypothesized that colorectal neoplastic sites provide a specific niche for SB resulting in sustained colonization, survival, and the

Table 3.3 Summary of studies on SB prevalence by colonic disease status

	No. of subjects				
Author (year)	Controls	Premalignant	Cancer	SB detection	Significant results
Klein et al. (1977)	105	25 (IBD)	63	Fecal culture	Cancer>controls
Burns et al. (1985)	216	62 (advanced polyps)	18	Fecal culture	Cancer>controls
Dubrow et al. (1991)	46	56 (polyps)		Fecal culture	No significant differences
Darjee and Gibb (1993)	16	–	16	Antibody titer	Cancer>controls
Potter et al. (1998)	13	–	11	Fecal culture	No significant differences
Teitelbaum and Triantafyllopoulou (2006)	34	47(IBD)		Fecal culture	No significant differences
Tjalsma et al. (2006)	8	4 (polyps)	12	Antibody patterns	Cancer/polyps>controls
Abdulamir et al. (2009)	50	14	60	Antibody titer	Cancer/adenoma>controls

establishment of a local tumour-associated (clinically silent) infection. Second, silent SB infection itself possibly promotes colorectal carcinogenesis, which has been supported by several experimental studies. Administration of SB or SB wall extracted antigens in rodents increases the formation of colorectal precursor lesions in a chemical carcinogenesis model (Ellmerich et al. 2000a). This was accompanied by increased expression of proliferative markers and enhanced interleukin IL-8 production in normal colonic mucosa of SB-injected animals. SB wall antigens are capable of adhering to various types of human cells, including GI-epithelial, endothelial and blood cells, as well as to extracellular matrix and induce IL-8 synthesis (Ellmerich et al. 2000b). In fact, increased IL-8 positive cells have been reported in SB seropositive human CRC cases compared with SB-seronegative cases (Abdulamir et al. 2009). IL-8 is a pro-inflammatory cytokine which also possesses mitogenic and angiogenic properties. It increases oxidative/ nitrosative stress and mediate the formation of carcinogenic compounds in gastrointestinal mucosa/ lumen (Federico et al. 2007; Vermeer et al. 2004; Hussain and Harris 2007). IL-8 also leads to cyclooxygenase (COX)-2 overexpression (Biarc et al. 2004). COX-2 driven prostaglandin synthesis stimulates cell proliferation, motility and metastatic potential, promotes angiogenesis, and induces local immunosuppression (Harris 2007; Mutoh et al. 2006). On the other hand, selective and non-selective COX-2 inhibitors reduce the incidence and prevalence of colorectal polyps (Steinbach et al. 2000; Logan et al. 2008). Importantly, increased COX-2 expression has been demonstrated in rodent infectious colorectal carcinogenesis models (Skinn et al. 2006; Newman et al. 2001; Balish and Warner 2002; Wang and Huycke 2007). The induction of COX-2 by SB in colon tissue has been reported for a rat model (Biarc et al. 2004) and may also occur in humans (Fig. 3.1). These enhanced COX-2 activities may also exert synergistic effects with other enzymes sharing substrates (e.g., CYP1 family) in metabolic activation of diet-derived carcinogens, such as polycyclic aromatic hydrocarbons (PAH) found in cooked meat (Wiese et al. 2001; Almahmeed et al. 2004). Such an enzyme, CYP1A1/B1, is indeed overexpressed in CRC and its precursors (McKay et al. 1993; Kumarakulasingham et al. 2005; Chang et al. 2005). This interplay is potentially important because meat consumption is one of the SB acquisition routes and because meat-derived PAH can induce intestinal CYP1A1/ B1 (Lampen et al. 2004). In addition, SB can induce matrix metalloproteinases (MMPs), e.g., MMP2 and MMP9 (Mungall et al. 2001), that play crucial roles in CRC growth and progression (Paduch et al. 2010; Sinnamon et al. 2008; Kim et al. 2009; Miyake et al. 2009). Furthermore, SB may also contribute to intra-colonic formation of potential carcinogens, e.g., nitroso-compounds (McKnight et al. 1999). The human large intestine contains a large amount of nitrogenous residues and nitrosating agents from dietary protein, and enzymatic activities of intestinal bacteria, e.g., streptococci, mediate these reactions (Hughes et al. 2001; Calmels et al. 1996, 1988). Intriguingly, consumption of red meat, a presumed route of SB acquisition, promotes colonic N-nitrosation via increasing supplies of colonic amine, nitrite and arginine (Hughes et al. 2001; Bingham et al. 1996, 2002; Silvester et al. 1997). Notably, large intestinal N-nitrosation does not occur in germ-free animals (Rowland et al. 1991).

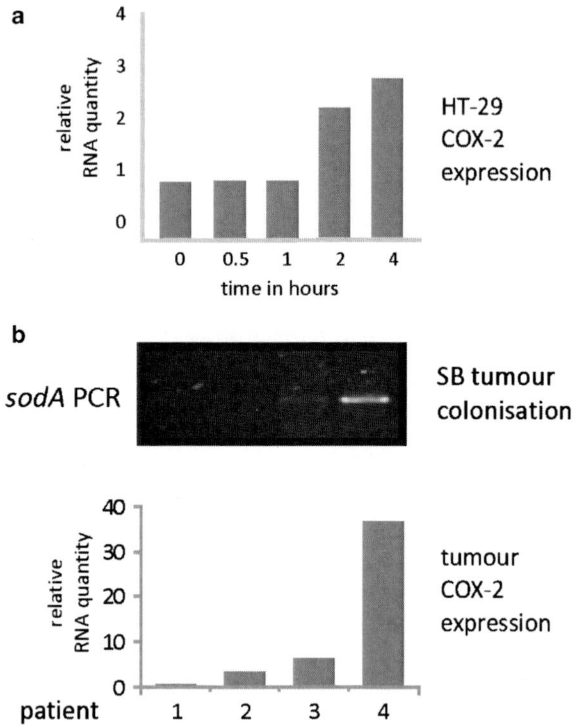

Fig. 3.1 COX-2 induction by SB. (**a**) The induction of COX-2 by SB was measured in HT-29 colorectal tumour cells *in vitro*. SB and HT-29 cells were co-incubated for 0.5, 1, 2 or 4 h. Subsequently RNA was extracted for real-time PCR procedures. The relative expression of COX-2 was determined by real time PCR using GAPDH as endogenous internal control and considered to be induced at values greater than 1.5. The *bar graph* shows that SB induces the expression of COX-2 after 2 and 4 h which is consistent with previously results in literature (Biarc et al. 2004). (**b**) The correlation of COX-2 expression and the presence of SB in tumour tissue from 4 CRC patients was determined in parallel. The presence of SB was monitored by a nested PCR on the SB *sodA* gene. The results are suggestive for a correlation of SB and COX-2 expression *in vivo*

3.5 SB Serology in CRC Patients

Although infections have been recognized as a major preventable cause of human cancer (Kuper et al. 2000), bacterial etiologies in sporadic CRC have not been established in humans. Notably, SB has indeed been recognized as an infectious agent that fulfills the criteria for inferring causality to the highest extent among the four agents evaluated as a potential cause of CRC in a recent review (Burnett-Hartman et al. 2008). However, to our knowledge there have been no epidemiologic studies properly designed to address this issue. The lack of good serological assays for SB infection may have been one of the reasons for scarcity of epidemiologic data. Darjee and Gibb (1993) were the first to monitor increased SB antibody responses in CRC patients by an ELISA approach. After that, Tjalsma et al. (2006, 2007) established an SB antibody profiling assay exploiting immunocapture time-of-flight mass spectrometry (IC-TOF MS) (Tjalsma et al. 2008), Abdulamir et al. (2009) also developed an ELISA to monitor SB antibodies in CRC patients and controls. As shown in Table 3.3, stronger associations observed by these approaches suggest that antibody assays may be a more powerful tool than fecal culture in assessing the associations between this bacterial infection and colorectal disease. Furthermore, as infectious agents in general induce a more pronounced immune response compared to tumour "self"-antigens, SB antigens could become instrumental in the immunodiagnosis of CRC (Tjalsma 2010).

3.6 SB and CRC Risk

To further investigate the exposure to SB in CRC patients, Boleij et al. (2010), developed an ELISA based on SB antigen RpL7/L12, previously assigned as a diagnostic antigen (Tjalsma et al. 2007). This assay was exploited for serological evaluation in Dutch (n = 209) and American (n = 112) populations. These analyses showed that an immune response against this bacterial antigen was increased in polyp patients and stage I/II CRC patients as compared to controls (Odds ratio (OR) 1.50, 95% Confidence Interval (CI) 0.48–4.62 in the Netherlands; OR 2.75, 95%CI 0.96–7.88 in the US) . Notably, increased anti-RpL7/L12 levels were not or only mildly detected in late stage colorectal cancer patients having lymph node or distant metastasis (Fig. 3.2). Increased anti-RpL7/L12 levels were not paralleled by increased antibody production to endotoxin, an intrinsic cell wall component of the majority of intestinal bacteria, which implicates that the humoral immune response against RpL7/L12 is not a general phenomenon induced by the loss of colonic barrier function. The age-adjusted OR for all colorectal tumours combined was very similar in the US (2.30 95% CI 1.06–5.00) and Netherlands (1.90, 95% CI 0.49–2.84). Even a relatively modest increase should be relevant for the progression of colon adenomas to carcinomas (accumulation of mutations), a process which can take over a decade to take place. In this respect, it is interesting to note that the ORs of 1.5 and 2.8 for early stage CRC were within the range of those calculated for the serological response to a panel of *Helicobacter pylori* antigens in patients with early stage gastric precancerous lesions (ORs ranging from 1 to 9) (Gao et al. 2009). Unfortunately, no data are yet available (August 2010) that correlate SB colonization of tumour tissue with the humoral immune response to SB antigens. Nevertheless, our preliminary studies suggest that tumour tissue provides a niche that allows increased SB colonization (Fig. 3.3). Altogether, these findings suggest that SB constitutes a risk factor for the development and/or progression of pre-malignant lesions into carcinomas. Importantly, cross-sectional and retrospective studies, including the current study and others, are not able to address the temporal relationship between an exposure and a disease outcome directly. Thus, future prospective studies are essential to elucidate the etiological roles of SB in colorectal carcinogenesis.

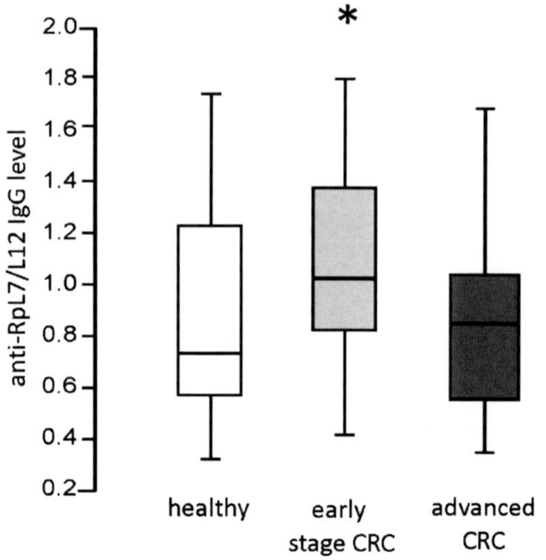

Fig. 3.2 Humoral immune response against SB antigen RpL7/L12. Serum anti-RpL7/L12 were determined in healthy control subjects (n=60), "early stage CRC" (polyp and local tumours; n=70) and "advanced CRC" (tumours with regional and distant metastases; n=50) by an ELISA (Tjalsma et al. 2007). The results are indicative for a moderate, but significant (*), increased exposure to this antigen during the early stages of CRC. Median levels, second and third quartile (*boxes*), ad ranges (*lines*) are indicated. Relative anti-RpL7/L12 IgG levels were expressed as arbitrary optical density units

3.7 Model for the Association of SB with CRC

Based on the current knowledge the following model for the association of SB with CRC can be envisaged (Fig. 3.4). Pre-malignant lesions are initiated by carcinogenic (dietary) factors that diffuse through the colonic mucus layer and induce mutations within the APC or B-catenin genes (Cho and Vogelstein 1992). These thereby immortalized epithelial cells are prone to the accumulation of other mutations and, as a side effect, the aberrant epithelial physiology disturbs the mucus layer covering the epithelial cells (Corfield et al. 2000) and makes it susceptible to bacterial infiltration. Such (pre-) malignant epithelial sites may also provide a selective bacterial microenvironment, for instance by the excretion of specific metabolites, recruitment of immune cells and/or production of selective anti-microbial substances. Bacteria, such as SB, which are unable to effectively colonize the healthy colon may have a competitive advantage in this microenvironment and survive for prolonged periods of time. Tumour infiltration of SB may exert inflammatory factors such as IL-8 and COX-2 and/or lead to increased levels of genotoxins and thereby promote intestinal carcinogenesis. These (pre-) malignant lesions also provide a portal of entry for SB which explains the increased anti-SB antibody titers and increased incidence of SB endocarditis in CRC patients. Late stage tumours entering the metastatic phase may change in such a way that bacterial survival on the tumour surface is diminished or

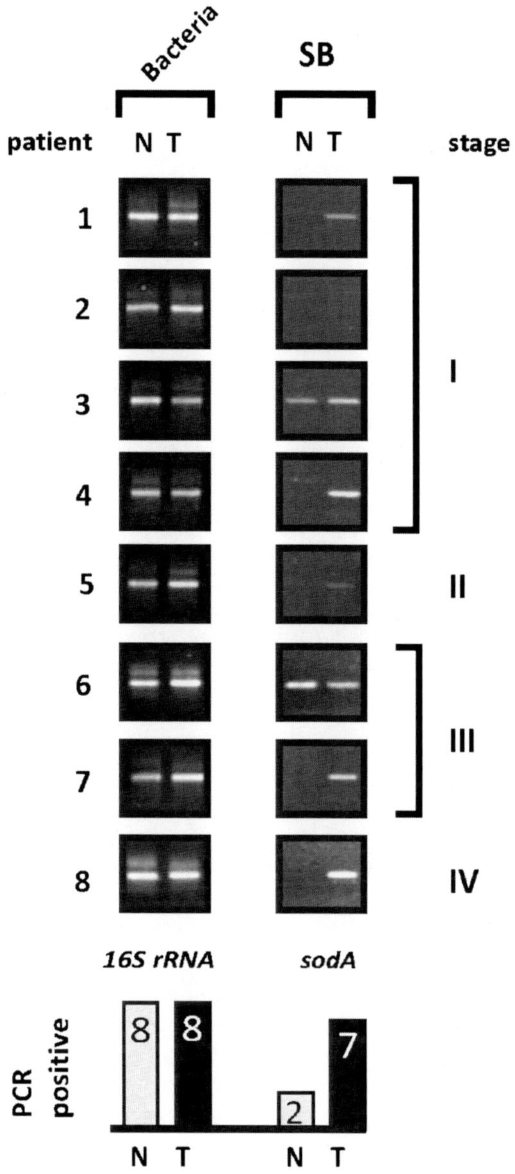

Fig. 3.3 SB detection in human colonic biopsy samples. The presence of SB in human biopsies from tumour tissue (T) and adjacent non-malignant mucosa (N) was monitored by a nested PCR on the SB *sodA* gene using biopsy-extracted DNA from 8 CRC patients as a template. CRC disease staging is indicated. A broad range 16S rRNA PCR was run in parallel to control for the presence of bacterial DNA and PCR inhibiting substances. The results are suggestive for a preferred colonization of tumour tissue by SB. The identity of the *sodA* PCR fragments was confirmed by nucleotide sequencing, which showed that all products had the highest degree of similarity with SGG (SB biotype I)

antibody expression due to bacterial interaction is reduced. The possibility that tumour progression may drive bacteria out of the cancerous tissue is similar to what has been reported for *H. pylori* during gastric cancer progression (Kang et al. 2006; Brenner et al. 2007). If true, this phenomenon may partly account for a wide range of the prevalence of SB reported for CRC patients that is comprised of various stages of the disease.

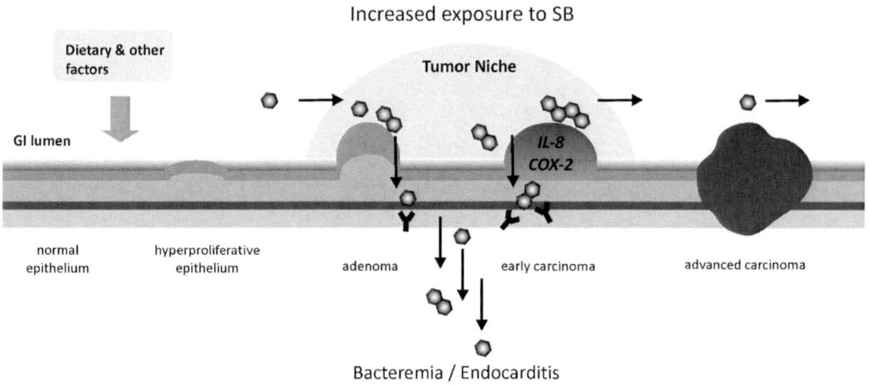

Fig. 3.4 Model for a temporal association between SB and CRC. The development of colorectal tumours is schematically depicted from *left* (healthy) to *right* (invasive and metastasizing carcinomas). Initiation of carcinogenesis is a multi-factorial process in which dietary factor play an important role. It may be envisaged that adenomas and early carcinomas provide a preferred niche for SB, which leads to subclinical infection and an increased exposure to SB which can be measured by serological assays. Moreover, this could explain the increased incidence of SB bacteremia and endocarditis in CRC patients as these (pre-) malignant lesions can form a portal of entry into the human body. In addition, SB may interfere with colon carcinogenesis for instance by the induction of IL-8 and COX-2, whereas tumour progression may drive SB out of advanced cancerous tissue (see text for details)

3.8 Conclusion

The clinical association between SB and CRC is widely acknowledged, and an SB infection is often regarded as an indication for full bowel examination in clinical practice. However, still little is known about the molecular mechanisms behind this association (Boleij et al. 2009a, b). The recent deciphering of the SGG (SB biotype I) genome revealed unique features among streptococci, probably related to its adaptation to the intestinal environment (Rusniok et al. 2010). For instance, SGG has the capacity to use a broad range of carbohydrates of plant origin, in particular to degrade polysaccharides derived from the plant cell wall. Its genome encodes a large repertoire of transporters and catalytic activities, like tannase, phenolic compounds decarboxylase, and bile salt hydrolase, which should contribute to the detoxification of the gut environment. Furthermore, SGG has the potential to synthesize all 20 amino acids and more vitamins than any other sequenced *Streptococcus* species (Rusniok et al. 2010). The surface properties (Fig. 3.5) of this bacterium might be implicated in resistance to innate immunity defenses, and glucan mucopolysaccharides, three types of pili, and collagen binding proteins may play a role in adhesion to tissues in the course of endocarditis. Recent *in vitro* studies revealed that SGG has a unique repertoire of virulence factors that may facilitate infection through (pre-)malignant colonic lesions and subsequently can provide SGG with a

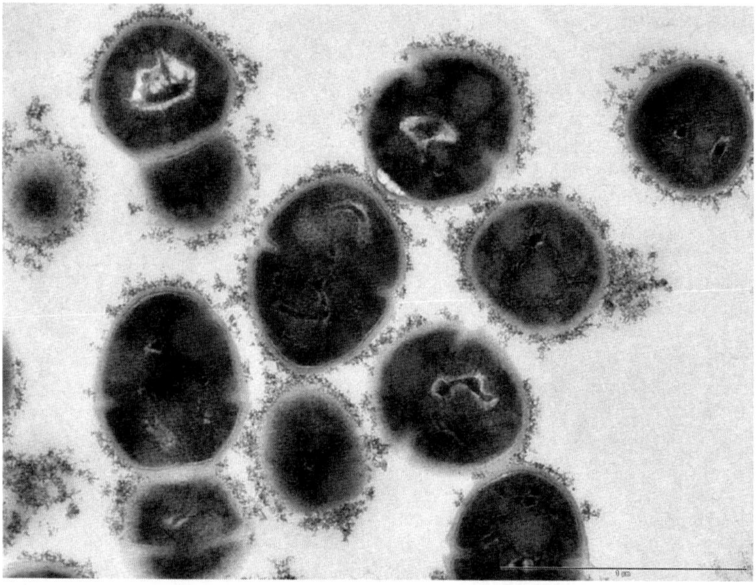

Fig. 3.5 SB surface structure. Electron microscopy picture of SGG cells (strain UCN34 (Rusniok et al. 2010)) showing the capsule of glucan mucopolysaccharides after polycationic ferritin labelling (Vanrobaeys et al. 1999), which may be important for immune evasion in the course of endocarditis (The picture was kindly provided by Philippe Glaser and Nadège Cayet, Unité de Génomique des Microorganismes Pathogènes, Institute Pasteur, Paris, France)

competitive advantage to evade the innate immune system and to form resistant vegetations at collagen-rich sites in susceptible CRC patients (Boleij et al. 2011a). However, many questions on the relationship between SGG and CRC remain to be answered. Therefore, future studies should answer to which extent polyps and tumours actually provide a niche for SB colonization, and if so, which factors are involved in the adherence to, and/or survival in, the tumour microenvironment and how this increased colonization promotes carcinogenesis. In addition, improved (ELISA) assays are desirable to address the relationship between SB exposure and CRC directly in prospective and retrospective studies. Together, these molecular and epidemiological studies are essential for the full elucidation of the etiological roles of SB in colorectal carcinogenesis.

Acknowledgements We thank all researchers for their valuable contributions to unravel the role of SB in CRC and we apologize for the fact that we could not mention all SB-related studies in this chapter. We especially thank Philippe Glaser, Albert Bolhuis, Julian Marchesi, Bas Dutilh, Dorine Swinkels and Rian Roelofs for their inspiring discussions on this subject. AB was supported by the Dutch Cancer Society (KWF; project KUN 2006–3591) and IK was supported in part by the US National Institutes of Health (Research Grant R01-CA93817).

References

Abdulamir AS, Hafidh RR, Mahdi LK et al (2009) Investigation into the controversial association of *Streptococcus gallolyticus* with colorectal cancer and adenoma. BMC Cancer 9:403

Alazmi W, Bustamante M, O'Loughlin C et al (2006) The association of *Streptococcus bovis* bacteremia and gastrointestinal diseases: a retrospective analysis. Dig Dis Sci 51:732–736

Almahmeed T, Boyle JO, Cohen EG et al (2004) Benzo[a]pyrene phenols are more potent inducers of CYP1A1, CYP1B1 and COX-2 than benzo[a]pyrene glucuronides in cell lines derived from the human aerodigestive tract. Carcinogenesis 25:793–799

Baba Y, Nosho K, Shima K et al (2010) PTGER2 overexpression in colorectal cancer is associated with microsatellite instability, independent of CpG island methylator phenotype. Cancer Epidemiol Biomarkers Prev 19:822–831

Balish E, Warner T (2002) Enterococcus faecalis induces inflammatory bowel disease in interleukin-10 knockout mice. Am J Pathol 160:2253–2257

Beck M, Frodl R, Funke G (2008) Comprehensive study of strains previously designated *Streptococcus bovis* consecutively isolated from human blood cultures and emended description of *Streptococcus gallolyticus* and *Streptococcus infantarius* subsp. coli. J Clin Microbiol 46:2966–2972

Biarc J, Nguyen IS, Pini A et al (2004) Carcinogenic properties of proteins with pro-inflammatory activity from *Streptococcus infantarius* (formerly *S. bovis*). Carcinogenesis 25:1477–1484

Bingham SA, Pignatelli B, Pollock JR et al (1996) Does increased endogenous formation of N-nitroso compounds in the human colon explain the association between red meat and colon cancer? Carcinogenesis 17:515–523

Bingham SA, Hughes R, Cross AJ (2002) Effect of white versus red meat on endogenous N-nitrosation in the human colon and further evidence of a dose response. J Nutr 132:3522S–3525S

Blaser NJ (2008) Understanding microbe-induced cancers. Cancer Prev Res 1:15–20

Boleij A, Schaeps RM, de Kleijn S et al (2009a) Surface-exposed histone-like protein a modulates adherence of *Streptococcus gallolyticus* to colon adenocarcinoma cells. Infect Immun 77:5519–5527

Boleij A, Schaeps RM, Tjalsma H (2009b) Association between *Streptococcus bovis* and colon cancer. J Clin Microbiol 47:516

Boleij A, Roelofs R, Schaeps RMJ et al (2010) Increased exposure to bacterial antigen RpL7/L12 in early stage colorectal cancer patients. Cancer 116:4014–4022

Boleij A, Muytjens CMJ, Bukhari CI, Cayet N, Glaser P, Hermans PW, Swinkels DS, Bolhuis A, Tjalsma H (2011a) Novel clues on the specific association of *Streptococcus gallolyticus* subsp *gallolyticus* with colorectal cancer. J Infect Dis 203:1101–1109

Boleij A, van Gelder MM, Swinkels DW, Tjalsma H (2011b) Clinical Importance of *Streptococcus gallolyticus* Infection Among Colorectal Cancer Patients: Systematic Review and Meta-analysis. Clin Infect Dis 53:870–878

Brenner H, Rothenbacher D, Weck MN (2007) Epidemiologic findings on serologically defined chronic atrophic gastritis strongly depend on the choice of the cutoff-value. Int J Cancer 121:2782–2786

Burnett-Hartman AN, Newcomb PA, Potter JD (2008) Infectious agents and colorectal cancer: a review of *Helicobacter pylori*, *Streptococcus bovis*, JC virus, and human papillomavirus. Cancer Epidemiol Biomarkers Prev 17:2970–2979

Burns CA, McCaughey R, Lauter CB (1985) The association of *Streptococcus bovis* fecal carriage and colon neoplasia: possible relationship with polyps and their premalignant potential. Am J Gastroenterol 80:42–46

Calmels S, Ohshima H, Bartsch H (1988) Nitrosamine formation by denitrifying and non-denitrifying bacteria: implication of nitrite reductase and nitrate reductase in nitrosation catalysis. J Gen Microbiol 134:221–226

Calmels S, Ohshima H, Henry Y et al (1996) Characterization of bacterial cytochrome cd(1)-nitrite reductase as one enzyme responsible for catalysis of nitrosation of secondary amines. Carcinogenesis 17:533–536

Chang H, Su JM, Huang CC et al (2005) Using a combination of cytochrome P450 1B1 and beta-catenin for early diagnosis and prevention of colorectal cancer. Cancer Detect Prev 29:562–569

Cho KR, Vogelstein B (1992) Genetic alterations in the adenoma–carcinoma sequence. Cancer 15:1727–1731

Clarridge JE 3rd, Attorri SM, Zhang Q et al (2001) 16 S ribosomal DNA sequence analysis distinguishes biotypes of *Streptococcus bovis*: *Streptococcus bovis* Biotype II/2 is a separate genospecies and the predominant clinical isolate in adult males. J Clin Microbiol 39:1549–1552

Corfield AP, Myerscough N, Longman R et al (2000) Mucins and mucosal protection in the gastrointestinal tract: new prospects for mucins in the pathology of gastrointestinal disease. Gut 47:589–594

Corredoira JC, Alonso MP, García JF et al (2005) Clinical characteristics and significance of *Streptococcus salivarius* bacteremia and *Streptococcus bovis* bacteremia: a prospective 16-year study. Eur J Clin Microbiol Infect Dis 24:250–255

Corredoira J, Alonso MP, Coira A et al (2008a) Characteristics of *Streptococcus bovis* endocarditis and its differences with *Streptococcus viridans* endocarditis. Eur J Clin Microbiol Infect Dis 27:285–291

Corredoira J, Alonso MP, Coira A et al (2008b) Association between *Streptococcus infantarius* (formerly S. bovis II/1) bacteremia and noncolonic cancer. J Clin Microbiol 46:1570

Coussens LM, Werb Z (2002) Inflammation and cancer. Nature 420:860–867

Curado MP, Edwards B, Shin HR et al (eds) (2007) Cancer incidence in five continents, vol IX, IARC Scientific Publications No. 160. IARC, Lyon

Darjee R, Gibb AP (1993) Serological investigation into the association between *Streptococcus bovis* and colonic cancer. J Clin Pathol 46:1116–1119

Del Valle L, Gordon J, Enam S et al (2002) Expression of human neurotropic polyomavirus JCV late gene product agnoprotein in human medulloblastoma. J Natl Cancer Inst 94:267–273

Dubé C, Rostom A, Lewin G et al (2007) U.S. Preventive Services Task Force. The use of aspirin for primary prevention of colorectal cancer: a systematic review prepared for the U.S. Preventive Services Task Force. Ann Intern Med 146:365–375

Dubrow R, Edberg S, Wikfors E et al (1991) Fecal carriage of *Streptococcus bovis* and colorectal adenomas. Gastroenterology 101:721–725

Ellmerich S, Djouder N, Schöller M et al (2000a) Production of cytokines by monocytes, epithelial and endothelial cells activated by *Streptococcus bovis*. Cytokine 12:26–31

Ellmerich S, Schöller M, Duranton B et al (2000b) Promotion of intestinal carcinogenesis by *Streptococcus bovis*. Carcinogenesis 21:753–756

Federico A, Morgillo F, Tuccillo C et al (2007) Chronic inflammation and oxidative stress in human carcinogenesis. Int J Cancer 121:2381–2386

Gao L, Weck MN, Michel A et al (2009) Association between chronic atrophic gastritis and serum antibodies to 15 *Helicobacter pylori* proteins measured by multiplex serology. Cancer Res 69:2973–2980

Ghali MB, Scott PT, Al Jassim RA (2004) Characterization of *Streptococcus bovis* from the rumen of the dromedary camel and Rusa deer. Lett Appl Microbiol 39:341–346

Giannitsioti E, Chirouze C, Bouvet A et al (2007) AEPEI Study Group. Characteristics and regional variations of group D streptococcal endocarditis in France. Clin Microbiol Infect 13:770–776

Gold JS, Bayar S, Salem RR (2004) Association of *Streptococcus bovis* bacteremia with colonic neoplasia and extracolonic malignancy. Arch Surg 139:760–765

Gonzlez-Quintela A, Martínez-Rey C, Castroagudín JF et al (2001) Prevalence of liver disease in patients with Streptococcus bovis bacteraemia. J Infect 42:116–119

Grivennikov SI, Greten FR, Karin M (2007) Immunity, inflammation, and cancer. Cell 140:883–899

Gupta A, Madani R, Mukhtar H (2010) *Streptococcus bovis* endocarditis, a silent sign for colonic tumour. Colorectal Dis 12:164–171

Harris RE (2007) Cyclooxygenase-2 (cox-2) and the inflammogenesis of cancer. Subcell Biochem 42:93–126

Herrera LA, Benítez-Bribiesca L, Mohar A et al (2005) Role of infectious diseases in human carcinogenesis. Environ Mol Mutagen 45:284–303

Hope ME, Hold GL, Kain R et al (2005) Sporadic colorectal cancer–role of the commensal microbiota. FEMS Microbiol Lett 244:1–7

Horner MJ, Ries LAG, Krapcho M et al (eds) (2009) SEER Cancer Statistics Review, 1975–2006. National Cancer Institute, Bethesda. http://seer.cancer.gov/csr/1975_2006/. Based on November 2008 SEER data submission, posted to the SEER web site, 2009

Hughes R, Cross AJ, Pollock JR et al (2001) Dose-dependent effect of dietary meat on endogenous colonic N-nitrosation. Carcinogenesis 22:199–202

Hussain SP, Harris CC (2007) Inflammation and cancer: An ancient link with novel potentials. Int J Cancer 121:2373–2380

IARC (1994) IARC monographs on the evaluation of carcinogenic risks to humans: vol 61. Schistosomes, liver flukes and Helicobacter pylori. IARC, Lyon

Ishitsuka T, Kashiwagi H, Konishi F (2001) Microsatellite instability in inflamed and neoplastic epithelium in ulcerative colitis. J Clin Pathol 54:526–532

Jass JR (2007) Classification of colorectal cancer based on correlation of clinical, morphological and molecular features. Histopathology 50:113–130

Jean SS, Teng LJ, Hsueh PR et al (2004) Bacteremic *Streptococcus bovis* infections at a university hospital, 1992–2001. J Formos Med Assoc 103:118–123

Kahveci A, Ari E, Arikan H et al (2010) *Streptococcus bovis* bacteremia related to colon adenoma in a chronic hemodialysis patient. Hemodial Int 14:91–93

Kang HY, Kim N, Park YS et al (2006) Progression of atrophic gastritis and intestinal metaplasia drives *Helicobacter pylori* out of the gastric mucosa. Dig Dis Sci 51:2310–2315

Kim YH, Kim MH, Kim BJ et al (2009) Inhibition of cell proliferation and invasion in a human colon cancer cell line by 5-aminosalicylic acid. Dig Liver Dis 41:328–337

Kim SY, Joo SI, Yi J et al (2010) A case of *Streptococcus gallolyticus* subsp. *gallolyticus* infective endocarditis with colon cancer: identification by 16 S ribosomal DNA sequencing. Korean J Lab Med 30:160–165

Klein RS, Recco RA, Catalano MT et al (1977) Association of *Streptococcus bovis* with carcinoma of the colon. N Engl J Med 297:800–802

Klein RS, Catalano MT, Edberg SC et al (1979) *Streptococcus bovis* septicemia and carcinoma of the colon. Ann Intern Med 91:560–562

Knudtson LM, Hartman PA (1993) Comparison of fluorescent gentamicin-thallous-carbonate and KF streptococcal agars to enumerate enterococci and fecal streptococci in meats. Appl Environ Microbiol 59:936–938

Kohonen-Corish MR, Daniel JJ, te Riele H et al (2002) Susceptibility of Msh2-deficient mice to inflammation-associated colorectal tumors. Cancer Res 62:2092–2097

Kumarakulasingham M, Rooney PH, Dundas SR et al (2005) Cytochrome p450 profile of colorectal cancer: identification of markers of prognosis. Clin Cancer Res 11:3758–3765

Kuper H, Adami H-O, Trichopoulos D (2000) Infections as a major preventable cause of human cancer. J Intern Med 248:171–183

Lampen A, Ebert B, Stumkat L et al (2004) Induction of gene expression of xenobiotic metabolism enzymes and ABC-transport proteins by PAH and a reconstituted PAH mixture in human Caco-2 cells. Biochim Biophys Acta 1681:38–46

Lee RA, Woo PC, To AP et al (2003) Geographical difference of disease association in *Streptococcus bovis* bacteraemia. J Med Microbiol 52:903–908

Li JH, Shi XZ, Lv S et al (2005) Effect of *Helicobacter pylori* infection on p53 expression of gastric mucosa and adenocarcinoma with microsatellite instability. World J Gastroenterol 11:4363–4366

Logan RF, Grainge MJ, Shepherd VC et al (2008) ukCAP Trial Group. Aspirin and folic acid for the prevention of recurrent colorectal adenomas. Gastroenterology 134:29–38

McKay JA, Murray GI, Weaver RJ et al (1993) Xenobiotic metabolising enzyme expression in colonic neoplasia. Gut 34:1234–1239

McKnight GM, Duncan CW, Leifert C et al (1999) Dietary nitrate in man: friend or foe? Br J Nutr 81:349–358

McMahon AJ, Auld CD, Dale BA et al (1991) *Streptococcus bovis* septicaemia associated with uncomplicated colonic carcinoma. Br J Surg 78:883–885

Miyake K, Shimada M, Nishioka M et al (2009) Downregulation of matrix metalloprotease-9 and urokinase plasminogen activator by TX-1877 results in decreased tumor growth and metastasis on xenograft model of rectal cancer. Cancer Chemother Pharmacol 64:885–892

Moellering RC Jr, Watson BK, Kunz LJ (1974) Endocarditis due to group D streptococci. Comparison of disease caused by *Streptococcus bovis* with that produced by the enterococci. Am J Med 57:239

Mungall BA, Kyaw-Tanner M, Pollitt CC (2001) In vitro evidence for a bacterial pathogenesis of equine laminitis. Vet Microbiol 79(3):209–223

Murray HW, Roberts RB (1978) *Streptococcus bovis* bacteremia and underlying gastrointestinal disease. Arch Intern Med 138:1097–1099

Mutoh M, Takahashi M, Wakabayashi K (2006) Roles of prostanoids in colon carcinogenesis and their potential targeting for cancer chemoprevention. Curr Pharm Des 12:2375–2382

Newman JV, Kosaka T, Sheppard BJ et al (2001) Bacterial infection promotes colon tumorigenesis in Apc(Min/+) mice. J Infect Dis 184:227–230

Nielsen SD, Christensen JJ, Laerkeborg A et al (2007) Molecular-biological methods of diagnosing colon-related *Streptococcus bovis* endocarditis. Ugeskr Laeger 169:610–611

Paduch R, Kandefer-Szerszeń M, Szuster-Ciesielska A et al (2010) Transforming growth factor-beta1 modulates metalloproteinase-2 and –9, nitric oxide, RhoA and alpha-smooth muscle actin expression in colon adenocarcinoma cells. Cell Biol Int 34(2):213–223

Parkin DM (2006) The global health burden of infection-associated cancers in the year 2002. Int J Cancer 118:3030–3044

Persing DH, Prendergast FG (1999) Infection, immunity, and cancer. Arch Pathol Lab Med 123:1015–1022

Pigrau C, Lorente A, Pahissa A et al (1988) *Streptococcus bovis* bacteremia and digestive system neoplasms. Scand J Infect Dis 20:459–460

Potter MA, Cunliffe NA, Smith M et al (1998) A prospective controlled study of the association of *Streptococcus bovis* with colorectal carcinoma. J Clin Pathol 51:473–474

Randazzo CL, Vaughan EE, Caggia C (2006) Artisanal and experimental Pecorino Siciliano cheese: microbial dynamics during manufacture assessed by culturing and PCR-DGGE analyses. Int J Food Microbiol 109:1–8

Reynolds JG, Silva E, McCormack WM (1983) Association of *Streptococcus bovis* bacteremia with bowel disease. J Clin Microbiol 17:696–697

Ribeiro ML, Godoy AP, Benvengo YH et al (2004) The influence of endoscopic procedures upon the contamination of *Helicobacter pylori* cultures. Arq Gastroenterol 41:100–103

Roberts RB (1992) Streptococcal endocarditis: the viridans and beta hemolytic streptococci. In: Kaye D (ed) Infective endocarditis. Raven Press, Ltd, New York, p 191

Rostom A, Dubé C, Lewin G et al (2007) U.S. Preventive Services Task Force. Nonsteroidal anti-inflammatorydrugs and cyclooxygenase-2 inhibitors for primary prevention of colorectal cancer: a systematic review prepared for the U.S. Preventive Services Task Force. Ann Intern Med 146:376–389

Rowland IR, Granli T, Bøckman OC et al (1991) Endogenous N-nitrosation in man assessed by measurement of apparent total N-nitroso compounds in faeces. Carcinogenesis 12:1395–1401

Ruoff KL, Miller SI, Garner CV et al (1989) Bacteremia with *Streptococcus bovis* and *Streptococcus salivarius*: clinical correlates of more accurate identification of isolates. J Clin Microbiol 27:305–308

Rusniok C, Couvé E, Da Cunha V et al (2010) Genome sequence of *Streptococcus gallolyticus*: insights into its adaptation to the bovine rumen and its ability to cause endocarditis. J Bacteriol 192:2266–2276

Schlegel L, Grimont F, Collins MD et al (2000) *Streptococcus infantarius* sp. nov., *Streptococcus infantarius* subsp. *infantarius* subsp. nov. and *Streptococcus infantarius* subsp. *coli* subsp. nov., isolated from humans and food. Int J Syst Evol Microbiol 50:1425–1434

Schlegel L, Grimont F, Ageron E et al (2004) Reappraisal of the taxonomy of the *Streptococcus bovis/Streptococcus equinus* complex and related species: description of *Streptococcus gallolyticus* subsp. *gallolyticus* subsp. nov., *S. gallolyticus* subsp. *macedonicus* subsp. nov. and *S. gallolyticus* subsp. *pasteurianus* subsp. nov. Indian J Med Res 119:252–256

Silvester KR, Bingham SA, Pollock JR et al (1997) Effect of meat and resistant starch on fecal excretion of apparent N-nitroso compounds and ammonia from the human large bowel. Nutr Cancer 29:13–23

Sinicrope FA (2007) Sporadic colorectal cancer: an infectious disease? Gastroenterology 132:797–801

Sinnamon MJ, Carter KJ, Fingleton B et al (2008) Matrix metalloproteinase-9 contributes to intestinal tumourigenesis in the adenomatous polyposis coli multiple intestinal neoplasia mouse. Int J Exp Pathol 89:466–475

Skinn AC, Vergnolle N, Zamuner SR et al (2006) *Citrobacter rodentium* infection causes iNOS-independent intestinal epithelial dysfunction in mice. Can J Physiol Pharmacol 84: 1301–1312

Steinbach G, Lynch PM, Phillips RK et al (2000) The effect of celecoxib, a cyclooxygenase-2 inhibitor, in familial adenomatous polyposis. N Engl J Med 342:1946–1952

Stone WL, Papas AM (1997) Tocopherols and the etiology of colon cancer. J Natl Cancer Inst 89:1006–1014

Teitelbaum JE, Triantafyllopoulou M (2006) Inflammatory bowel disease and *Streptococcus bovis*. Dig Dis Sci 51:1439–1442

Terzić J, Grivennikov S, Karin E et al (2010) Inflammation and colon cancer. Gastroenterology 138:2101–2114

Thian TS, Hartman PA (1981) Gentamicin-thallous-carbonate medium for isolation of fecal streptococci from foods. Appl Environ Microbiol 41:724–728

Tjalsma H (2010) Identification of biomarkers for colorectal cancer through proteomics-based approaches. Exp Rev Proteomic 7:879–895

Tjalsma H, Schöller-Guinard M, Lasonder E et al (2006) Profiling the humoral immune response in colon cancer patients: diagnostic antigens from *Streptococcus bovis*. Int J Cancer 119:2127–2135

Tjalsma H, Lasonder E, Schöller-Guinard M et al (2007) Shotgun immunoproteomics to identify disease-related bacterial antigens: application to human colon cancer. Proteomic Clin Appl 1:429–434

Tjalsma H, Schaeps RMJ, Swinkels DW (2008) Immunoproteomics: from biomarker discovery to diagnostic applications. Proteomic Clin Appl 2:167–180

Vanrobaeys M, De Herdt P, Charlier G et al (1999) Ultrastructure of surface components of *Streptococcus gallolytics* (*S. bovis*) strains of differing virulence isolated from pigeons. Microbiology 145:335–342

Vaska VL, Faoagali JL (2009) *Streptococcus bovis* bacteraemia: identification within organism complex and association with endocarditis and colonic malignancy. Pathology 41:183–186

Vermeer IT, Henderson LY, Moonen EJ et al (2004) Neutrophil-mediated formation of carcinogenic N-nitroso compounds in an in vitro model for intestinal inflammation. Toxicol Lett 154:175–182

Wang X, Huycke MM (2007) Extracellular superoxide production by *Enterococcus faecalis* promotes chromosomal instability in mammalian cells. Gastroenterology 132:551–561

Wentling GK, Metzger PP, Dozois EJ et al (2006) Unusual bacterial infections and colorectal carcinoma–Streptococcus bovis and Clostridium septicum: report of three cases. Dis Colon Rectum 49:1223–1227

Wiese FW, Thompson PA, Kadlubar FF (2001) Carcinogen substrate specificity of human COX-1 and COX-2. Carcinogenesis 22:5–10

Zarkin BA, Lillemoe KD, Cameron JL et al (1990) The triad of *Streptococcus bovis* bacteremia, colonic pathology, and liver disease. Ann Surg 211:786–791

Chapter 4
Chlamydial Disease: A Crossroad Between Chronic Infection and Development of Cancer

Carlo Contini and Silva Seraceni

Abstract *Chlamydia* is an intracellular bacterium implicated as potentially oncogenic for its tendency to cause chronic and persistent infections. These organisms have been frequently associated with several types of cancer including cervical dysplasia and cancer by *C. trachomatis*, lung cancer and cutaneous T-cell lymphoma by *C. pneumoniae* and a number of non-gastrointestinal MALT lymphomas such as ocular adnexal lymphoma by *C. psittaci*, suggesting a potential role. *C. trachomatis*, which causes ocular-genital infections in humans, was recently demonstrated at molecular and cultural level in patients with ocular cancer, thus implying also for this bacterium a role in the pathogenesis of the above malignancy. The pathophysiological processes and molecular mechanisms that lead to the development of chronic inflammatory disease, persistence, and ultimately cancer, still need to be clarified. This chapter describes the pathogenetic aspects of Chlamydial infections favouring the onset of chronic diseases and cancers as well as the diagnostic and clinical features in relation to *Chlamydia* species involved. The potential application of bacteria-eradicating therapy would certainly represent an exciting challenge for the next few years.

Keywords *Chlamydia* • *Chlamydophila pneumoniae* • *Chlamydia pneumoniae* • *Chlamydophila psittaci* • *Chlamydia psittaci* • *Chlamydia trachomatis* • Cancer • MZL • MALT lymphoma • Non-Hodgkin's lymphoma • Ocular adnexal lymphoma • Apoptosis • Hsp • TETR-PCR • PCR • RT-PCR • Cell culture • PBMC • Doxycycline • Macrolides • Quinolones

C. Contini (✉) • S. Seraceni
Department of Clinical and Experimental Medicine, Section of Infectious Diseases, University of Ferrara, via F. di Mortara, 23, Ferrara I-44100, Italy
e-mail: cnc@unife.it

A.A. Khan (ed.), *Bacteria and Cancer*, DOI 10.1007/978-94-007-2585-0_4,

Abbreviations

AM	Alveolar macrophages
EB	Elementary bodies
EBV	Epstein-Barr virus
HBV	Hepatitis B virus
HPV	Human papilloma virus
Hsp	Heat Shock Proteins
LGV	Lymphogranuloma venereum
MIP	Macrophage Infectivity potentiator lipoprotein
MOMP	Major outer membrane protein
MS	Multiple Sclerosis
MZL	Marginal zone B-cell lymphomas
NHL	Non-Hodgkin's lymphomas
OAL	Ocular adnexal lymphoma
PAMP	Pathogen-associated molecular patterns
PBMC	Peripheral blood mononuclear cells
PCNSL	Primary central nervous system marginal zone B cell lymphoma
PID	Pelvic inflammatory disease
PRR	Pattern recognition receptors
RB	Reticulate bodies
STD	Sexually transmitted disease
T3S	Type III secretion
TETR-PCR	Time-release polymerase chain reaction
TLR	Toll-like receptors

4.1 Introduction

Infectious agents play an important role in the aetiology of certain human malignancies, and are thought to be responsible for around 20% of the worldwide cancer burden (Parkin 2001).

Much of the burden of cancer incidence, morbidity, and mortality occurs in the developing world (up to 27%), with a large body of evidence regarding the role of viruses such as human papilloma virus (HPV), hepatitis B virus (HBV) and Epstein-Barr virus (EBV) in the complex processes of carcinogenesis of the cervix, stomach and liver (Jemal et al. 2010).

In addition to viral agents implicated in carcinogenesis, a theory of possible association between bacterial infection and cancer has been proposed in early nineteenth century (Lax 2005).

Helicobacter pylori infection has shown to cause gastric cancer, chronic carriers of *Salmonella typhi* are at more risk of developing gallbladder or hepatobiliary carcinoma than non-carriers, Mycoplasma like organisms have been suggested to be associated with Hodgkin's disease (Johnson et al. 1996). Moreover, many bacteria that cause persistent infections produce toxins that disrupt cellular signalling, alter the regulation of cell growth, induce inflammation or directly damage DNA.

Toxins may also mimic carcinogens and tumour promoters and might represent a paradigm for bacterially induced carcinogenesis. The question however remains quite controversial especially with regard to certain species of bacteria for oncogenic properties.

In recent years, other bacterial agents have been increasingly recognised as important pathogenetic factors for various malignant tumours including non-Hodgkin's lymphomas (NHL) (Groves et al. 2000; Slater 2001). Especially, in marginal zone lymphomas of mucosa-associated lymphoid tissue (MALT), the chronic antigenic stimulation and the action of infectious oncogenes somehow related to these agents (bacterial or viral) has been suggested to evoke host immune responses and to promote and sustain clonal B-cell expansion (Wyatt and Rathbone 1988).

In this setting, one of the bacterial agents which in recent years has gained attention is the bacterium *Chlamydia* implicated as potentially oncogenic for its tendency to cause chronic and persistent infections. This chapter aims to analyze the role that *Chlamydiae* play in chronic and persistent diseases as well as in development of cancer and ocular adnexal lymphoma (OAL).

4.2 *Chlamydiae*

4.2.1 Characteristics and Developmental Cell Cycle

The *Chlamydiae* are ubiquitous gram negative, aerobic, obligate intracellular bacteria once considered viruses, which grow in eukaryotic cells and are responsible for wide range of human diseases.

Chlamydiae were taxonomically categorised into their own order *Chlamydiales*, with one family, *Chlamydiaceae*, and a single genus, *Chlamydia*, which included four species: *C. trachomatis*, *C. pneumoniae*, *C. psittaci*, and *C. pecorum*.

C. trachomatis and *C. pneumoniae* are common pathogens in humans, but the routes of transmission, susceptible populations, and clinical presentations differ markedly. Genetic mapping of these two *Chlamydia* species, has confirmed little similarity as 70 genes in *C. trachomatis* do not exist in *C. pneumoniae* (Kalman et al. 1999). *C. psittaci* and *C. pecorum* occur mainly in animals, although *C. psittaci* may be also implicated in human respiratory diseases.

In 1999, a new taxonomic classification proposed to rename some *Chlamydiaceae* including *C. pneumoniae* and *C. psittaci* in *Chlamydophila pneumoniae* and *Chlamydophila psittaci* (Everett et al. 1999). However, the proposal to change the taxonomic nomenclature has not been universally accepted and both names are currently in use by different authors. The revised taxonomic classification currently contains at least 4 distinct families (*Chlamydiaceae, Simkaniaceae,*

Parachlamydiaceae and Waddliaceae) based on >90% 16S rRNA identity and a common developmental cycle.

The intracellular growth cycle of the *Chlamydiae* is complex and several growth options are possible, depending on the host-cell type, the particular environmental conditions in the host cell and the nature of tissue that is being affected. *Chlamydiae*, have a characteristic biphasic growth cycle within a eukaryotic host cell, during which infectious elementary bodies (EBs, 0.3–0.6 μm diameter) differentiate into the metabolically active but non infective reticulate bodies (RBs, 0.6–1 μm diameter) that divide by binary fission within the host derived vacuoles named Chlamydial inclusions. After 48–72 h, RBs multiply by binary fission and reorganize into EBs which are released after host cell lysis.

In vitro, this orderly alternation between EB and RB in life cycle development usually take place in 72 h, ranging from 36 to 96 h to complete, depending on each species, and in the number of inclusions per host cell (from one in *C. trachomatis* infected cell, to several inclusions for *C. pneumoniae* and *C. psittaci* infected cell).

Under *in vitro* conditions, RBs block division and maintain a stable association with the infected cell and become the aberrant or persistent bodies with enlarged forms, altered gene expression profile and multiple nucleoids instead of undergoing rapid replication and differentiating into infectious EBs (Fig. 4.1). During persistent growth, aberrant RBs continue chromosome replication but fail to divide (Gérard et al. 2001). These events constitute the basis of clinical persistence leading to chronic sequelae.

Although the life cycle of *Chlamydiae* is well characterized by microscopy, the signals that trigger interconversion of the morphologically distinct forms are not completely known (Beatty et al. 1994; Dautry-Varsat et al. 2005). However, EBs are no longer considered as inert organisms. The discovery that EBs can translocate stored proteins into the host under distinct signalling pathways is further evidence that the entry process results from a dialogue between the bacteria and the host, although many features including EB protein attachment to target cells remain to be clarified or discovered (Dautry-Varsat et al. 2005; Wuppermann et al. 2008).

4.2.2 Chlamydia Strategies for Evading Host Immune Response

Chlamydia pathogenesis depends on the cell population invaded, the initiation of the replicative genetic state of the pathogen and the efficiency of the release of effector molecules into the host cell.

A number of mechanisms can be considered to explain the evasion of host immune response. As many other intracellular bacteria (Brinkmann et al. 1987), endocystosed *Chlamydiae* are in fact sequestered within a host derived phagosome during the intracellular phase of developmental cycle. Their intracellular location largely protects them from antibody and complement attack.

Cell mediated immunity is the predominant component in controlling *Chlamydia* infection, even if *Chlamydia* antibodies may play a significant role in controlling the infection at a later stage of the disease (Zhong 2009). Moreover, studies using animal models have shown that both the IgA secreting B cells and IFNγ-producing

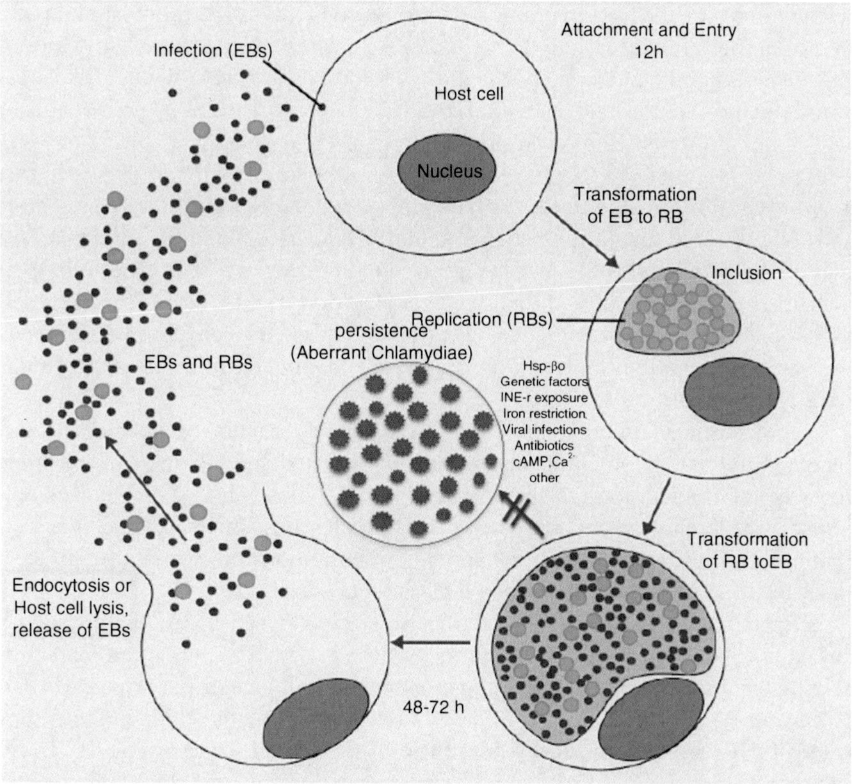

Fig. 4.1 Chlamydial cell life cycle development and aberrant Chlamydial bodies (persistence). The *in vitro* developmental cycle is fundamental to all *Chlamydia* and varies only in timing (from 48 to 72 to 96 h, depending on the species). Factors including nutrient depletion, viral infections (HSV), iron restriction, amino acids and certain antibiotics (penicillin, D-Cycloserin, Chloramphenicol, Sulfonamides) can induce *in vitro Chlamydia* persistent stage. Among cytokines, INF-γ directly inhibits bacterial growth and induces *Chlamydia* persistence by inhibition of 60 and 12 kDa cystein protein and tryptophan depletion which can stop the expression of MOMP that in turn stop the progress of RB division and RB conversion into EBs and formation of aberrant *Chlamydiae* (Modified and adapted from Behrens-Baumann (2007))

$CD4^+$ Th1 T cells are the most important adaptive immunity mechanisms in course of infection, although other immune components also play some roles (Morrison et al. 2000).

Despite these powerful host defence mechanisms, acute infection (if not treated) can activate inflammation, inducing the production of a wide variety of inflammatory cytokines, (IL-1, IL-6, IL-8 and TNFα) and can persist in some infected hosts (Rasmussen et al. 1997). In fact, *Chlamydia* species have shown a tendency to cause persistent infections that may also play a role in oncogenesis. In this regard, the induced inflammatory responses cannot only fail to effectively clear the infection but also contribute to inflammatory pathologies (Stephens 2003).

The failure by the host to eradicate the disease involves the establishment of a state of chronic infection in which *Chlamydiae*, after internalization into mononuclear cells, enter into a state of quiescence (cryptic body) with intermittent periods of replication and characterized by antigenic variation, production of Heat Shock Proteins (Hsps) and proinflammatory cytokines (capable of evading host defences) which trigger tissue damage (Stratton and Mitchell 1997). In this regard, Hsp-60, an ubiquitous and evolutionarily conserved chaperonin, normally sequestered inside the cell, particularly into mitochondria, and can elicit an immune response in humans which although directed against the microbial molecule but also reacts with endogenous Hsp-60 (Pockley 2003). During cell stress conditions, as well as during carcinogenesis, this chaperonin becomes exposed on the cell surface and/or is secreted from cells into the extracellular space and circulation.

Quantification of circulating Hsp-60 has recently become a potential useful marker of infection for clinicians in patients affected by a variety of diseases. However, interpretation of its values should be carefully evaluated, as a correlation between chaperonin levels and disease is difficult to establish. Hsp-60 is also a ligand of Toll-like receptors (TLR) and its expression on cell membrane surface's correlates with apoptotic phenomena (Ohashi et al. 2000).

During *Chlamydia* invasion and intracellular growth, sensors of the host innate immunity (pattern recognition receptors, PRR) can detect the infection by recognizing microbial components (pathogen associated molecular patterns, PAMPs). *Chlamydia* PAMPs such as Hsp-60 and Macrophage Infectivity Potentiator lipoprotein (MIP) are recognized by host PRR TLR4 and TLR2 respectively. These host receptors selectively recognize a broad spectrum of microbial components and endogenous molecules released by injured tissue (Bulut et al. 2002; Bas et al. 2008). In particular, TLR4 and TLR2 have been reported to be essential mediators of *C. pneumoniae* related host cell activation and defence (Kern et al. 2009).

MIP or other lipoproteins could be released from EB surface and RBs, and retain inside tissues where they might activate resident cells and perpetuate inflammatory response even after the eradication of live bacteria with antibiotic therapy (Bas et al. 2008). In general, PRRs, upon ligand binding, can lead to activation of various inflammatory signalling pathways including NF-κB, NF-IL-6 and MAP kinases.

Type III secretion (T3S) apparatus is another mechanism which seems central to the biology of the *Chlamydiaceae,* as it mediates the translocation of bacterial toxins to the cytosol of infected cells. It happens in several important gram-negative bacterial pathogens (Peters et al. 2007). It consist a molecular injection system protruding from the outer membrane that appears to be expressed and functional in acute as well as in chronic infection and may represent a prominent virulence factor.

A major role of T3S may also involve, ensuring growth and development of the pathogen by modifying apoptosis signals or some other transcriptional regulation important for *Chlamydia* survival.

4.2.3 *Chlamydia* Persistence and Chronic Infection

The exact role of persistent stage in the *Chlamydia* developmental cycle as well as the molecular mechanisms allowing persistence remains to be elucidated. *In vivo* studies of *Chlamydia* persistence are hampered by genotype definition and viability of organism. However, characterization of *in vitro* persistent phase of *Chlamydiae* and multiple lines of *in vivo* evidence, suggest that *Chlamydiae* persist in an altered form during chronic disease (Hogan et al. 2004).

Persistence has long been recognized as a major factor in the pathogenesis of *Chlamydia* disease. It has been described as a viable but non-cultivable growth stage resulting in a long-term relationship with the infected host cell that may not necessarily manifest as clinically recognizable disease.

It is distinct from unapparent infections, which may or may not involve evident *Chlamydia* growth and refers to an atypical, intracellular and metabolically less active state that is difficult to resolve not only by the host-defence system, but also by antibiotic therapy.

Unlike the re-infections believed to be the result of exposure to a *Chlamydia* serotype different from the initial, persistent infections are due to the same type of pathogen genotype entered into a metabolic quiescent and non-infectious form and responsible of three to ten recurrences which can last many years (Dean and Powers 2001).

In vitro studies have shown that several factors including nutrient depletion, cytokines, iron restriction, amino acids, Ca^{++}, and certain antibiotics can induce the Chlamydial persistent stage (Beatty et al. 1994; Raulston 1997; Dreses-Werringloer et al. 2000) (Fig. 4.1). In particular IFN-γ directly inhibits bacterial growth (Sueltenfuss and Pollard 1963) and induces *Chlamydia* persistence by allowing 60 and 12 kDa cystein protein inhibition and tryptophan depletion which can stop the expression of late proteins such as major outer membrane protein (MOMP) that in turn stop the progress of RB division and RB conversion into EBs leading to aberrant Chlamydial RBs. (Sardinia et al. 1988; Beatty et al. 1994). In this context, a complete transcriptome analysis of *C. trachomatis* serovar D growth in HeLa cells exposed to IFN-γ, demonstrated the up-regulation of many genes involved in active metabolic processes in the aberrant RBs, including those involved in DNA repair and recombination, protein translation, and phospholipid utilization (Belland et al. 2003).

Moreover, separate studies at transcriptional level have demonstrated a down regulation of *C. trachomatis* MOMP in HeLa cells and an upregualtion of *C. pneumoniae* MOMP in response to IFN-γ stimulation (Mathews et al. 2001; Molestina et al. 2002). This underlines the different roles played by MOMP in the two species. Also, HSP-60–1/groEL was found to be expressed predominantly during acute phase growth of *C. trachomatis* serovar K and that the Hsp-60-copy 2/Ct-604 gene transcript/protein was increased in iron-induced persistent cultures (Gérard et al. 2004).

Antibiotics such as penicillin and quinolones (such as ciprofloxacin and ofloxacin) have shown to favour persistence instead of resolving infection because of inducing aberrant but viable particles which may explain therapy failure (Dreses-Werringloer et al. 2000). *C. trachomatis* exposure to penicillin leads to enlarged and aberrant RBs, the so called "penicillin forms" that return to normal growth after penicillin removal. On the other hand, although Chlamydial RBs are killed by macrolide treatment (azithromycin), residual antigens can persist for more than 28 days continuing to harbour inflammatory responses (Wyrick and Knight 2004).

In vivo studies have shown that the presence of Chlamydial antigens and nucleic acids even in absence of cultivable organisms is indicative of persisting organisms probably as result of immunologic stimulation during chronic disease. Chlamydia rRNA demonstration may provide evidence for inapparent Chlamydial infections (Cheema et al. 1991).

All different species of *Chlamydia* have tendency to cause persistent infections that may play a role in chronic diseases (inflammation and scarring with significant damage to the host) and oncogenesis.

A chronic *C. pneumoniae* infection increases the expression of its own Hsp-60 which continuously stimulates host immune cells to produce anti-Hsp-60 antibodies inside but also outside cells. Thus the immune response to microbial Hsp-60 may gradually lead or contribute to autoimmunity to host self antigens and consequently, to the development of chronic diseases such as asthma, stroke, degenerative joint diseases, atherosclerosis, coronary heart disease (Kuo et al. 1995; Saikku et al. 1988; Elkind et al. 2000). All attempts to eradicate *C. pneumoniae* in patients with cardiovascular diseases have failed because the persistent state is completely refractory to antibiotic treatment (O'Connor et al. 2003). This problem should be solved in cell cultures and animal experiments before new human intervention trials.

Chlamydiae can ultimately favour the onset of cancer (Chaturvedi et al. 2010) or lymphoma. In this regard, a strong relationship between *C. psittaci* and its potential carcinogenic role in OAL development has been established (Ferreri et al. 2004) owing in part to their mitogenic *in vitro* activity (Rasanen et al. 1986; Byrne and Ojcius 2004), *in vivo* polyclonal lymphoproliferation and resistance to apoptosis in infected cells (Hanna et al. 1979; Miyairi and Byrne 2006).

Previous studies have also revealed that in *C. trachomatis* infection, the cytosolic levels of Hsp-60 *in vivo* gradually increase during carcinogenetic steps, from normal tissue to dysplasia to fully developed carcinoma in various organs (Cappello et al. 2009).

4.2.4 Chlamydia and Apoptosis

Cell death by apoptosis is an active and important defence mechanism against invading pathogens. Apoptosis has a direct role in many infectious diseases, especially those caused by viruses, intracellular protozoans and intracellular bacteria (Byrne and Ojcius 2004). For many of these pathogens, the apoptotic signalling starts from the pathogen and not by the host cell.

In this regard, *Chlamydia* inhibits apoptotic signalling cascades during productive growth as part of its intracellular survival strategy (Miyairi and Byrne 2006) in order to maintain the integrity of the host cell for the completion of its intracellular growth (Zhong 2009). This is in part due to the proteolysis of host proteins for ensuring its own intracellular replication while maintaining the integrity of the infected host cells for long periods of time. *Chlamydia* also inhibits apoptosis during persistent growth or in phagocytes, but induces apoptosis in T cells, which suggests that apoptosis has an immunomodulatory role in Chlamydial infections.

The anti-apoptotic activity has shown to be prolonged during *Chlamydia* persistence. This strengthens the hypothesis that active *Chlamydia* metabolism maintains host cell integrity and contributes to intracellular survival (Perfettini et al. 2002).

The circumstances that dictate whether the *Chlamydiae* inhibit or activate host-cell death reflect important pathogenic considerations, including whether an acute or chronic infection is in progress and whether intracellular *Chlamydia* growth is programmed to go through a productive infectious cycle or is stalled under non-productive growth conditions.

It is possible that apoptotic activity is controlled to some extent by the intracellular growth status of the *Chlamydiae*, which can be influenced by any or all of these considerations (Byrne and Ojcius 2004) and by strain.

While for *C. pneumoniae*, active inhibition of apoptosis occurs in epithelial cells, macrophages and neutrophils, for *C. trachomatis* and *C. psittaci* the anti-apoptotic activity has been demonstrated mainly in epithelial cells later in their developmental cycle (Miyairi and Byrne 2006; van Zandbergen et al. 2004).

It is not known exactly how pro-apoptotic and antiapoptotic effects correlate with the wide spectrum of clinical manifestations and Chlamydial diseases? Although a *Chlamydia* induced apoptotic activity has been hypothesised during acute manifestation of disease, whereas inhibition of apoptosis in chronic disease states (Byrne and Ojcius 2004).

Chronic infection and clinical persistence are closely related. Inhibition of apoptosis could represent a mechanism that has evolved to establish a chronic infection. Several lines of evidence suggest that to provoke chronic infection, *C. trachomatis* could adopt several strategies. One of these consists of being silent, resulting in asymptomatic infections that cannot be diagnosed at that time. This promotes bacterial progression, even to the most internal tissues. In addition, *C. trachomatis* MOMP displays variable immunodominant antigenic epitopes. Variations in these epitopes explain the absence of strain specific immunity and multiple re-infections by different serovars or by the same mutated serovar are still possible (Millman et al. 2001).

For these reasons, even if the initial infection is resolved, re-infections are possible and can lead to auto-pathological immune response induction (Beatty et al. 1994). Although re-infections occur, the refinement of Chlamydial diagnostic methods will allow us to establish whether *C. trachomatis* can persist.

4.3 *Chlamydiae* spp. and their Role in Cancer and in Chronic Diseases

4.3.1 C. pneumoniae

Although all *Chlamydiaceae* share the same unique developmental cycle but they show differences in tissue tropism, infectivity, inclusion morphology and cycle length.

C. pneumoniae is a gram-negative bacillus and a compulsory intracellular parasite which causes respiratory infection in more than 50% of adults. It is a widespread pathogen with serum positivity in more than 50% of the general population (Kuo et al. 1995).

Chronic infections by *C. pneumoniae* are associated with small and squamous lung cancer and cutaneous T-cell lymphoma (Littman et al. 2005; Abrams et al. 2001). Elevated *C. pneumoniae* antibody titers have been observed in individuals with lung cancer. Persons with elevated anti *C. pneumoniae* IgA antibody titers have up to a twofold increased risk for small cell carcinomas and adenocarcinomas of the lung (Littman et al. 2005). This risk is increased in male smoker patients with chronic *C. pneumoniae* infection (Kocazeybek 2003).

Other recent finding have shown a significant association between elevated *Chlamydia* Hsp-60 seropositivity and the risk for lung cancer, supporting an etiologic role of *C. pneumoniae* in lung carcinogenesis (Chaturvedi et al. 2010). A more recent meta-analysis involving 12 studies using data from the electronic databases PubMed, Embase, Web of Science and CNKI, has suggested that *C. pneumoniae* infection is associated with an increased risk of lung cancer, and higher IgA titres may be a better predictor of lung cancer risk (Zhan et al. 2011).

Accumulating evidence suggests that immunological events constitute in part the basis of the carcinogenic activity of *Chlamydia*. Previous *in vitro* studies have in fact demonstrated that superoxide oxygen radicals, TNF-α, IL1β and IL8 released by alveolar macrophages (AM) from healthy volunteers, play an essential role, contributing to lung tissue and DNA damage that may result in carcinogenesis (Redecke et al. 1998). Thus, AM play an important role as a target cell in *C. pneumoniae* infection because it can establish and maintain a productive infection *in vitro* and elicit a marked inflammatory response to the microorganism.

C. pneumoniae is also potent inducer of the proinflammatory cytokines TNF-α, IL-1β, and IL-6 in human monocytic cells, which may contribute to cancer development.

In this regard, this pathogen has the ability to activate PBMCs *in vitro*, as demonstrated by a cytokine response, and to grow inside these cells (as shown in PBMC circulating and in co-cultured PBMC, Fig. 4.2).

Monocytes and macrophages are known to be carriers of the infection from the respiratory site to other sites in the body and cause both local and systemic infection (Bodetti and Timms 2000). In fact, as happens in various chronic diseases,

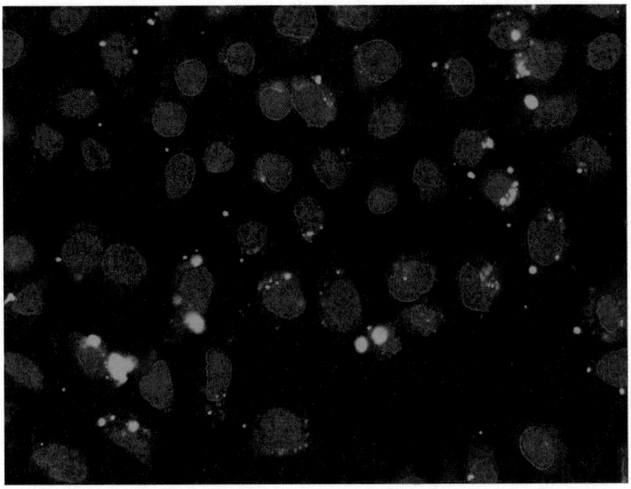

Fig. 4.2 Fresh PBMCs containing *C. pneumoniae* co-cultured on Hep-2 cells and immunostained with a fluorescein isothiocyanate-labelled anti-*Chlamydia* monoclonal antibody (Washington Research Foundation, Seattle, USA). *C. pneumoniae* show *bright apple-green* fluorescence (Magnification × 400)

this pathogen may be transported by infected monocytes and participate in the maintenance of local immunological response and inflammation, delaying its clearance (Airenne et al. 1999; Sessa et al. 2007; Cirino et al. 2006; Contini et al. 2008, 2010b).

C. pneumoniae can also replicate in lymphocytes suggesting that these cells may be an important host cell for dissemination of organisms and may alter lymphocyte functions and certain immune mechanisms in infected individuals (Haranaga et al. 2001). *C. pneumoniae* can thus survive in different immune cells, and use the immune cells as carriers for potentially cause chronic/persistent infections due to host inability for complete elimination of the pathogen (Yucesan and Sriram 2001; Stratton and Sriram 2003).

As for other *Chlamydiae*, *C. pneumoniae* has recently been connected to a range of chronic systemic diseases with a subsequent significant tissue damage including chronic inflammatory lung conditions (Teig et al. 2005), inflammatory arthritis (Gérard et al. 2000; Contini et al. 2010b) as well as cardiovascular and peripheral vascular diseases (Watson and Alp 2008; Mussa et al. 2006).

Although evidence exists for a direct role of *C. pneumoniae* in atherogenesis (Campbell and Kuo 2004) but this still remains to be confirmed (Ieven and Hoymans 2005). Especially, if *Chlamydiae* initiate atherosclerotic injury, whether it facilitates progression of existing plaques or merely colonizes the lesions, is not known.

C. pneumoniae is also considered a neurotropic pathogen since it is able to cross the blood-brain barrier and to infect the CNS. Monocytes may traffic *C. pneumoniae* across the blood-brain barrier, shed the organism in the CNS and induce neuroinflammation (Contini et al. 2010b). A growing body of evidence concerns the involvement of this pathogen in chronic neurological disorders and particularly in Alzheimer's disease and Multiple Sclerosis (MS).

The main obstacles that have so far presented to support a definitive role of *C. pneumoniae* in these diseases are represented by the fact that no methods exist to safely and confidently diagnose chronic infection, and because Chlamydial chronic infections are characterized by the inaccessibility of the "Chlamydial persistent state" to conventional anti Chlamydial agents.

Recent molecular, ultrastructural, and cultural advances have provided evidence that *C. pneumoniae* is viable and metabolically active in different biological compartments such as cerebrospinal fluid and PBMC from subgroups of MS patients, suggesting an association between this pathogen and MS disease (Sriram et al. 2005; Fainardi et al. 2008; Contini et al. 2010a). Hence, *C. pneumoniae* may induce a chronic persistent brain infection acting as a cofactor in the development of MS.

4.3.2 Other Chlamydiae

Among the other *Chlamydia* species involved in development of cancer, *C. trachomatis* can cause cervical dysplasia and cancer (Schachter et al. 1975) including ovarian cancer (Quirk and Kupinski 2001) and can also lead to vulvar carcinogenesis (Olejek et al. 2009).

C. trachomatis also causes chronic, clinically unapparent infections of the upper genital tract that may result in significant damage to the reproductive organs (Manavi 2006). In this regard, most *C. trachomatis* infections are asymptomatic or paucisymptomatic remaining undetected and thus untreated for a prolonged period with the possibility of developing chronic infections because of spread via monocytes and cause local and systemic infections.

Acute urogenital infections can also progress to persistent infection, which in turn may initiate a pathogenic process leading to chronic diseases, including pelvic inflammatory disease, ectopic pregnancy, tubal factor infertility, and *Chlamydia* induced arthritis in individuals who are genetically predisposed to this condition (Girschick et al. 2008). Importantly, *C. trachomatis* has been shown to be fully viable and metabolically active in both the acute and chronic, persistent infection state.

Infections by *C. psittaci* are predominantly associated with chronic conjunctivitis and have been recently linked with OALs of MALT type (Ferreri et al. 2004). This aspect will be discussed later in this chapter.

Besides OAL, *C. psittaci* infection was associated with non gastrointestinal MALT lymphomas (lung, thyroid and salivary gland, skin) and autoimmune precursor lesions (Hashimoto thyroiditis and Sjögren syndrome), suggesting possible involvement of *C. psittaci* induced antigen driven MALT lymphoma genesis (Aigelsreiter et al. 2011).

More recently, *C psittaci* was identified with various methods including PCR in an Italian patient with primary central nervous system marginal zone B cell lymphoma (PCNSL), a tumour mainly characterized by diffuse large B-cell histology and aggressive clinical course (Ponzoni et al. 2011).

4.4 Role of *Chlamydia* in MALT Lymphomas

MALT is the third commonest form of Non Hodgkin Lymphoma (NHL) and its incidence has risen over the last two decades (Müller et al. 2005). MALT lymphomas constitute a group of low-grade extranodal B-cell neoplasms classified as specific subtypes of NHL that share similar clinical, pathologic, immunologic, and molecular features and arise in the context of pre-existing prolonged lymphoid proliferation in mucosal sites. MALT NHL is classified as a peripheral B-cell neoplasm in the group of marginal zone B-cell lymphomas (MZLs) which accounts for 7–8% of adult NHL and comprises three subtypes: nodal, extranodal (MALT type), and splenic (Harris et al. 1994).

Despite overlapping morphologic features, nodal and extranodal MZLs demonstrate distinct clinical behaviour. Nodal MZL behaves as a conventional indolent and incurable low grade lymphoma, often widely disseminated early in the disease course. Extranodal MALT lymphomas, so called because arise outside the lymph nodes and spleen, remain localized for an extended duration, lack poor prognostic features, and have a superior 5-year overall survival rate (81% versus 56%) and failure-free survival rate (65% versus 28%) compared with its nodal counterpart (Nathwani et al. 1999).

The lymphoid tissue in which MALT lymphomas arise normally may be present at the site of origin (e.g., Waldeyer's ring or intestinal Peyer's patches) or may be acquired in the setting of chronic antigenic stimulation because of persistent inflammation or infection, or autoimmune disorder.

MALT lymphomas occur most commonly in the gastroenteric tract, but may affect a variety of extranodal sites including stomach and eye (Thieblemont et al. 1995; Lee et al. 2005).

Gastric MALT lymphoma, with its typical dependence on *H. pylori* infection, is the most common and the most extensively studied model for MALT lymphoma genesis. For this pathogen, all Koch postulates linking a specific organism to a disease have been fulfilled (Isaacson and Du 2004; Du 2007; Verma et al. 2008). Other examples of lymphoma genesis concern Epstein–Barr virus (EBV), a worldwide disseminated virus with carcinogenic activity (Magrath 1990).

MALT lymphomas with similar pathology and associated to infectious agents other than *H. pylori* and EBV have been found. Although, their role in MALT lymphoma genesis is not completely fulfilling Koch postulates. Such condition have been found in a variety of primary sites including the salivary gland, thyroid, lung, thymus, breast and ocular adnexa, and can develop in response to chronic and persistent stimuli during infection or autoimmune process (Verma et al. 2008)

In this setting, although for certain bacterial agents including *Campylobacter jejuni* and *Borrelia burgdorferi* (Du 2007), the role in lymphoma genesis is not definitely established, but for others, such as *Chlamydia*, a potential role has been established in MALT lymphomas and OAL. This microorganism, found at molecular and ultrasctuctural level, has been the target for unconventional therapy with encouraging results (Ferreri et al. 2004; Ponzoni et al. 2010).

OALs are the most common lymphomas of the ocular adnexa (Jenkins et al. 2000), occurring principally in the conjunctiva, orbital soft tissue and lachrymal apparatus. They include lesions affecting the conjunctiva, orbit, lachrymal glands, and eyelids. They constitutes a subtype of NHL (approximately 12%) of all MALT lymphomas, accounting for 60–80% of all OAL cases (Isaacson and Du 2004; Guidoboni et al. 2006).

OALs are mostly seen in the fifth to seventh decade of life (median age, 65 years), with a female predominance and the majority (85–90%) of patients present with localized disease (stage I). The slow evolution of symptoms, especially in conjunctival lymphomas explains the long median interval between onset of symptoms and time of diagnosis ranging from 1 month to 10 years (median, 7 months). One of the main reasons of the slow evolution may lie in the lack of connective tissue stroma of neoplastic lymphocyte population which is able to model surrounding tissue without causing particular irritation.

The most frequent site of origin is the orbit (40%), followed by conjunctiva (35–40%), lachrymal gland (10–15%), and eyelid (10%). Most patients with conjunctival lymphoma have visible, characteristic "salmon-pink patch" lesions reflecting tumour vascularity associated with conjunctival swelling, redness and irritation. Lack of symptoms is also possible in a large number of cases.

Orbital lymphoma most frequently presents with painless diffuse ill-defined or circumscribed mass that may be associated with proptosis, diplopia or motility disturbance. Only a few patients, primarily with orbital lymphoma, complain of pain or irritation.

For lymphoma arising in the lachrymal sac, swelling is one of the predominant symptoms. These are often non-specific, and might be misdiagnosed as chronic dacryocystitis with nasolacrimal duct obstruction.

MALT lymphoma may also be associated with chronic conjunctivitis, resembling what happens for gastric MALT lymphoma, which is associated with *H. pylori* derived chronic gastritis.

Several structural chromosomal abnormalities have been demonstrated in MALT lymphoma. Their frequency varies with the anatomic site of the lymphoma (Streubel et al. 2004).

The aetiology of OAL is currently unclear, although much attention has been focused on determining whether OAL is caused by an autoimmune disorder, chronic antigenic stimulation or both. Now it is becoming evident that infectious agents underlying chronic eye infection, as *Chlamydia*, herpes simplex and adenovirus may play a role in ocular lymphoma genesis in certain geographic regions. In general, *C. psittaci* and *H. pylori* may contribute to the pathogenesis of conjunctival

MALT lymphoma, whereas *Chlamydia* (*C. psittaci* and *C. pneumoniae*) and HCV can be considered the putative organisms that may play a role in the aetiology of orbital MALT lymphoma (Verma et al. 2008).

The chronic antigen stimulation hypothesis claims that a specific infectious agent initiates a reactive lymphoid infiltrate in the normally sterile ocular adnexal tissues and leads to a B-cell clonal expansion and proliferation. At this stage, genetic alterations and microenvironment may sustain the growth independent of the infectious agent. In this regard several genetic abnormalities have been reported in OALs of MALT type. Most common are trisomy of chromosome 3 and chromosome 18 in up to 68% and 57% of patients, respectively (McKelvie 2010).

Recently, homozygous deletions of the chromosomal band 6q23, involving the tumour necrosis factor α induced protein 3 (TNFAIP3, A20) gene, a negative regulator of NF-kB, were described in OALs, suggesting a role for A20 gene as a tumour suppressor in this disease (Honma et al. 2008). A20 inactivation may thus, represent a common mechanism for constitutive NF-kB activation, which may contribute to lymphoma genesis by stimulating cell proliferation and survival (Novak et al. 2009).

4.4.1 *Chlamydia spp. and Ocular Adnexal MALT Lymphoma*

The first report to provide sero-epidemiological evidence on the association between Chlamydial infections and malignant lymphomas was by Anttila et al. 1998. Lymphoma patients were found an approximate ten-fold risk to have both *C. pneumoniae* and *C. trachomatis* specific immune complexes present in their circulation. The risk was considerably higher in males than in females.

In the past few years several studies have reported more consistently the possible role of *Chlamydiae* in the development of OAL. Although the large number of papers reported in the literature concern *C. psittaci*, there is evidence that other *Chlamydiae* such as *C. pneumoniae* and *C. trachomatis* can somehow be implicated in the pathogenesis of OAL.

4.4.1.1 Data Supporting *C. psittaci* as Potential Infectious Agent in the Development of OAL

C. psittaci is an obligate and intracellular gram negative bacterium with seven known genotypes (A–F and E/B) that can be transmitted to humans. *C. psittaci* is the etiologic agent of psittacosis, a human infection caused by exposure to infected birds. *C. psittaci* can infect 465 avian species in 30 avian orders (Smith et al. 2005). Additionally, *C. psittaci* has been isolated in secretions of infected cats with upper respiratory tract infections of seasonal conjunctivitis, thus being an additional source of human infections (Sykes et al. 1999).

This microorganism has a well known tendency to cause persistent infections (Lietman et al. 1998). In particular, *C. psittaci* related follicular conjunctivitis may favour the development of OAL through a chronic antigenic stimulation (Lietman et al. 1998). The association of OAL with *C. psittaci* has been reported mostly in subjects coming from rural areas, following a prolonged contact with household animals (Ferreri et al. 2004).

The first evidence of such an association came from a study performed by Ferreri on an Italian cohort of patients, where *C. psittaci* DNA was detected by PCR in ocular lesions from 80% of patients with OAL and in none of non-neoplastic orbital lesions used as controls (Ferreri et al. 2004). In addition, PBMCs extracted from patients with OALs and harbouring *C. psittaci* DNA, were also positive (43%) for *Chlamydia* DNA, whereas controls were negative. In some cases, PBMC samples were positive for *C. psittaci* DNA more than 5 years after lymphoma diagnosis, with a concurrent relapse of the disease, suggesting that the natural history of OAL is associated with the persistence of the *C. psittaci* infection, due to chronic antigenic stimulation and further supporting an etiologic role of this bacterium (Ferreri et al. 2005).

The same group also reported the success in eradicating, (partially or completely) the disease after *C. psittaci* specific antibiotic therapy with tetracycline using standard dosing regimens for 3 weeks. This therapeutic approach has revealed to be fast, cheap and well tolerated for treating OALs even for aged and heavily pretreated patients with conventional more aggressive therapy (Ferreri et al. 2004, 2008a).

The assessment of *C. psittaci* DNA in patient's PBMCs was thus considered a potential and useful tool to evaluate disease eradication and even detect reinfection or reactivation. The genotypic and phenotypic differences among the strains of the pathogen should also be taken into account while performing this assessment (Rosado et al. 2006).

Finally, *C. psittaci* elementary bodies in the PBMCs were found viable and infectious and capable to grow *in vitro* and to be isolated from patients with OAMZL (Ferreri et al. 2008b).

Taken together, the fact that *C. psittaci* can be isolated and grown *in vitro* from biological samples of OAL patients improves the evidence level supporting this bacteria–lymphoma association and fulfils the second Koch's postulate hitherto reserved for *H. pylori* for gastric lymphoma (Du 2007; Falco Jover et al. 1999).

Other evidences suggesting the potential implication of *C. psittaci* in OAL come from Yoo et al. (2007) who found a high correlation between *C. psittaci* and OAL in Korean patients with a detection rate of bacteria DNA in about 80% and from Aigelsreiter et al. (2008) who detected *C. psittaci* DNA in 53.8% of MALT lymphomas from Austrian patients.

These evidences strengthen that *C. psittaci* and bacterium driven antigens may have a causal role in the development of OALs and in sustaining the disease over time, although often clinically unapparent.

4.4.1.2 Data not Supporting the Exclusive Role of *C. psittaci* in Development of OAL

Although OAL of MALT type was found in 40% of patients with prolonged contacts with household animals and a history of chronic conjunctivitis (Ferreri et al. 2004), the link between transmission from infected animals, establishment of an asymptomatic *C. psittaci* persistence and development of OAL must be definitively confirmed. In fact, Koch postulates linking *H. pylori* to gastric MALT lymphoma have been completely fulfilled, but there is much less evidence for a relationship between *C. psittaci* and OAL and results so far obtained are often conflicting (Grünberger et al. 2006; Daibata et al. 2006; Du 2007; Husain et al. 2007; Yakushijin et al. 2007; Decaudin et al. 2008; McKelvie 2010).

Since the first descriptions, many groups worldwide have investigated the possible association between *C. psittaci* and OAL occurrence, but concluded with discordant results in different countries and variable response to antibiotic treatment. The possibility that *C. psittaci* may be variably associated with OAL in various geographic regions, the genetic background of different populations, the varying epidemiological risk factors among different geographic areas, the fact that *C. psittaci* plays a causative role only in some OAL patients, may affect the incidence of these lymphomas.

To confirm what is still obscure regarding pathogenesis of OAL and the immunostimuli predisposing to this malignancy, the history of *C. psittaci* and prevalence of positive detection rates of *C. psittaci* in patients with OALs was examined. There are regions such as Italy and Korea where the prevalence of the microorganism is greater than 75%, others like Germany, Netherlands, UK, China, Eastern U.S. and Cuba, where the prevalence is included between 10% and 50%, while other countries such as France, Japan etc. have prevalence below 10% (Chanudet et al. 2006; Gracia et al. 2007; Decaudin et al. 2008; McKelvie 2010).

A suggestive study was done on OAL patients from Kenya, where although the underdeveloped conditions and the rather high plethora of infectious agents which could establish an excellent pabulum for the cancerous transformation and the development of the OAL, no association was found between OAL and *C. psittaci*. This suggests that other mechanisms in addition to infectious agents may play a role, including the genetic background of different populations and the genetic polymorphism that may be responsible for resistance or susceptibility to a particular infectious agent (Carugi et al. 2010). Whether this non-association between infectious agents and OAL can also be applied to other African countries or regions rests to be further investigated.

Currently, the scientific debate still continues. Two recent meta-analysis showed not only a striking variability in the prevalence of *C. psittaci* across different geographical regions and even between studies from the same geographic regions, but also the uneven effectiveness of antibiotic therapy for the treatment of OAL, essentially linked to the short follow-up (Husain et al. 2007) and also to the fact that the regression of lymphoma after antibiotic administration, is often observed in those patients not harbouring *C. psittaci* DNA, suggesting that other undiscovered doxycycline sensitive organisms may also be involved in the pathogenesis of disease. However, accumulating evidence seems to rule out the possible involvement of agents commonly associated with chronic eye diseases. No alternative pathogens have been found using PCR with universal pan bacterial primer (Husain et al. 2007; Matthews et al. 2008) as demonstrated in OAL patients from South Florida, in whom *C. psittaci* DNA was never isolated using the same techniques employed by Ferreri (Rosado et al. 2006).

In the light of these findings it might be assumed that the variability of the prevalence rated worldwide may be affected not only by geographical differences but also by the different sensitivity of the methods used for sample collection, PCR analysis and sequence diversity in the target region of *C. psittaci* (Vargas et al. 2006).

A large study involving an adequate number of cases from several countries and conducted with the same methodological diagnostic approach in order to exclude false-positive results, would be desirable.

In this context, however, the molecular approach based on touchdown enzyme time-release polymerase chain reaction (TETR-PCR) designed to simultaneously detect *Chlamydia* DNA proved to be the best method for the molecular detection of *C. psittaci* DNA and able to discriminate between *C. psittaci* and *C. pneumoniae* (Ponzoni et al. 2008).

Recently, a number of different techniques including immunohistochemistry, electron microscopy, laser capture microdissection and highly sensitive TETR-PCR employed to assess *Chlamydia* prevalence in both nodal and extranodal lymphomas. These techniques have detected *C. psittaci* in a consistent number of lymphomas arising in different organs from ocular adnexa, thus suggesting the possible contribution of this bacterium in lymphoma development and the possible adoption of the same antibiotic therapy as in OALs (Ponzoni et al. 2008). Table 4.1 shows the molecular techniques so far employed for detecting *Chlamydiae* in clinical samples.

The question remains whether all OAL patients should be treated for a possible infectious aetiology. In general, a diagnostic biopsy with immunophenotyping and genetic analysis to determine the grade of the lymphoma combined with PCR analysis to determine the presence of an infective agent should be recommended. If *C. psittaci* infection is found, a trial of oral doxycycline (100 mg twice daily for 3 weeks) is worthwhile. At present, despite the ocular side effects, localized radiotherapy still remains the standard first-line treatment for stage I low-grade OAL.

4.4.2 Role of *C. pneumoniae*

The first report on the possible association between *C. pneumoniae* and OAL of MALT type came from Shen and colleagues, reporting on a patient from Hong Kong with bilateral orbital MALT lymphoma. *C. pneumoniae* DNA, but not of

Table 4.1 Molecular protocols employed for detecting *Chlamydiae* in clinical specimens obtained from patients with OAL

Author	Patients no.	Specimen type	Method	Genes target (PCR product)	No. of positive patients /% *Chlamydia* detection
Ferreri et al. (2004)	40	Biopsy tissue[b] (formalin-fixed, paraffin-embedded) PBMC	Multiplex (TETR-PCR)[c] Sequencing	Amplified DNA from the end 16s rRNA and the beginning of 16S-23S spacer region primer pairs for *C. pneumoniae* and *C. trachomatis* (in 16s rRNA, 197 bp, 315 bp), for *C. psittaci* one primer in 16s rRNA and one in 16S-23S spacer region (111 bp)	32/80, *C. psittaci* (confirmed with sequencing in three patients)
Ferreri et al. (2005)	9	Biopsy tissue (formalin-fixed, paraffin-embedded) PBMC	Multiplex (TETR-PCR)	16s rRNA and 16S-23S spacer region (*C. psittaci* 111 bp)	9/100, *C. psittaci* 7/ 71, in biopsies before therapy
Rosado et al. (2006)	57	Biopsy tissue (fresh or formalin-fixed, paraffin-embedded)	TETR-PCR[c, d]	16s rRNA and 16S-23S spacer region (*C. psittaci* 111 bp)	0/–, *C. psittaci*
Chan et al. (2006)	8	Microdissected tumour cells from formalin-fixed, paraffin-embedded tissue	TETR-PCR[c]	16s rRNA and 16S-23S spacer region (*C. psittaci* 111 bp, *C. pneumoniae* 197 bp, *C. trachomatis* 315 bp) using ^{32}P-labelled primers	1/12.5, *C. pneumoniae*
Shen et al. (2006)	1	Microdissected tumour cells	Multiplex (TETR-PCR)[d]	16s rRNA and 16S-23S spacer region (*C. psittaci* 111 bp, *C. pneumoniae* 197 bp, *C. trachomatis* 315 bp)	1/100, *C. pneumoniae*

(continued)

Table 4.1 (continued)

Author	Patients no.	Specimen type	Method	Genes target (PCR product)	No. of positive patients /% *Chlamydia* detection
Mulder et al. (2006)[a]	19	Biopsy tissue (formalin-fixed, paraffin-embedded)	TETR-PCR	16s rRNA and 16S-23S spacer region (*C. psittaci* 111 bp)	0/–, *C. psittaci*
			Real-time PCR[e]	OmpA (*C.psittaci,* 82 bp)	
Daibata et al. (2006)[a]	21	Biopsy tissue (frozen or formalin-fixed, paraffin-embedded)	TETR-PCR	16s rRNA and 16S-23S spacer region (*C. psittaci* 111 bp)	0/–, *C. psittaci*
			n-PCR	16s rRNA (*C. psittaci*, 127 bp)	
			PCR	OmpA (*C. psittaci,* 248 bp)	
Liu et al. (2006)	17	Biopsy tissue (formalin-fixed, paraffin-embedded)	TETR-PCR Southern blot	16s rRNA and 16S-23S spacer region (*C. psittaci* 111 bp)	0/–, *C. psittaci*
Chanudet et al. (2006)[a]	142	Biopsy tissue (formalin-fixed, paraffin-embedded)	TETR-PCR	16s rRNA and 16S-23S spacer region (*C. psittaci* 111 bp)	31/22, *C. psittaci*
			Sequencing	New primer for 16s rRNA (*C. pneumoniae,* 73 bp and *C. trachomatis,* 116 bp)	
Ferreri et al. (2006)	27	Biopsy tissue (formalin-fixed, paraffin-embedded)	Multiplex (TETR-PCR)	16s rRNA and 16S-23S spacer region (*C. psittaci* 111 bp)	11/41, *C. psittaci*
		PBMC	Sequencing		
Zhang et al. (2007)	28	Biopsy tissue (formalin-fixed, paraffin-embedded)	TETR-PCR	16s rRNA and 16S-23S spacer region (C. *psittaci* 111 bp)	0/–, *C. psittaci*
			Real time PCR	16s rRNA (*C. psittaci* 148 bp)	
Yoo et al. (2007)	33	Biopsy tissue (formalin-fixed, paraffin-embedded)	TETR-PCR[c, d]	16s rRNA and 16S-23S spacer region (*C. psittaci*, 111 bp, *C pneumoniae* 197 bp, *C. trachomatis* 315 bp)	26/78, *C. psittaci* 3/9, *C. pneumoniae* 0/–, *C. trachomatis*

Husain et al. (2007)	458	Biopsy tissue PBMC	TETR-PCR	According to Madico et al. (2000) only (8 studies) and to Madico et al. (2000) plus other authors (3 studies[a])	92/23, *C. psittaci*
Ferreri et al. (2008a)	6	Biopsy tissue PBMC	TETR-PCR[c,d]	16s rRNA and 16S-23S spacer region (*C. psittaci* 111 bp)	6/100, *C. psittaci*
Matthews et al. (2008)	49	Biopsy tissue (fresh and frozen)	PCR[f,g] ALH-PCR[h]	16s RNA /V1-V2 hypervariable region (1,400 and 312–314 bp)	0/–, *C. psittaci*
			TETR-PCR[i]	ITS	
			Cloning Sequencing	16s rRNA and 16S-23S spacer region (*C. psittaci* 111 bp)	
Aigelsreiter et al. (2008)	13	Macrodissected formalin-fixed, paraffin-embedded tissue	Real time PCR[d, j]	16s rRNA and 16S–23S spacer region (C. *psittaci* 111 bp) *C pneumoniae* (227 bp)	7/54, *C. psittaci*
Ferreri et al. (2008b)	20	Biopsy tissue (formalin-fixed, paraffin-embedded) conjunctival swabs PBMC (fresh after culture in J774A.1 cells)	TETR-PCR[c, d] Immunostaining	16s rRNA and 16S–23S spacer region (C. psittaci, 111 bp, *C pneumoniae* 197 bp, *C. trachomatis* 315 bp,)	15/75 *C. psittaci*
			PCR[k, l] and RFLP	16s rRNA gene of *Chlamydiae* (*C. psittaci*, 269 bp RFLP with enzyme MseI)	
			Sequencing	Hsp60 (universal target 555 bp, *C. psittaci* 174 bp)	

(continued)

Table 4.1 (continued)

Author	Patients no.	Specimen type	Method	Genes target (PCR product)	No. of positive patients /% *Chlamydia* detection
Ponzoni et al. (2008)	35	Biopsy tissue (formalin-fixed, paraffin-embedded)	TETR-PCR PCR[l] n-PCR[m] Sequencing	16s rRNA and 16S-23S spacer region (*C. psittaci*, 111 bp, *C pneumoniae* 197 bp, *C. trachomatis* 315 bp,) Hsp60 (universal target 555 bp, *C. psittaci* 174 bp)	26/74, *C.psittaci* (OAL) 4/11, 1/3, *C. pneumoniae* and *C. trachomatis,* respectively (not OAL)
			IHC, IFS, EM, laser-capture microdissection	OmpA (*C. psittaci,* 207 bp and *C. pneumoniae* 333 bp)	
Contini et al. (2009)	1	Biopsy tissue PBMC (fresh and after co-culture in Hep2 cells)	n-PCR and RT-PCR[n,o,p,q] TETR-PCR	16s rRNA (*C. pneumoniae* 270 bp), 16s rRNA (*C. psittaci* 126 bp, *C. pneumoniae* 221 bp, *C. trachomatis* 412 bp)	1/100, *C. trachomatis* (confirmed by sequencing)
			Sequencing	OmpA (*C. pneumoniae* 488 bp, *C. psittaci* 494 bp, *C. trachomatis* 527 bp,) Hsp60 (*C. trachomatis*: ct110 114 bp, ct755 155 bp, ct604 161 bp) 16s rRNA and 16S-23S spacer region (*C psittaci* 111 bp, *C. pneumoniae* 197 bp, *C. trachomatis* 315 bp)	

Carugi et al. (2010)	39 (nine African, 30 Italian)	Biopsy tissue (formalin-fixed, paraffin-embedded)	TETR-PCR Sequencing FISH	16S rRNA and 16S–23S spacer region (*C. psittaci* 111 bp)	0/–, *C. psittaci* African patient 5/17, *C. psittaci* Italian patient

[a]Referred to studies which have performed PCR according to Madico procedure and other authors

[b]Refered to Ocular Adenxal Lymphoma specimen (*OAL*); Peripheral Blood Mononuclear Cells (*PBMC*); Touchdown Enzyme Time-Release Polymerase Chain Reaction (*TETR-PCR*) often designed to simultaneously detect *C. psittaci, C. trachomatis* and *C. pneumoniae* DNA; nested-PCR (*n-PCR*); Amplicon Length Heterogeneity Polymerase Chain Reaction (*ALH-PCR*); Reverse Transcriptase PCR (*RT-PCR*); Fluorescent *In Situ* Hybridization (*FISH*)

[c]Madico et al. (2000)

[d]Ferreri et al. (2004)

[e]Heddema et al. (2006)

[f]Lecuit et al. (2004)

[g]Suzuki et al. (1998)

[h]Cardinale et al. (2004)

[i]Rosado et al. (2006)

[j]Raggam et al. (2005)

[k]Ossewaarde and Meijer (1999)

[l]Hill et al. (2005)

[m]Tong and Sillis (1993)

[n]Mahony et al. (2000)

[o]Messmer et al. (1997)

[p]Cunningham et al. (1998)

[q]Gérard et al. (2004)

C. psittaci, was identified in biopsy of the lesion in a Chinese patient with bilateral orbital of MALT lymphoma (Shen et al. 2006). More recently, another case confirming evidence of molecular signature of *C. pneumoniae* was detected in a Chinese patient from Hong Kong with bilateral recurrent orbital MALT lymphoma who had a negative serology against *C. pneumoniae* and in whom neither *C. psittaci* nor *C. trachomatis* and *H. pylori* genes were found (Chan et al. 2006).

Although, there are few reports describing *C. pneumoniae* associated with MALT lymphomas, there is evidence of how these pathogens may relate to differences in the geographic incidence of their infections. *H. pylori* infection is higher in France, *C. psittaci* in Italy, and *C. pneumoniae* in Southeast Asia. These differences may also relate to the genotypic and phenotypic differences among the strains of the pathogen (Rosado et al. 2006).

However, the relatively high seroprevalence of *C. pneumoniae* (ranging from 40% to 90%) in some geographic areas, including Japan, Taiwan, and Hong Kong, limits the value of this serological index (Wang et al. 1993). Molecular identification may thus represent an interesting approach for the diagnosis of *Chlamydia* infection, notably for *C. pneumoniae*.

4.4.3 Role of *C. trachomatis*

4.4.3.1 Epidemiology and Clinical Features

Although, *C. psittaci* is the most detected pathogen in OALs, but the association between *C. psittaci* and OAL is not consistent around the world. *C. trachomatis* is an obligate intracellular gram negative bacterium which causes ocular-genital infections in humans with its highest prevalence amongst young men and women. Infection with this agent can be asymptomatic in up to 80% of women which can make diagnosis and detection difficult (Manavi 2006). The asymptomatic character of most genital Chlamydial infections makes this pathogen the major cause of bacterial sexually transmitted infections worldwide.

C. trachomatis has several serovariants based on the features of their major membrane proteins. Serovars A, B, Ba and C of biovar trachoma infect the conjunctival epithelium and lead to ocular infections that can progress to trachoma. In developing countries, these serovars remain endemic.

Trachoma is a chronic conjunctivitis in which, repeated Chlamydial infections, constant exposure to Chlamydial antigens and the concomitant presence of other bacterial infections constitute the basis of the disease. The transmission of infection occurs primarily within the family in situations of overcrowding, poor hygiene, the abundance of flies that can mediate the transmission of infection. To date, trachoma is a major cause of preventable blindness and is believed to be endemic in 56 countries (Polack et al. 2005).

It is estimated that approximately 1.3 million people are blind from this disease and probably a further 1.8 million have low vision. Overall, it is the eighth commonest blinding disease (Resnikoff et al. 2008). Owing to its special impact in resource poor nations, it has been recently placed on the WHO's priority list for intervention.

The serotypes D-K of *C. trachomatis* are most frequently involved in genital infections, with a prevalence of 16–38% in women belonging to populations at risk for sexually transmitted infections and 30–50% for male non-gonococcal urethritis. Genital infections by *C. trachomatis* infection are among the most common causes of reproductive tract diseases and infertility.

The incidence of Chlamydial infection in women increased from 79 per 100,000 in 1987 to 467 per 100,000 in 2003 (CDC 2004). Women with Chlamydial infection in the lower genital tract may develop an ascending infection that causes acute salpingitis with or without endometritis, also known as pelvic inflammatory disease (PID). PID can also be caused by genital *Mycoplasmas, endogenous vaginal flora, aerobic Streptococci* and *Neisseria gonorrhoeae*.

The dominant cause of PID in developed countries is considered to be genital *C. trachomatis* infection. Symptoms tend to have a subacute onset and usually develop during menses or in the first 2 weeks of the menstrual cycle. Women with *C. trachomatis* associated PID, tubal infertility, and ectopic pregnancy have high titers of serum antibodies to *Chlamydia* Hsp-60 (Toye et al. 1993). Twenty percent of women who develop PID become infertile, 18% develop chronic pelvic pain, and 9% have a tubal pregnancy (Manavi 2006).

PID has also been associated with increased risk of ovarian cancer (Risch and Howe 1995). The Centers for Disease Control and Prevention (CDC) recommend that physicians maintain a low threshold for diagnosing PID and that empiric treatment be initiated in women at risk of sexually transmitted disease (STD) who have uterine, adnexal, or cervical motion tenderness with no other identifiable cause.

The majority of men with urethral infection (non-gonococcal urethritis or post-gonococcal), whether symptomatic or not, have purulent urethral discharge typically low at presentation. The subjective symptoms are burning and dysuria. The inflammatory process is generally limited to anterior urethra, while a small percentage of cases may extend back to the urethra, prostate, according to some, in rare cases epididymis. Serovars D to K also infect respiratory epithelial cells and cause infant pneumonitis.

Serovars L1 and L3 cause lymphogranuloma venereum (LGV), a STD that occurs mostly in tropical countries. Rare cases of import are observed in Western countries. However, in the UK, it mainly occurs in men who have sex with men (Nieuwenhuis et al. 2004).

C. trachomatis, has also been implicated as a major organism that can trigger reactive arthritis (Girschick et al. 2008) which develops in a small percentage of individuals with Chlamydial infection and includes the triad of urethritis (cervicitis in women), conjunctivitis, and painless mucocutaneous lesions.

4.4.3.2 *C. trachomatis* and Development of OAL

Data concerning the possible involvement of *C. trachomatis* in the development of OAL seem fragmented and not supported by strong evidence of the literature. The MEDLINE and PubMed databases were searched for publications regarding the association between *C. trachomatis* and OAL as well as the efficacy of antibiotics employed for the treatment of OAL. The literature search spanned the period between January 2004 (when what to our knowledge was the first report of the possible association between *C. trachomatis* and OAL was published) and November 2010.

Only two papers were found. One of them featured a combination of asymptomatic MALT-type lymphoma in the left eye with chronic and bilateral conjunctivitis with giant follicles (Adult Inclusion Conjunctivitis) occurred in a 18 year old male patient suspected to be due to *C. trachomatis* on the basis of positive results for *Chlamydia* antigen assay (Yeung et al. 2004). However, no molecular and no other microbiological investigation was done to know what type of *Chlamydia* was involved. Moreover, conjunctivitis was unresponsive to doxycycline treatment.

The second paper concerns a case observed by our research group (Contini et al. 2009) concerning a 53-year Italian woman suffering from chronic HBV hepatitis, under treatment with antiviral nucleos(t)ide analogues, (initially lamivudine, 100 mg) once day continuously, then lamivudine plus adefovir (10 mg). No viral mutations were detected in the last 2 years, but quantification of HBV viral load yet showed detectable DNA viraemia and ALT were elevated and fluctuating.

The patient presented to the Ophthalmology clinic with an indolent natural history of bilateral OAL previously treated with chemotherapy. At that time, total body CT scan showed no lymph node enlargement and tissue involvement. However, microscopic and immunocytochemical analysis of osteomedullary biopsy and the flow cytometry analysis of bone marrow cells obtained from a bone marrow aspirate documented 12% neoplastic cells of clonal origin. The multicolour flow cytometry analysis excluded the presence of tumour cells in the peripheral blood. Following chemotherapy, a complete disappearance of clonal lymphoid cells was detected.

Because of the possible link between OAL and *Chlamydia* infection, she was investigated for *Chlamydiae* and other pathogens. Serological screening for *Chlamydia* did not detect anti-Chlamydial antibodies. Microbiological and serological analyses carried out for a number of bacteria including *Mycoplasma* and *Ureaplasma urealyticum,* proved all negative. Blood and urine cultures were negative. Synovial fluid cultures as well as urogenital swabs for detecting *C. trachomatis* did not grow inclusion bodies. Screening for *C. trachomatis* were reported to be negative but without any documentation. The screening by us made was negative. In any case, the patient never had urogenital symptoms.

PBMCs as well as lymphoma specimens were tested by molecular and cell culture assay (Hep-2 cell lines) based on additional centrifugation and extension of culture time (Contini et al. 2008; Contini et al. 2010a).

C. trachomatis 16S rRNA, ompA, together with different Hsp-60 encoding genes were demonstrated using highly sensitive nested PCR and reverse transcriptase

a

Primer Sequence (5'-3')	*C. tachomatis* Sequence Alignment	PCR protocol Program (35 cycles) /Product size (bp)
Ct604 (HsP60s) TTTATCTAGCGCAGTGGGTT **Ct604 (Hsp60r)** TATTCTACCGTTCCTGTTCC	**Sample** GTAGGGGGAAGCATATCCAAAATCCCACACAGGCATGCTCTTCGGGCATGCCAGNTATGT C. t. GTAAGGAGAAGCATATCCAAAATCCCACACA-GCATGCTCTTC-GGCATGCCAG-TATGT **Sample** TTCAGATTCTTCATGTTCTTCTACGATATAGTAACCGTTCTTCCCTCCATAACCCACTGC C. t. TTCAGATTCTTCATGTTCTTCTACGATATAGTAACCGTTCTTCCCTCCATAACCCACTGC **Sample** GCTAGATAAAA C. t. GCTAGATAAAA	95° C 40 sec 61° C 40 sec 72° C 40 sec 161 bp 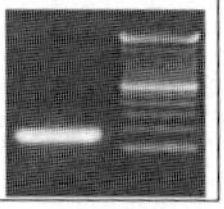

b

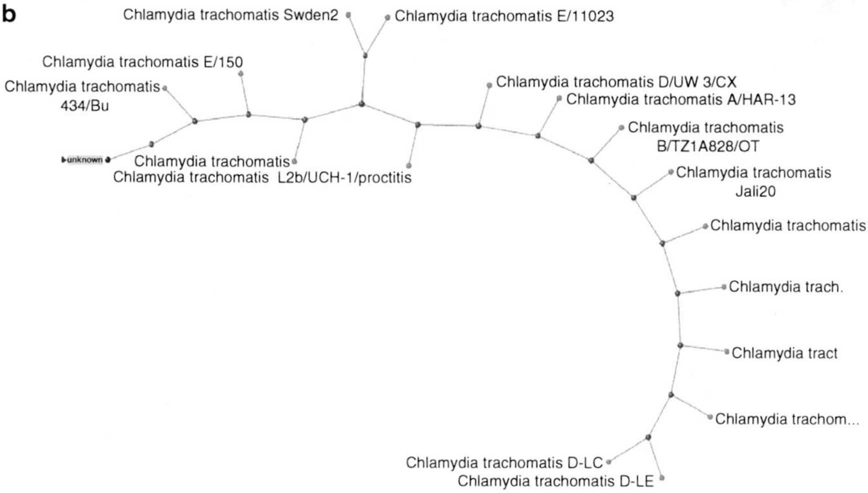

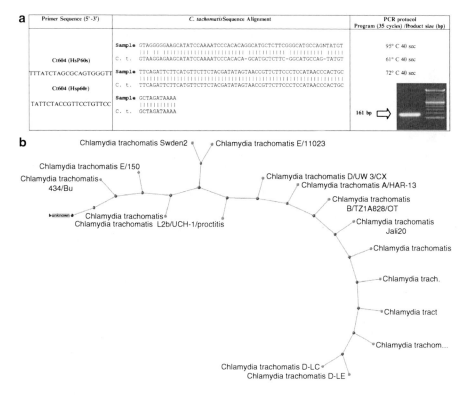

Fig. 4.3 Sample PCR gene product and phylogenetic tree (*panel A* and *B*, respectively) of *C. trachomatis* (C.t) analyzed by BLAST system showing the homology degree (99%) with Ct Hsp 60 (GroEL2) gene (GeneBank accession n. AY447002.1)

PCR (RT-PCR) in fresh lymphoma lesions and PBMC specimens co-cultured in optimized Hep-2 cell lines (Gérard et al. 2004; Contini et al. 2004, 2008; Pruckler et al. 1999). In particular, positive products of 412 bp and 527 bp corresponding to *C. trachomatis* 16S rRNA and ompA gene, respectively, and three HsP-60-encoding genes (Ct-110, Ct-604, Ct-755) corresponding to 114–161 bp were detected by PCR and RT-PCR in both clinical specimens. Sequencing of PCR products (ABI PRISM1 DNA Sequencing) detected a strict homology with *C. trachomatis*, as shown by BLAST analysis (Fig. 4.3).

The patient was given doxycycline 100 mg, twice a day, for 4 weeks without receiving any concomitant antiblastic or corticosteroid therapy in the period from the start of antibiotic therapy which was safe and well tolerated.

C. trachomatis PCR DNA and mRNA transcripts were negative in PBMCs after the conclusion of antibiotic treatment and a significant reduction of the ocular lesions was observed (Fig. 4.4). *C. trachomatis* eradication was monitored at molecular level by assessing patient's PBMCs by blinded investigators 6 months later and after 9 months of follow-up. An ultrasonography performed at 6 months

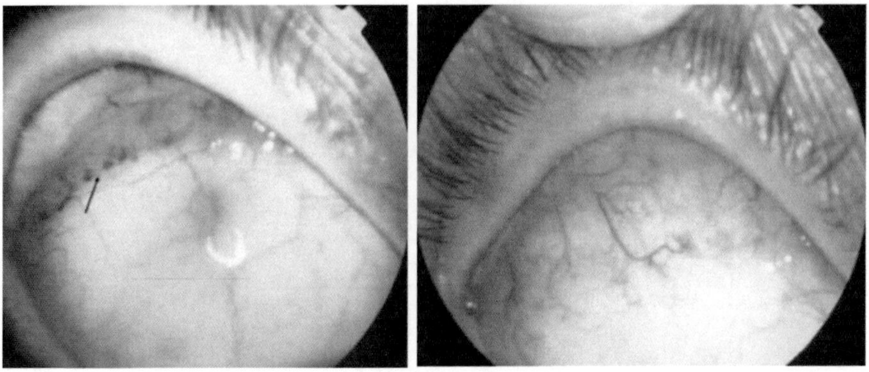

Fig. 4.4 Right conjunctival lymphoma with typical salmon colour with diffuse margins (*left, arrow*) and complete regression after doxycycline treatment (*right side*) (Reproduced from Contini et al. (2009)). This material is reproduced with permission of John Wiley & Sons, Inc.

from therapy, showed a complete disappearance of the lesion described. At 12 month from doxycycline assumption, *C. trachomatis* was no longer detectable in the patient's PBMCs.

One year after, the patient relapsed and a new ocular biopsy did detect OAL characterized by an indolent lesion with salmon red patch appearance in the bulbar conjunctiva of the right eye. Histological diagnosis was again a marginal zone B-cell lymphoma of MALT-type. PCR identified again all three genes products associated to Hsp-60 gene in absence of 16S rRNA and ompA gene, whereas RT-PCR revealed an amplification signal corresponding only to the single Ct-604 portion of Hsp-60 cDNA which continued to be positive after the suspension of doxycycline, in absence of OAL.

Doxycycline was readministered as earlier. A marked reduction of the eye lesion was observed. A new evaluation of patient's PBMCs 1 month after starting therapy did detect again the identical molecular profile. Thus, an antibiotic regimen based on clarithromycin plus levofloxacin (1 g per day for 1 month) was started. The PBMC RT-PCR was negative for Hsp-60 gene in absence of DNA (data not published).

Currently, the patient, at a distance of about 2 years from the lymphomatous manifestations, has no complaints and molecular tests are negative in PBMCs. Ultrasound examinations and CT scans of the eye are also negative. The PBMC follow up is performed every 6 months and so far there have not been observed signatures of *C. trachomatis* DNA.

4.4.3.3 Antibiotic Treatment and Significance of Molecular Recovery of *C. trachomatis* in Patients with OAL

The above experience expands the knowledge about the aetiology of OAL and suggests the following considerations:

First, the finding of *C. trachomatis* in peripheral blood and in eye bioptic samples from a patient suffering from OAL of MALT type indicates that also this pathogen is another doxycycline-sensitive organism that may be involved in the pathogenesis of OAL.

As previously reported, tetracycline has shown to be a valid alternative to chemotherapy and radiation without causing the same toxic side effects. However, it was eradicating at initial diagnosis but not at relapse.

If other *Chlamydia* strains are implicated and may justify their later resistance to doxycycline, is not known. There is not enough knowledge about the either of isolation of strains that display resistance to recommended antibiotics or of mechanisms of possible resistance in isolates from patients failing treatment. On the other hand, an evaluation of antimicrobial resistance and treatment outcome in *C. trachomatis* infection is complicated by the lack of standardized tests, as well as by the fact that *in vitro* resistance does not correlate with clinical outcome.

Second, although clinical studies on the employment of macrolides and quinolones alone or in association have not been largely performed, these compounds could be effectively employed in clinical settings when doxycycline doesn't work as expected.

The rationale for the use of macrolide clarithromycin is that this drug is among the first line agents for the treatment of uncomplicated genital infections by *C. trachomatis* other than tetracyclines, quinolones and penicillins.

Clarithromycin has also shown to exhibit antitumoral activity in murine cancer models and also in patients with MZL of the gastrointestinal tract (Matsumoto and Iida 2005). Conversely, ciprofloxacin and ofloxacin have shown to be not eradicating the organism from host cells *in vitro*, but rather to induce a state of *Chlamydia* persistence characterized by the presence of noncultivable, but fully viable bacteria and the development of aberrant inclusions (Dreses-Werringloer et al. 2000). Despite this, however, the association of both compounds has been effective in eradicating *C. trachomatis* Hsp-60 cDNA in clinical specimens as demonstrated by RT-PCR (Contini et al. 2009).

Third, the finding of Chlamydial DNA alone does not necessarily correspond to an established, replicating infection. Hence, the molecular approach to demonstrate the presence and viability of *C. trachomatis* by identifying rRNA primary transcripts is strongly justified. Therefore, molecular methods directed to detection of both DNA and mRNA on fresh biopsy specimens and PBMCs could be useful not only for diagnosis but also for monitoring the Chlamydial genes related to bacterial persistence and to assess the effectiveness of therapy in these patients. In this setting, culture system has revealed to be highly efficacious and sensitive, as active transcription of DNA from the organism was found in a persistent and metabolically active state in co-cultured PBMCs.

Finally, the absence of detection of multiple or single genes and in particular of the Ct 604 portion of Hsp-60 cDNA after clinical recovery, could represent a useful marker to monitor the progress of OAL. This however, remains an investigational approach which must be carefully assessed by properly designed trials.

The underlying HBV infection from which the patient suffered, is another important factor which might have contributed to favour a condition of continuation of the persistent antigenic stimuli and to trigger a B-cell expansion into a lymphoma.

Patients with HCV seropositivity have shown increased risk of developing B-cell non-Hodgkin lymphomas, particularly with an extranodal localization in the liver, spleen, salivary glands and OAL (De Vita et al. 1997).

Up to now, few studies have investigated the association between chronic HBV infection and NHL and OALs. Chronic HBV infected patients have shown to be nearly three times more likely to develop NHL than comparison patients suggesting the possibility that HBV may also play in addition to its established role in the induction of hepatocellular carcinoma, an etiologic role in the induction of B-NHL (Ulcickas Yood et al. 2007).

The possible contribution of HBV as well as HCV in lymphoma genesis is still a subject of discussion.

4.5 Conclusion

Infection from a variety of microorganisms must be considered as an important risk factor for cancer in humans. Approximately 20% of cancers worldwide are linked to viruses, bacteria, and parasites (Samaras et al. 2010) and the number of malignancies associated with specific infectious disease agents continues to grow.

Chlamydiae are intracellular bacteria growing in eukaryote cells with a tendency to cause persistent infection. *In vitro Chlamydia* persistence is well established and is characterised by altered growth and common ultrastructural traits. However, *in vitro* procedures currently available do not adequately reflect the complex system of events that occur *in vivo* and which lead to persistence, although there have been observations of altered morphological forms similar to those observed *in vitro*, for detection of *Chlamydia*.

A long list of questions awaits a definitive answer about the exact role of *Chlamydia* in the context of cancer development. Although many advances have been made in better definition of the growth cycle, signalling pathways together with biochemical and microbiological aspects of bacterial pathogenesis, the underlying pathophysiological processes that lead to the development of chronic inflammatory disease and persistence have not been defined. Furthermore, the specific molecular mechanisms inducing cancer development by this pathogen have not yet been fully clarified.

Chlamydia has been associated with several types of cancer (cervical dysplasia and cancer, by *C. trachomatis*, lung cancer and cutaneous T-cell lymphoma by *C. pneumoniae* and non gastrointestinal MALT-type lymphomas including OAL, by *C. psittaci*).

The high prevalence of *C. psittaci* in patients with OAL has suggested a potential role, although it has been documented an evident geographical variability and response to antibiotic treatment among different studies possibly related to the different techniques employed, the lack of stratification of response rates on histological subtypes of OAL, genetic and phenotypic differences of *C. psittaci* strains in different countries and the reduced time of follow-up of these patients.

In addition, the regression of lymphoma after antibiotic administration was often observed in patients not harbouring *C. psittaci* DNA, suggesting that other undiscovered doxycycline-sensitive organisms may also be involved in the pathogenesis of disease.

The molecular and cultural evidence of *C. trachomatis* demonstrated for the first time in a female patient with OAL, seems to suggest that this pathogen may also contribute to pathogenesis of such lymphoma.

Prospective studies on a larger number of cases, longer follow-up time and above all, the use of a panel of techniques directed to detection of both *Chlamydia* DNA and mRNA transcripts combined with appropriate cell cultures would be useful not only for screening and diagnosis but also for monitoring genes related to bacterial persistence and to assess the effectiveness of therapy in these patients.

The potential application of bacteria-eradicating therapy at local and systemic level may ultimately result in safer and more efficient therapeutic option for patients affected by these malignancies.

Their management will necessarily take in account a close collaboration between experts in ophthalmology, infectious diseases and haematology, but would certainly represent an exciting challenge for the next few years.

References

Abrams JT, Balin BJ, Vonderheid EC (2001) Association between Sézary T cell-activating factor, *Chlamydia pneumoniae*, and cutaneous T cell lymphoma. Ann N Y Acad Sci 941:69–85

Aigelsreiter A, Leitner E, Deutsch AJ et al (2008) *Chlamydia psittaci* in MALT lymphomas of ocular adnexals: the Austrian experience. Leuk Res 32(8):1292–1294

Aigelsreiter A, Gerlza T, Deutsch AJ et al (2011) *Chlamydia psittaci* infection in nongastrointestinal extranodal MALT lymphomas and their precursor lesions. Am J Clin Pathol 135(1):70–75

Airenne S, Surcel HM, Alakärppä H et al (1999) *Chlamydia pneumoniae* infection in human monocytes. Infect Immun 67(3):1445–1449

Anttila TI, Lehtinen T, Leinonen M et al (1998) Serological evidence of an association between chlamydial infections and malignant lymphomas. Br J Haematol 103(1):150–156

Bas S, Neff L, Vuillet M et al (2008) The proinflammatory cytokine response to *Chlamydia trachomatis* elementary bodies in human macrophages is partly mediated by a lipoprotein, the macrophage infectivity potentiator, through TLR2/TLR1/TLR6 and CD14. J Immunol 180(2):1158–1168

Beatty WL, Morrison RP, Byrne GI (1994) Persistent *Chlamydiae:* from cell culture to a paradigm for Chlamydial pathogenesis. Microbiol Rev 58(4):686–699

Behrens-Baumann W (2007) Chlamydial diseases of the eye. A short overview. Ophthalmologe 104(1):28–34

Belland RJ, Zhong G, Crane DD et al (2003) Genomic transcriptional profiling of the developmental cycle of *Chlamydia trachomatis*. Proc Natl Acad Sci USA 100(14):8478–8483

Bodetti TJ, Timms P (2000) Detection of *Chlamydia pneumoniae* DNA and antigen in the circulating mononuclear cell fractions of humans and koalas. Infect Immun 68(5):2744–2747

Brinkmann V, Remington JS, Sharma SD (1987) Protective immunity in toxoplasmosis: correlation between antibody response, brain cyst formation, T-cell activation, and survival in normal and B-cell-deficient mice bearing the H-2 k haplotype. Infect Immun 55(4): 990–994

Bulut Y, Faure E, Thomas L et al (2002) Chlamydial heat shock protein 60 activates macrophages and endothelial cells through Toll-like receptor 4 and MD2 in a MyD88-dependent pathway. J Immunol 168(3):1435–1440

Byrne GI, Ojcius DM (2004) *Chlamydia* and apoptosis: life and death decisions of an intracellular pathogen. Nat Rev Microbiol 2(10):802–808

Campbell LA, Kuo CC (2004) *Chlamydia pneumoniae*: an infectious risk factor for atherosclerosis? Nat Rev Microbiol 2(1):23–32

Cappello F, Conway de Macario E, Di Felice V et al (2009) *Chlamydia trachomatis* infection and anti-Hsp60 immunity: the two sides of the coin. PLoS Pathog 5(8):e1000552

Cardinale M, Brusetti L, Quatrini P et al (2004) Comparison of different primer sets for use in automated ribosomal intergenic spacer analysis of complex bacterial communities. Appl Environ Microbiol 70(10):6147–6156

Carugi A, Onnis A, Antonicelli G et al (2010) Geographic variation and environmental conditions as cofactors in *Chlamydia psittaci* association with ocular adnexal lymphomas: a comparison between Italian and African samples. Hematol Oncol 28(1):20–26

Center for Disease Control and Prevention, Atlanta (2004) Sexually transmitted disease surveillance 2003 supplement. Center for Disease Control and Prevention, Atlanta

Chan CC, Shen D, Mochizuki M et al (2006) Detection of *Helicobacter pylori* and *Chlamydia pneumoniae* genes in primary orbital lymphoma. Trans Am Ophthalmol Soc 104:62–70

Chanudet E, Zhou Y, Bacon CM et al (2006) *Chlamydia psittaci* is variably associated with ocular adnexal MALT lymphoma in different geographical regions. J Pathol 209(3):344–351

Chaturvedi AK, Gaydos CA, Agreda P et al (2010) *Chlamydia pneumoniae* infection and risk for lung cancer. Cancer Epidemiol Biomarkers Prev 19(6):1498–1505

Cheema MA, Schumacher HR, Hudson AP (1991) RNA-directed molecular hybridization screening: evidence for inapparent Chlamydial infection. Am J Med Sci 302(5):261–268

Cirino F, Webley WC, West C et al (2006) Detection of *Chlamydia* in the peripheral blood cells of normal donors using *in vitro* culture, immunofluorescence microscopy and flow cytometry techniques. BMC Infect Dis 6:23

Contini C, Cultrera R, Seraceni S et al (2004) Cerebrospinal fluid molecular demonstration of *Chlamydia pneumoniae* DNA is associated to clinical and brain magnetic resonance imaging activity in a subset of patients with relapsing-remitting multiple sclerosis. Mult Scler 10(4):360–369

Contini C, Seraceni S, Castellazzi M et al (2008) *Chlamydophila pneumoniae* DNA and mRNA transcript levels in peripheral blood mononuclear cells and cerebrospinal fluid of patients with multiple sclerosis. Neurosci Res 62(1):58–61

Contini C, Seraceni S, Carradori S et al (2009) Identification of *Chlamydia trachomatis* in a patient with ocular lymphoma. Am J Hematol 84(9):597–599

Contini C, Seraceni S, Cultrera R et al (2010a) *Chlamydophila pneumoniae* infection and its role in neurological disorders. Interdiscip Perspect Infect Dis [Epub Feb 21:273573]

Contini C, Grilli A, Badia L et al (2010b) Detection of *Chlamydophila pneumoniae* in patients with arthritis: significance and diagnostic value. Rheumatol Int Apr 10 [Epub ahead of print]

Cunningham AF, Johnston SL, Julious SA et al (1998) Chronic *Chlamydia pneumoniae* infection and asthma exacerbations in children. Eur Respir J 11(2):345–349

Daibata M, Nemoto Y, Togitani K et al (2006) Absence of *Chlamydia psittaci* in ocular adnexal lymphoma from Japanese patients. Br J Haematol 132(5):651–652

Dautry-Varsat A, Subtil A, Hackstadt T (2005) Recent insights into the mechanisms of *Chlamydia* entry. Cell Microbiol 7(12):1714–1722

De Vita S, Sacco C, Sansonno D et al (1997) Characterization of overt B-cell lymphomas in patients with hepatitis C virus infection. Blood 90(2):776–782

Dean D, Powers VC (2001) Persistent *Chlamydia trachomatis* infections resist apoptotic stimuli. Infect Immun 69(4):2442–2447

Decaudin D, Dolcetti R, De Cremoux P et al (2008) Variable association between *Chlamydophila psittaci* infection and ocular adnexal lymphomas: methodological biases or true geographical variations. Anticancer Drugs 19(8):761–765

Dreses-Werringloer U, Padubrin I, Jürgens-Saathoff B et al (2000) Persistence of *Chlamydia trachomatis* is induced by ciprofloxacin and ofloxacin *in vitro*. Antimicrob Agents Chemother 44(12):3288–3297

Du MQ (2007) MALT lymphoma: recent advances in aetiology and molecular genetics. J Clin Exp Hematop 47(2):31–42

Elkind MS, Lin IF, Grayston JT et al (2000) *Chlamydia pneumoniae* and the risk of first ischemic stroke: the Northern Manhattan Stroke Study. Stroke 31(7):1521–1525

Everett KD, Bush RM, Andersen AA (1999) Emended description of the order *Chlamydiales*, proposal of *Parachlamydiaceae* fam. nov. and *Simkaniaceae* fam. nov., each containing one monotypic genus, revised taxonomy of the family *Chlamydiaceae*, including a new genus and five new species, and standards for the identification of organisms. Int J Syst Bacteriol 49(Pt 2):415–440

Fainardi E, Castellazzi M, Seraceni S et al (2008) Under the microscope: focus on *Chlamydia pneumoniae* infection and multiple sclerosis. Curr Neurovasc Res 5(1):60–70

Falco Jover G, Martınez Egea A, Sanchez Cuenca J et al (1999) Regression of primary gastric B-cell mucosa-associated lymphoid tissue (MALT) lymphoma after eradication of *Helicobacter pylori*. Rev Esp Enferm Dig 91(8):541–548

Ferreri AJ, Guidoboni M, Ponzoni M et al (2004) Evidence for an association between *Chlamydia psittaci* and ocular adnexal lymphoma. J Natl Cancer Inst 96(8):586–594

Ferreri AJ, Ponzoni M, Guidoboni M et al (2005) Regression of ocular adnexal lymphoma after *Chlamydia psittaci*-eradicating antibiotic therapy. J Clin Oncol 23(22):5067–5073

Ferreri AJ, Ponzoni M, Guidoboni M et al (2006) Bacteria-eradicating therapy with doxycycline in ocular adnexal MALT lymphoma: a multicenter prospective trial. J Natl Cancer Inst 98(19):1375–1382

Ferreri AJ, Dognini GP, Ponzoni M et al (2008a) *Chlamydophila psittaci*-eradicating antibiotic therapy in patients with advanced-stage ocular adnexal MALT lymphoma. Ann Oncol 19(1):194–195

Ferreri AJ, Dolcetti R, Dognini GP et al (2008b) *Chlamydophila psittaci* is viable and infectious in the conjunctiva and peripheral blood of patients with ocular adnexal lymphoma: results of a single-center prospective case–control study. Int J Cancer 123(5):1089–1093

Gérard HC, Schumacher HR, El-Gabalawy H et al (2000) *Chlamydia pneumoniae* present in the human synovium are viable and metabolically active. Microb Pathog 29(1):17–24

Gérard HC, Krausse-Opatz B, Wang Z et al (2001) Expression of *Chlamydia trachomatis* genes encoding products required for DNA synthesis and cell division during active versus persistent infection. Mol Microbiol 41(3):731–741

Gérard HC, Whittum-Hudson JA, Schumacher HR et al (2004) Differential expression of three *Chlamydia trachomatis* hsp60-encoding genes in active vs. persistent infections. Microb Pathog 36(1):35–39

Girschick HJ, Guilherme L, Inman RD et al (2008) Bacterial triggers and autoimmune rheumatic diseases. Clin Exp Rheumatol 26(1):S12–S17

Gracia E, Froesch P, Mazzucchelli L et al (2007) Low prevalence of *Chlamydia psittaci* in ocular adnexal lymphomas from Cuban patients. Leuk Lymphoma 48(1):104–108

Groves FD, Linet MS, Travis LB et al (2000) Cancer surveillance series: Non-Hodgkin's lymphoma incidence by histologic subtype in the United States from 1978 through 1995. J Natl Cancer Inst 92(15):1240–1251

Grünberger B, Hauff W, Lukas J et al (2006) 'Blind' antibiotic treatment targeting *Chlamydia* is not effective in patients with MALT lymphoma of the ocular adnexa. Ann Oncol 17(3):484–487

Guidoboni M, Ferreri AJ, Ponzoni M et al (2006) Infectious agents in mucosa-associated lymphoid tissue-type lymphomas: pathogenic role and therapeutic perspectives. Clin Lymphoma Myeloma 6(4):289–300

Hanna L, Schmidt L, Sharp M et al (1979) Human cell-mediated immune responses to Chlamydial antigens. Infect Immun 23(2):412–417

Haranaga S, Yamaguchi H, Friedman H et al (2001) *Chlamydia pneumoniae* infects and multiplies in lymphocytes *in vitro*. Infect Immun 69(12):7753–7759

Harris NL, Jaffe ES, Stein H et al (1994) A revised European-American classification of lymphoid neoplasms: a proposal from the International Lymphoma Study Group. Blood 84(5):1361–1392

Heddema ER, Beld MG, de Wever B et al (2006) Development of an internally controlled real-time PCR assay for detection of *Chlamydophila psittaci* in the LightCycler 2.0 system. Clin Microbiol Infect 12(6):571–575

Hill JE, Goh SH, Money DM et al (2005) Characterization of vaginal microflora of healthy, nonpregnant women by chaperonin-60 sequence-based methods. Am J Obstet Gynecol 193(Pt 1):682–692

Hogan RJ, Mathews SA, Mukhopadhyay S et al (2004) Chlamydial persistence: beyond the biphasic paradigm. Infect Immun 72(4):1843–1855

Honma K, Tsuzuki S, Nakagawa M et al (2008) TNFAIP3 is the target gene of chromosome band 6q23.3–q24.1 loss in ocular adnexal marginal zone B cell lymphoma. Genes Chromosomes Cancer 47(1):1–7

Husain A, Roberts D, Pro B et al (2007) Meta-analyses of the association between *Chlamydia psittaci* and ocular adnexal lymphoma and the response of ocular adnexal lymphoma to antibiotics. Cancer 110(4):809–815

Ieven MM, Hoymans VY (2005) Involvement of *Chlamydia pneumoniae* in atherosclerosis: more evidence for lack of evidence. J Clin Microbiol 43(1):19–24

Isaacson PG, Du MQ (2004) MALT lymphoma: from morphology to molecules. Nat Rev Cancer 4(8):644–653

Jemal A, Center MM, DeSantis C et al (2010) Global patterns of cancer incidence and mortality rates and trends. Cancer Epidemiol Biomarkers Prev 19(8):1893–1907

Jenkins C, Rose GE, Bunce C et al (2000) Histological features of ocular adnexal lymphoma (REAL classification) and their association with patient morbidity and survival. Br J Ophthalmol 84(8):907–913

Johnson L, Wirotsko E, Wirotsko W et al (1996) *Mycoplasma*-like organisms in Hodkin's disease. Lancet 347(9005):901–902

Kalman S, Mitchell W, Marathe R et al (1999) Comparative genomes of *Chlamydia pneumoniae* and *C. trachomatis*. Nat Genet 21(4):385–389

Kern JM, Maass V, Maass M (2009) Molecular pathogenesis of chronic *Chlamydia pneumoniae* infection: a brief overview. Clin Microbiol Infect 15(1):36–41

Kocazeybek B (2003) Chronic *Chlamydophila pneumoniae* infection in lung cancer, a risk factor: a case-control study. J Med Microbiol 52(Pt 8):721–726

Kuo CC, Jackson LA, Campbell LA et al (1995) *Chlamydia pneumoniae* (TWAR). Clin Microbiol Rev 8(4):451–461

Lax AJ (2005) Opinion: bacterial toxins and cancer–a case to answer? Nat Rev Microbiol 3(4):343–349

Lecuit M, Abachin E, Martin A et al (2004) Immunoproliferative small intestinal disease associated with *Campylobacter jejuni*. N Engl J Med 350(3):239–248

Lee JL, Kim MK, Lee KH et al (2005) Extranodal marginal zone B-cell lymphomas of mucosa-associated lymphoid tissue-type of the orbit and ocular adnexa. Ann Hematol 84(1):13–18

Lietman T, Brooks D, Moncada J et al (1998) Chronic follicular conjunctivitis associated with *Chlamydia psittaci* or *Chlamydia pneumoniae*. Clin Infect Dis 26(6):1335–1340

Littman AJ, Jackson LA, Vaughan TL (2005) *Chlamydia pneumoniae* and lung cancer: epidemiologic evidence. Cancer Epidemiol Biomarkers Prev 14(4):773–778

Liu YC, Ohyashiki JK, Ito Y et al (2006) *Chlamydia psittaci* in ocular adnexal lymphoma: Japanese experience. Leuk Res 30(12):1587–1589, Letter to Editor

Madico G, Quinn TC, Boman J et al (2000) Touchdown enzyme time release-PCR for detection and identification of *Chlamydia trachomatis, C. pneumoniae*, and *C. psittaci* using the 16S and 16S—23S spacer rRNA genes. J Clin Microbiol 38(3):1085–1093

Magrath I (1990) The pathogenesis of Burkitt's lymphoma. Adv Cancer Res 55:133–270

Mahony JB, Chong S, Coombes BK et al (2000) Analytical sensitivity, reproducibility of results, and clinical performance of five PCR assays for detecting *Chlamydia pneumoniae* DNA in peripheral blood mononuclear cells. J Clin Microbiol 38(7):2622–2627

Manavi K (2006) A review on infection with *Chlamydia trachomatis*. Best Pract Res Clin Obstet Gynaecol 20(6):941–951

Mathews S, George C, Flegg C et al (2001) Differential expression of ompA, ompB, pyk, nlpD and Cpn0585 genes between normal and interferon-gamma treated cultures of *Chlamydia pneumoniae*. Microb Pathog 30(6):337–345

Matsumoto T, Iida M (2005) Extra-gastric lymphoma of MALT type and *H. pylori* eradication. Nippon Rinsho 63(Suppl 1):308–311

Matthews JM, Moreno LI, Dennis J et al (2008) Ocular adnexal lymphoma: no evidence for bacterial DNA associated with lymphoma pathogenesis. Br J Haematol 142(2):246–249

McKelvie PA (2010) Ocular adnexal lymphomas: a review. Adv Anat Pathol 17(4):251–261

Messmer TO, Skelton SK, Moroney JF et al (1997) Application of a nested, multiplex PCR to psittacosis outbreaks. J Clin Microbiol 35(8):2043–2046

Millman KL, Tavare S, Dean D (2001) Recombination in the ompA gene but not the omcB gene of *Chlamydia* contributes to serovar-specific differences in tissue tropism, immune surveillance, and persistence of the organism. J Bacteriol 183(20):5997–6008

Miyairi I, Byrne GI (2006) *Chlamydia* and programmed cell death. Curr Opin Microbiol 9(1):102–108

Molestina RE, Klein JB, Miller RD et al (2002) Proteomic analysis of differentially expressed *Chlamydia pneumoniae* genes during persistent infection of HEp-2 cells. Infect Immun 70(6):2976–2981

Morrison SG, Su H, Caldwell HD et al (2000) Immunity to murine *Chlamydia trachomatis* genital tract reinfection involves B cells and CD4(+) T cells but not CD8(+) T cells. Infect Immun 68(12):6979–6987

Mulder MM, Heddema ER, Pannekoek Y et al (2006) No evidence for an association of ocular adnexal lymphoma with *Chlamydia psittaci* in a cohort of patients from the Netherlands. Leuk Res 30(10):1305–1307

Müller AM, Ihorst G, Mertelsmann R et al (2005) Epidemiology of non-Hodgkin's lymphoma (NHL): trends, geographic distribution, and etiology. Ann Hematol 84(1):1–12

Mussa FF, Chai H, Wang X et al (2006) Chlamydia pneumoniae and vascular disease: an update. J Vasc Surg 43(6):1301–1307

Nathwani BN, Anderson JR, Armitage JO et al (1999) Marginal zone B-cell lymphoma: a clinical comparison of nodal and mucosa-associated lymphoid tissue types. Non-Hodgkin's Lymphoma Classification Project. J Clin Oncol 17(8):2486–2492

Nieuwenhuis RF, Ossewaarde JM, Götz HM et al (2004) Resurgence of lymphogranuloma venereum in Western Europe: an outbreak of *Chlamydia trachomatis* serovar l2 proctitis in The Netherlands among men who have sex with men. Clin Infect Dis 39(7):996–1003

Novak U, Rinaldi A, Kwee I et al (2009) The NF-{kappa}B negative regulator TNFAIP3 (A20) is inactivated by somatic mutations and genomic deletions in marginal zone lymphomas. Blood 113(20):4918–4921

O'Connor CM, Dunne MW, Pfeffer MA et al (2003) Azithromycin for the secondary prevention of coronary heart disease events: the WIZARD study: a randomized controlled trial. JAMA 290(11):1459–1466

Ohashi K, Burkart V, Flohe S et al (2000) Cutting edge: heat shock protein 60 is a putative endogenous ligand of the Toll-like receptor-4 complex. J Immunol 164(2):558–561

Olejek A, Kozak-Darmas I, Kellas-Sleczka S et al (2009) *Chlamydia trachomatis* infection in women with lichen sclerosus vulvae and vulvar cancer. Neuro Endocrinol 30(5):671–674

Ossewaarde JM, Meijer A (1999) Molecular evidence for the existence of additional members of the order *Chlamydiales*. Microbiology 145(Pt 2):411–417

Parkin DM (2001) Global cancer statistics in the year 2000. Lancet Oncol 2(9):533–543

Perfettini JL, Darville T, Dautry-Varsat A et al (2002) Inhibition of apoptosis by gamma interferon in cells and mice infected with *Chlamydia muridarum* (the mouse pneumonitis strain of *Chlamydia trachomatis*). Infect Immun 70(5):2559–2565

Peters J, Wilson DP, Myers G et al (2007) Type III secretion a` la Chlamydia. Trends Microbiol 15(6):241–251

Pockley AG (2003) Heat shock proteins as regulators of the immune response. Lancet 362(9382):469–476

Polack S, Brooker S, Kuper H et al (2005) Mapping the global distribution of trachoma. Bull World Health Organ 83(12):913–919

Ponzoni M, Ferreri AJ, Guidoboni M et al (2008) *Chlamydia* infection and lymphomas: association beyond ocular adnexal lymphomas highlighted by multiple detection methods. Clin Cancer Res 14(18):5794–5800

Ponzoni M, Ferreri AJ, Doglioni C et al (2010) Unconventional therapies in ocular adnexal lymphomas. Expert Rev Anticancer Ther 10(9):1341–1343

Ponzoni M, Bonetti F, Poliani PL et al (2011) Central nervous system marginal zone B-cell lymphoma associated with *Chlamydophila psittaci* infection. Hum Pathol 42(5):738–742

Pruckler JM, Masse N, Stevens VA et al (1999) Optimizing culture of *Chlamydia pneumoniae* by using multiple centrifugations. J Clin Microbiol 37(10):3399–3401

Quirk JT, Kupinski JM (2001) Chronic infection, inflammation, and epithelial ovarian cancer. Med Hypotheses 57(4):426–428

Raggam RB, Leitner E, Berg J et al (2005) Single-run, parallel detection of DNA from three pneumonia-producing bacteria by real-time polymerase chain reaction. J Mol Diagn 7(1):133–138

Rasanen L, Lehto M, Jokinen I et al (1986) Polyclonal antibody formation of human lymphocytes to bacterial components. Immunology 58(4):577–581

Rasmussen SJ, Eckmann L, Quayle AJ et al (1997) Secretion of proinflammatory cytokines by epithelial cells in response to *Chlamydia* infection suggests a central role for epithelial cells in Chlamydial pathogenesis. J Clin Invest 99(1):77–87

Raulston JE (1997) Response of *Chlamydia trachomatis* serovar E to iron restriction *in vitro* and evidence for iron-regulated Chlamydial proteins. Infect Immun 65(11):4539–4547

Redecke V, Dalhoff K, Bohnet S et al (1998) Interaction of *Chlamydia pneumoniae* and human alveolar macrophages: infection and inflammatory response. Am J Respir Cell Mol Biol 19(5):721–727

Resnikoff S, Pascolini D, Mariotti SP et al (2008) Global magnitude of visual impairment caused by uncorrected refractive errors in 2004. Bull World Health Organ 86(1):63–70

Risch HA, Howe GR (1995) Pelvic inflammatory disease and the risk of epithelial ovarian cancer. Cancer Epidemiol Biomarkers Prev 4(5):447–451

Rosado MF, Byrne GE Jr, Ding F et al (2006) Ocular adnexal lymphoma: a clinicopathological study of a large cohort of patients with no evidence for an association with *Chlamydia psittaci*. Blood 107(2):467–472

Saikku P, Leinonen M, Mattila K et al (1988) Serological evidence of an association of a novel *Chlamydia*, TWAR, with chronic coronary heart disease and acute myocardial infarction. Lancet 2(8618):983–986

Samaras V, Rafailidis PI, Mourtzoukou EG et al (2010) Chronic bacterial and parasitic infections and cancer: a review. J Infect Dev Ctries 4(5):267–281

Sardinia LM, Segal E, Ganem D (1988) Developmental regulation of the cysteine-rich outer-membrane proteins of murine *Chlamydia trachomatis*. J Gen Microbiol 134(4):997–1004

Schachter J, Hill EC, King EB et al (1975) Chlamydial infection in women with cervical dysplasia. Am J Obstet Gynecol 123(7):753–757

Sessa R, Di Pietro M, Schiavoni G et al (2007) Measurement of *Chlamydia pneumoniae* bacterial load in peripheral blood mononuclear cells may be helpful to assess the state of Chlamydial infection in patients with carotid atherosclerotic disease. Atherosclerosis 195(1):e224–e230

Shen D, Yuen HK, Galita DA et al (2006) Detection of *Chlamydia pneumoniae* in a bilateral orbital mucosa-associated lymphoid tissue lymphoma. Am J Ophthalmol 141(6):1162–1163

Slater DN (2001) *Borrelia burgdorferi*–associated primary cutaneous B-cell lymphoma. Histopathology 38(1):73–77

Smith KA, Bradley KK, Stobierski MG et al (2005) Compendium of measures to control *Chlamydophila psittaci* (formerly *Chlamydia psittaci*) infection among humans (psittacosis) and pet birds. J Am Vet Med Assoc 226(4):532–539

Sriram S, Ljunggren-Rose A, Yao SY et al (2005) Detection of chlamydial bodies and antigens in the central nervous system of patients with multiple sclerosis. J Infect Dis 192(7):1219–1228
Stephens RS (2003) The cellular paradigm of Chlamydial pathogenesis. Trends Microbiol 11(1):44–51
Stratton CW, Mitchell WM (1997) The immunopathology of Chlamydial infections. Antimicrob Infect Dis Newsl 16:89–94
Stratton CW, Sriram S (2003) Association of *Chlamydia pneumoniae* with central nervous system disease. Microbes Infect 5(13):1249–1253
Streubel B, Simonitsch-Klupp I, Müllauer L et al (2004) Variable frequencies of MALT lymphoma-associated genetic aberrations in MALT lymphomas of different sites. Leukemia 18(10):1722–1726
Sueltenfuss EA, Pollard M (1963) Cytochemical assay of interferon produced by duck hepatitis virus. Science 139(3555):595–596
Suzuki M, Rappe MS, Giovannoni SJ (1998) Kinetic bias in estimates of coastal picoplankton community structure obtained by measurements of small-subunit rRNA gene PCR amplicon length heterogeneity. Appl Environ Microbiol 64(11):4522–4529
Sykes JE, Anderson GA, Studdert VP et al (1999) Prevalence of feline *Chlamydia psittaci* and feline herpesvirus 1 in cats with upper respiratory tract disease. J Vet Intern Med 13(3):153–162
Teig N, Anders A, Schmidt C et al (2005) *Chlamydophila pneumoniae* and *Mycoplasma pneumoniae* in respiratory specimens of children with chronic lung diseases. Thorax 60(11):962–966
Thieblemont C, Berger F, Coiffier B (1995) Mucosa-associated lymphoid tissue lymphomas. Curr Opin Oncol 7(5):415–420
Tong CY, Sillis M (1993) Detection of *Chlamydia pneumoniae* and *Chlamydia psittaci* in sputum samples by PCR. J Clin Pathol 46(4):313–317
Toye B, Laferrière C, Claman P et al (1993) Association between antibody to the Chlamydial heat-shock protein and tubal infertility. J Infect Dis 168(5):1236–1240
Ulcickas Yood M, Quesenberry CP Jr, Guo D et al (2007) Incidence of non-Hodgkin's lymphoma among individuals with chronic hepatitis B virus infection. Hepatology 46(1):107–112
Van Zandbergen G, Gieffers J, Kothe H et al (2004) *Chlamydia pneumoniae* multiply in neutrophil granulocytes and delay their spontaneous apoptosis. J Immunol 172(3):1768–1776
Vargas RL, Fallone E, Felgar RE et al (2006) Is there an association between ocular adnexal lymphoma and infection with *Chlamydia psittaci*? The University of Rochester experience. Leuk Res 30(5):547–551
Verma V, Shen D, Sieving PC et al (2008) The role of infectious agents in the etiology of ocular adnexal neoplasia. Surv Ophthalmol 53(4):312–331
Wang JH, Liu YC, Cheng DL et al (1993) Seroprevalence of *Chlamydia pneumoniae* in Taiwan. Scand J Infect Dis 25(5):565–568
Watson C, Alp NJ (2008) Role of Chlamydia pneumoniae in atherosclerosis. Clin Sci 114(8):509–531
Wuppermann FN, Mölleken K, Julien M, Jantos CA, Hegemann JH (2008) Chlamydia pneumoniae GroEL1 protein is cell surface associated and required for infection of HEp-2 cells. J Bacteriol 190(10):3757–3767, Epub 2008 Feb 29
Wyatt JI, Rathbone BJ (1988) Immune response of the gastric mucosa to *Campylobacter pylori*. Scand J Gastroenterol Suppl 142:44–49
Wyrick PB, Knight ST (2004) Pre-exposure of infected human endometrial epithelial cells to penicillin *in vitro* renders *Chlamydia trachomatis* refractory to azithromycin. J Antimicrob Chemother 54(1):79–85
Yakushijin Y, Kodama T, Takaoka I et al (2007) Absence of Chlamydial infection in Japanese patients with ocular adnexal lymphoma of mucosa-associated lymphoid tissue. Int J Hematol 85(3):223–230
Yeung L, Tsao YP, Chen PY et al (2004) Combination of adult inclusion conjunctivitis and mucosa-associated lymphoid tissue (MALT) lymphoma in a young adult. Cornea 23(1):71–75

Yoo C, Ryu MH, Huh J et al (2007) *Chlamydia psittaci* infection and clinicopathologic analysis of ocular adnexal lymphomas in Korea. Am J Hematol 82(9):821–823

Yucesan C, Sriram S (2001) *Chlamydia pneumoniae* infection of the central nervous system. Curr Opin Neurol 14(3):355–359

Zhan P, Suo LJ, Qian Q et al (2011) *Chlamydia pneumoniae* infection and lung cancer risk: meta-analysis. Eur J Cancer 47(5):742–747

Zhang GS, Winter JN, Variakojis D et al (2007) Lack of an association between *Chlamydia psittaci* and ocular adnexal lymphoma. Leuk Lymphoma 48(3):577–583

Zhong G (2009) Killing me softly: Chlamydial use of proteolysis for evading host defenses. Trends Microbiol 17(10):467–474

Chapter 5
Salmonella typhi and Gallbladder Cancer

Catterina Ferreccio

Abstract This chapter reviews the current epidemiological indicators suggesting that *Salmonella typhi* could be a causal agent of gallbladder cancer (GBC). Both GBC and *S. typhi* occur in poor areas and are frequent in some Latin American (Chile, Mexico) and Asian (India, Korea) countries. In the developed world, typhoid or GBC are seen mainly among travelers or immigrants. While most people are susceptible to *S. typhi* infection, only 3% of infected become chronic carriers of *S. typhi* in their biliary tract. Gallstone carriers have a 6–15 times higher risk of becoming *S. typhi* carriers (Hofmann and Chianale, J Infect Dis, 167(4):993–994, 1993). *S. typhi* chronic carriers have 3–200 times higher risk of GBC than non-carriers, and 1–6% lifetime risk to develop GBC (Caygill et al., Lancet 343(8889):83–84, 1994; Eur J Cancer Prev 4(2):187–193, 1995). *S. typhi* associated relative risk of GBC is higher than that for gallstones (GS) alone; GS carriers have OR from 3 to 6 and less than 1% lifetime risk to develop GBC (Hsing et al., Biliary tract cancer. In: Schottenfeld D, Fraumeni JF (eds) Cancer epidemiology and prevention, 3rd edn. Oxford University Press, New York, pp. 787–800, 2006; Pandey, J Surg Oncol 93(8):640–643, 2006; Diehl, JAMA 250(17):2323–2326, 1983). The major steps in gallbladder carcinogenesis would be: *S. typhi* acute infection, *S. typhi* persistence as a chronic infection for decades, latency of 10 years to progress from metaplasia or dysplasia to pre cancer, and 5 years latency for invasion. Mechanistic research is needed to understand *S. typhi-GBC* association and provide tools for its control. Nevertheless, with the current epidemiological evidence much suffering could be prevented by: *(i)* health education, basic sanitation and food hygiene, to interrupt *S. typhi* transmission; *(ii)* treatment of chronic carriers, *(iii)* promotion of healthy diet and lifestyles to prevent gallstone formation, and *(iv)* cholecystectomy of gallstone disease in high risk areas to prevent invasive GBC.

C. Ferreccio (✉)
Departamento de Salud Pública/Escuela de Medicina, Pontificia Universidad Católica de Chile, Marcoleta 434, Santiago, Chile
e mail: cferrec@med.puc.cl

A.A. Khan (ed.), *Bacteria and Cancer*, DOI 10.1007/978-94-007-2585-0_5,

Keywords Gallbladder cancer • Cancer causes • Digestive cancer • Biliary tract cancer • Gallstones • Cholesterol gallstone • *Salmonella enterica* • *Salmonella typhi* • Enteric infections • Typhoid • Typhoid fever • Enteric fever • Amerindians • Chronic infection • Chronic carrier • Vi antibodies

Abbreviations

CDT	Cytolethal distending toxin
GB	Gallbladder
GBC	Gallbladder cancer
GS	Gallstone
GSD	Gallstone disease
HR	Hazard ratio
OR	Odds ratio
SMR	Standard mortality ratio
TF	Typhoid fever

5.1 Introduction

S. typhi is a human adapted pathogen causing self limited acute infections, 80% of which are asymptomatic; among symptomatic patients, 3% become chronic carriers of *S. typhi* in the biliary tract. Various studies have described *S. typhi* persistence in GB associated with GBC decades after the primo infection (Welton et al. 1979; Vaishnavi et al. 2005; Caygill et al. 1994, 1995; Mellemgaard and Gaarslev 1988; Lai et al. 1992; el-Zayadi et al. 1991; Andia et al. 2008). The major steps in GB carcinogenesis would be: *S. typhi* infection, certain host (gallstones) and bacterial (microfilms) factors that facilitate *S. typhi* persistence over decades, still other factors (endogenous or exogenous) that facilitate progression to dysplasia or precancerous lesions, and invasion. It is not known if this chronic infection requires live *S. typhi* or a dormant or L form of *S. typhi*. Other than *S. typhi, Salmonella paratyphi* A and B cause a similar clinical picture, but little is known about their potential for GBC carcinogenicity, and since these types are less frequent than *S. typhi,* they will not be considered in this chapter. We will first review *S. typhi* acute and chronic infection, then go over the epidemiology of GBC, ending with a discussion of the elements that make plausible a causal role of *S typhi* in at least some GBC cases.

5.2 *Salmonella typhi* Infection, Transmission and Acquisition

S. enterica serotype *typhi* is a member of the family *Enterobacteriaceae*, it has some antigens that induce antibodies used in clinical diagnoses of acute (somatic O and flagellar H) or chronic infection (capsular Vi). The Vi capsular antigen of *S. typhi* is

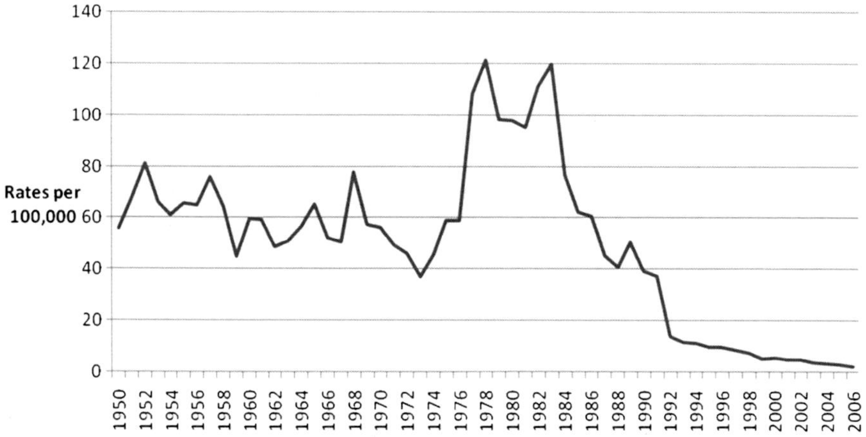

Fig. 5.1 Typhoid Fever Incidence in Chile 1950–2006 (*Source*: Personal elaboration from Chilean Ministry of Health Statistics)

also present in some strains of *S. enterica* serotypes *hirschfeldii (paratyphi C)* and *dublin*, and *Citrobacter freundii* (Parry et al. 2002). Humans are the reservoir and natural host of *S. typhi*. The fecal-oral route has been the main route of acquisition of *S. typhi* infection. *S. typhi* has been isolated in various food items, rivers, irrigation waters, and sewage, where it can survive several days (Sears et al. 1984, 1986). An infected person can transmit *S. typhi* directly by contaminated hand-mouth (as in children) or hand-food (as food-handlers) routes; these are called the short cycle of transmission, in contrary to the long cycle, in which the bacteria gets to the host after an environmental transit, typically sewage contaminated water sources. TF was an important killer in Europe and North America in the nineteenth century, but its transmission and occurrence declined sharply with the availability of basic sanitation for most of that population. The decline of TF began much earlier than any specific treatment (like Cloramphenicol and other antibiotics) or preventive measure (like Vi vaccines) were available, confirming the environmental nature of this infection. In developing countries, the most noticeable impact on TF was described in Chile, where it had been highly endemic since the firsts reports in the 1930s, associated with the wide use of sewage contaminated irrigation water. It affected the populated urban areas reaching its peak in the mid 1970s in the Metropolitan Region, with rates of 121/100,000 cases per population and lethality below 1%. TF dropped dramatically in Chile in 1990, being an unexpected result of the strong environmental and public health interventions including, police enforced prohibition of sewage contaminated irrigation waters, health education, food handlers control, which took place in response to the Cholera epidemic that was affecting the neighboring country of Peru that year (Fig. 5.1) (Laval and Ferreccio 2007). This panic driven intervention demonstrated what most epidemiologists knew at that time, that the provision of safe drinking water was not sufficient to stop transmission, a safe sewage disposal system was necessary. In most developing countries the final disposal of sewage is usually neglected, thus fecal contamination of food and secondary contamination of water sources is frequent.

5.2.1 Acute S. typhi Infection: Typhoid Fever (TF)

5.2.1.1 Natural History and Clinical Presentation

Once ingested, *S. typhi* reaches the small intestine, adhere to mucosal cells and invades the mucosa. It is transported to the underlying lymphoid tissue and to the mononuclear phagocyte system in lymphoid follicles, liver, and spleen, where *Salmonella* multiplies within the phagocytes. After an incubation period of 7–14 days, bacteria are released into the bloodstream to invade the liver, spleen, bone marrow, gallbladder and the intestine. This second invasion causes fever, malaise, headache, anorexia, nausea, abdominal discomfort, dry cough, and myalgia, lasting 2 weeks if uncomplicated. Up to 10% of convalescing patients with untreated typhoid excrete *S. typhi* in the feces for up to 3 months, 1–4% become long-term carriers, excreting the organism for more than 1 year. Up to 25% of long-term carriers have no history of typhoid. Chronic carriage is more common among women and the elderly and in patients with cholelithiasis (Parry et al. 2002). Antibiotic (Cloramphenicol) treatment for TF was introduced in 1948, decreasing significantly its lethality. In the last decades, however, there has been an emergence of resistance to chloramphenicol and other antimicrobial agents, which is causing concern (Parry et al. 2002). Current case-fatality rate is 1–10%, with an average of 3%. Nevertheless, the available data on the natural history of TF, as well as the estimates of its incidence and severity, is insufficient to estimate the real burden of the disease (Crump et al. 2008).

Risk Factors for Typhoid Fever

The main route of transmission in areas with high TF incidence is via the environment, by fecal contamination of water and food. In areas of intermediate incidence, the short cycle becomes more relevant, i.e. by the hand to mouth or hand to food routes. In areas of very low incidence, most cases are imported (infection acquired abroad). Poverty is a common factor for both environmental pollution and insufficient water and food handling controls. The short cycle is mainly caused by silent chronic carriers of *S. typhi*. Some genetic polymorphisms have been associated to the evolution of *S. typhi* infection (Kothari et al. 2008). There are no other vectors or animal reservoirs of *S. typhi*.

5.2.1.2 Worldwide Distribution of Typhoid Fever

Incidence in Endemic Areas

According to Crump et al. (2004), the current high risk (rates > 100/100,000 cases/year) areas for TF are South-Central Asia and South-East Asia, while medium incidence (rates 10–100/100,000 cases/year) areas include the rest of Asia, Africa, Latin America and the Caribbean, and Oceania (except Australia and New Zealand); in Africa the disease is probably under-reported. Europe, North America and the rest of the western world present low incidence (rates <10/100,000 cases/year) of TF. Crump also estimated that in 2000 there were 22 million cases of TF and 217,000 deaths. The incidence can vary widely depending on the study methods, as summarized by Kothari et al. (2008). A study in Pakistan in 2006 estimated an incidence rate of 170/100,000 using blood cultures, rising to 710/100,000 based on serological testing (Siddiqui et al. 2006).

Age Specific Incident Infection

Most cases occur in school-age children, with a sharp increase at 5–9 years of age, associated with an increase in the number of contacts once entering school, and higher levels of exposure at school, where children eat and share food. Incidence rates are high until age 20; at 20–34 years of age incidence of TF decreases, and after 35 years of age it drops even further, reaching a very low incidence at 65 years of age or older. Recent reviews suggest an inverse correlation between TF incidence and mean age of cases. In moderate incidence areas TF peak age is 5–19 years, while in highly endemic areas the peak incidence is at 1–5 years of age (Kothari et al. 2008; Crump et al. 2008). Most infections are asymptomatic, particularly in children under 2 years of age who do not develop the typical TF presentation, but a benign bacteremia (Ferreccio et al. 1984). It is not known if these early infections represent a higher risk of GBC at older ages nor if the asymptomatic infections present a similar rate of becoming chronic or latent infections than the symptomatic ones.

Seasonality

In many countries, TF is more frequent in the summer seasons, associated with a higher environmental transmission rate in that period. When the incidence is low, the age curve, as well as its seasonality, tends to flatten, and most cases are associated to food handlers with small outbreaks that affect all susceptible that have been exposed (Laval and Ferreccio 2007).

Recent Trends

In the last 20 years, there have been reports of the isolation of *S. typhi* strains resistant to Ampicillin, Cloramphenicol and Trimethoprim-sulfamethoxazole, estimated to be 12% in 1985–94 and reaching 17% in 1997. In USA, there have also been reports of increasing rates of isolation of strains resistant to Nalidix acid, a proxy for resistance to fluoroquinolones, from 19% in 1999 to 54% in 2006 with most cases occurring in travelers from the Indian subcontinent (Lynch et al. 2009).

In Asia, there have been reports of an increase in the isolation of *S. paratyphi A*, reaching 50% of the cases in some areas. Its clinical characteristics are indistinguishable from those of *S. typhi* (Crump and Mintz 2010).

5.2.2 *Chronic S. typhi Infection*

5.2.2.1 Prevalence of *S. typhi* Chronic Carriers

The chronic carriage rate varies markedly by age, from 0.1% in children to 13.5% in adults, and at age 50 the prevalence doubles that of age 20, whilst carriage is at least twice as high among women (Cvjetanovic et al. 1971). A study of the New York State register showed that only 0.3% of cases under 20 years old became chronic carriers, doubling the rate of carriage every 10-year interval, reaching 2.1%, 4.4%, 8.8% and 10.1% at ages 20–30, 30–40, 40–50 and 50–60 respectively (Ames and Robins 1943). During the acute phase, there was no difference in the rate of fecal recovery of *S. typhi* among males and females (Ames and Robins 1943). The rate of carriage is higher among those having TF concurrently with cholelithiasis (Levine et al. 1982). The prevalence of chronic carriers in India, based on Vi serology, was 1.8% among blood bank donors, in contrast with 7–12% among patients with various gastroenterological conditions (Vaishnavi et al. 2005). This is comparable with the figures estimated by Levine et al. (1982) for the Chilean population of Santiago in 1980, where chronic *S. typhi* carriage ranged from 0.1% to 1.0% in men and from 3.8% to 5.6% among women. Levine estimated that, there were 22,000 female and 4,000 male chronic carriers of *S. typhi*, with a prevalence of 6.9 per 1,000 inhabitants (Levine et al. 1982, 1984). The isolation of *S. typhi* was higher in acute cholecystitis (6–9%) than in chronic cholecystitis (1.5–3.6%) (Nervi et al. 1984; Ristori et al. 1982a, b; Levine et al. 1984). Levine's estimates were conservative because they were based on bacterial culture in bile. A follow-up of 45 carriers of *S. typhi* demonstrated that the bacteria persisted in 22% of them after gallbladder removal (Ristori et al. 1982b), indicating that in those cases *S. typhi* had a niche, probably in another part of the biliary tract.

5.2.2.2 Laboratory Methods to Identify Chronic Carriers

The most used method to identify chronic carriers of *S. typhi* has been the antibodies detection against its Vi capsular antigen. These tests have evolved improving their accuracy. Since 1930, the test was based on a Vi-rich *S. typhi* strain antigen and the detection technique was a direct bacterial agglutination, which caused both false negatives and false positives results (Lanata et al. 1983). In 1972, Wong and Feeley described a method based on a highly purified Vi antigen prepared from *Citrobacter freundii* and used a passive haemagglutination assay (Lanata et al. 1983). This new test was evaluated in both endemic and non-endemic areas, demonstrating that Vi serology was a moderately adequate screening technique to identify chronic carriers of *S. typhi* (sensitivity 75%, specificity 92%) (Lanata et al. 1983). In 1987, the technique was improved by the development of an ELISA test adapted to assess the relative occurrence of Vi-specific IgG, IgM, and IgA in the carrier state, resulting in a better test performance (IgG sensitivity 86% and specificity 95%), even in epidemic settings (Losonsky et al. 1987). The performance of IgM and IgA antibodies in detecting chronic carriers was poorer in comparison to IgG. The sensitivity of IgM-specific Vi was 37%, while IgA was present in similar proportion (72%) in chronic carriers and acute TF cases (Losonsky et al. 1987). Thus the most sensitive means to date for identifying chronic carriers had been the ELISA test to measure Vi antibody levels, which are usually much higher in chronic carriers than in acute infection or convalescence (Losonsky et al. 1987). Similar results were found in USA (Lin et al. 1988).

Recently, Tewari et al. (2010) compared Widal test, Vi antigen, culture and PCR among GBC and gallstone controls. He did not find any significant difference in the serological tests, but PCR in bile or tissue were positive only among GBC (3.7% and 33.3% respectively), suggesting that PCR is more specific for identifying chronic carriers (Tewari et al. 2010). Nath et al. (2008) compared GBC, gallstone patients, healthy population and corpses with regards to various methods of identifying *S. typhi* infection (Nath et al. 2008). Overall, the culture was positive in only 3.8% in GBC, 0.9% in gallbladder benign diseases and 0% in corpses. Vi antibodies were significantly higher among GBC than benign gallbladder disease or the healthy controls with 38%, 14% and 9% respectively (Nath et al. 2008). Antibodies against antigen O or antigen H were also higher among GBC cases than the two other groups (23%, 12% and 13% for antigen O respectively and 21%, 8% and 10% for antigen H). Also specific PCR against *S. typhi* flagellin was significantly higher among GBC than benign gallbladder disease or corpses without gallbladder disease (67%, 43% and 8%, respectively). Interestingly, in the control groups he found that women had higher *S. typhi* chronic infection than men at every age, confirming the higher female susceptibility to become *S. typhi* carriers. Male and female Vi antibodies were 0% and 10.2% under age 30, 20% and 24% at 30–60 years of age (Nath et al. 2008).

5.2.2.3 Determinants of Chronic Persistence of *S. typhi*

In some cases *S. typhi* will persist in the bile duct and feces and Vi antibodies will be higher than in acute infection or in convalescence. Subjects with gallstone disease have an odds ratio of 15 for becoming a chronic carrier compared with people without gallbladder disease (Hofmann and Chianale 1993). Females have two to three times higher risk of becoming carriers (Levine et al. 1982), but most women with GS who develop TF are able to clear the infection in a few months. *S. typhi* can be cleared completely by the cell mediated immune system and most cases become resistant to a new infection. Nevertheless, new episodes of TF years after an acute infection have been described, which could represent a new infection or a reactivation of a latent infection. *S. typhi* has not been studied enough to be able to determine whether the new infection is caused by the same or a new bacteria. Also, the traditional techniques to identify *S. typhi*, based on bacterial cultures, were very insensitive and there may have been many more asymptomatic chronic carriers than estimated.

In Chile, among 874 patients with confirmed TF, those who became chronic carriers had a higher frequency of erythrocyte ABH non-secretor phenotype than non-carriers of *S typhi*. This finding suggests that the carrier state may be genetically determined and that the ABH carbohydrates secreted in the bile may interfere with *S. typhi* adherence to the bile tract (Hofmann and Chianale 1993).

5.2.2.4 Mechanism of *S. typhi* Persistence in the Gallbladder

Salmonella is highly resistant to bile. The bacteria adhere to specific parts of GS suggesting a receptor which could either be cholesterol or calcium bilirrubinate (Prouty et al. 2002). It has been shown that *S. typhi* has the capability of forming a microfilm around the stones, where it is protected from the host bile and from antibiotics. Bile is the signal for the bacteria to form a microfilm (Prouty et al. 2002).

5.2.2.5 Characteristics of *S. typhi* That Could Explain Its Oncogenicity

The presence of other still unknown factors is necessary for a chronic GB infection with *S. typhi* to evolve into cancer. Characteristics of the infection itself may influence risk, for instance, *S. typhi* type or co-infection with other microorganism agents. Recent studies have demonstrated that *S. typhi* has certain characteristics that may explain its association with GBC. In 2010, Nath reviewed the mechanisms that could be behind *S. typhi* carcinogenicity. He proposed that *S. typhi* lives in the liver and is excreted intermittently into the gallbladder, where its metabolites and mutagens are concentrated more than ten times. He also discussed various carcinogens produced by *S. typhi,* such as bacterial glucuronidase, secondary bile acids and nitroso compounds. He emphasized the cytolethal distending toxin (CDT), a genotoxin with immunomodulatory capability which reaches the nucleus of the infected cell causing DNA damage (Nath et al. 2010). Thus, today there is enough mechanistic data to consider *S. typhi* as a plausible causal agent of GBC.

In the last few years there have been numerous reports of genetic and phylogenetic studies using new high throughput techniques which have permitted to identify *S. typhi* haplotypes and which open the field for more molecular epidemiological studies (Baker et al. 2010). It has also been speculated that the well known capability of bacteria to interchange DNA with other microorganisms, as well as with humans (lateral gene transfer), could be playing a role in the carcinogenic process (Khan and Shrivastava 2010).

5.2.3 Prevention and Control of Typhoid Fever

5.2.3.1 Environmental Hygiene

The main prevention strategy is environmental hygiene. It goes to the original cause of the problem and has many other positive externalities. Nevertheless, it requires significant initial resources which may not be available in areas where they are most needed. Thus, other individual based prevention strategies may be considered.

5.2.3.2 Individual Based Interventions Including *S. typhi* Vaccine, Control of Food Handlers, Eradication of Chronic Carriers

There are two widely used vaccines, the TY21 oral vaccine and the purified Vi parenteral vaccine. They both provide a protection of approximately 70%. The oral vaccine is given in four doses to children over age 6 with a booster every 5 years. The Vi vaccine is one intramuscular dose with a recommended booster every 2 years.

To decrease the transmission by silent chronic carriers the most efficient is health education. It is impractical and non-feasible to conduct a study of chronic carriers, either population-based detection or among food handlers. For this last group, what is usually done is to conduct an enteric disease prevention program through education and control of food handling practices. Additionally, in many places food handlers must get a *S. typhi* vaccine. To control localized outbreaks of TF, health authorities must search for a chronic carrier and when found she/he must be separated from food handling until eradication of ST has been documented.

5.3 Gallbladder Cancer

5.3.1 Epidemiology and Worldwide Distribution

The best available information of GBC incidence worldwide comes from the IARC (IARC 2009). In every country reported, the highest risk is found among women, thus we will use age standardized incidence rates per 100,000 women for worldwide comparisons. In Africa, the GBC incidence rates in women range from 0.4 to 0.7 in Uganda and Zimbabwe to 10.0 in Algeria. In Central and South America, rates range from around 3 in Argentina, Brazil and Costa Rica, increasing to around 7 in Peru and Ecuador, and peaking in 27.3 in Chile, being the highest ever reported worldwide. In North America, Canada provinces are mostly below 2, increasing to 5 in the Northwestern Territories, and in USA all states report rates below 2. In Asia, China reports rates from 2 to 5 in Shanghai, in India rates range from 0.8 in Madras to 8.6 in New Delhi, Japan reports rates from 4.5 to 6.5, Korea reports rates from

6 to 8, Kuwait, Thailand and Turkey's rates are all around 2, and Pakistan reports rates close to 5. In Europe, the lowest rates, around 1, are in UK, most countries report rates between 2 and 3. Oceania reports rates from 1 in Hawaii to 2.6 in Australia's Northern Territory (Table 5.1). Rates in the province of Valdivia in Chile are higher than the most cited previous IARC report of population-based cancer registries worldwide (Ferlay et al. 2004). The last IARC incidence data included for the first time the Valdivia data (GBC incidence 27.3 and 8.6 per 100,000 women and men) (Curado et al. 2007). A particularly high risk groups in Valdivia are Mapuche women from low socioeconomic status with an incidence rate of 269.2. These women are probably concentrating various risk factors (Bertran et al. 2010). Chile also ranks first in terms of GBC mortality among females, with an age standardized mortality rate of 12.7, followed by the Czech Republic and the Republic of Korea with 4.5 and 4.4 respectively. Among males, Korea, Chile and Japan occupy first place with rates of 6.5, 6.2 and 5.1 respectively (IARC 2009) (Fig. 5.2).

5.3.2 Clinical Presentation and Survival

Most of GBC cases (68%) present at a very advanced stage (stage IV) of the disease, with silent jaundice, weight loss and palpable abdominal tumour. The most frequent symptom is abdominal pain (83%), in 16% of cases the diagnosis is incidental during gallstone surgery, and in 9% of cases the diagnosis is post mortem. Nearly 80% of cases present GS (Bertran et al. 2010).

The 5-year survival is only 10.3% for population based series in high risk areas (Bertran et al. 2010), which is very similar to the 11–14% survival in hospital based series including high and low risk areas (Gabrielli et al. 2010). These figures are within the range of those reported worldwide, including USA (Konstantinidis et al. 2009), Japan (Kayahara et al. 2008), Mexico (Gomez-Roel et al. 2007) and India (Batra et al. 2005). In all scenarios, the main factor for survival is stage at diagnosis.

5.3.3 Natural History of GBC

Most (80%) GBC in high risk areas arise in gallbladders with GS. The prevalence of an incidental finding of GBC in gallbladder surgical specimens was around 4% in various studies in Mexico and Chile (Albores-Saavedra et al. 1980; Roa et al. 2006, 2009; Goldin and Roa 2009). In addition to the 4.0% of carcinoma, they found 0.8% of dysplasia. Most (92%) invasive GBC presented chronic inflammation of the gallbladder, 81% had dysplasia, 69% had cancer *in situ* and 66% had metaplasia in the tissue adjacent to the cancer. They also coincide in that the majority (72%) of mucous carcinoma was undetectable macroscopically. From the pathologist's view, the natural history of GBC begins with cholecystitis, continues with the development of dysplasia 10 years later, and ends in GBC 5–10 years after that

Table 5.1 Incidence rates of GBC in women worldwide

Age standardized (world) incidence rate (per 100,000) of gallbladder cancer, selected registries

		Male	Female
Africa	Algeria, Setif	2.1	10.0
	Tunisia, Centre, Sousse	1.8	3.1
	Egypt, Gharbiah	1.2	1.0
	Zimbabwe, Harare: African	0.7	0.7
	Uganda, Kyadondo Country	0.5	0.4
Gallbladder America, Central And South	Chile, Valdivia	12.3	27.3
	Ecuador, Quito	4.5	7.4
	Brazil, Goiania	3.1	3.5
	Brazil, Cuiaba	1.1	2.0
	France, La Martinique	1.1	2.2
America, North	USA, California Los Angeles: Korean	5.9	3.6
	USA, California Los Angeles: Japanese	2.6	1.4
	USA, California Los Angeles: Hispanic White	2.4	3.9
	USA, California Los Angeles: Filipino	2.3	2.0
	USA, Georgia: Black	0.9	1.1
	USA, District of Columbia: White	0.8	0.9
Asia	Japan, Yamagata Prefecture	8.4	6.0
	India, New Delhi	3.9	8.6
	Bahrain: Bahraini	0.8	0.9
	India, Nagpur	0.8	0.8
	Korea, Daegu	10.4	7.5
	Philippines, Manila	1.2	1.3
	Oman: Omani	0.8	1.1
	Thailand, Songkhla	0.8	1.2
	Malaysia, Sarawak	0.6	0.9
Europe	Czech Republic	4.4	6.0
	Italy, Varese Province	4.1	3.1
	Germany, Mecklenburg-Western Pomerania	3.7	4.9
	Spain, Navarra	3.1	3.0
	Slovak Republic	3.6	5.9
	France, Loire-Atlantique	1.2	1.0
	Belgium, Flanders	1.0	1.1
	UK, England, North Western	0.9	0.9
Oceania	Australia, Northern Territory	2.7	2.6
	New Zealand	1.3	1.4
	USA, Hawaii: Chinese	1.2	1.3

Source: Personal elaboration from International Agency for Research on Cancer (International Agency for Research on Cancer (IARC) 2009).

(Roa et al. 2009). Mean age of dysplasia cases was 46.3, early cancer 57.5, advanced cancer 59 and metastatic cancer 61 (Roa et al. 1996). The vast majority (98%) of the histological types of GBC correspond to adenocarcinomas, moderately or poorly differentiated (Albores-Saavedra et al. 1980; Roa et al. 2006, 2009; Goldin and Roa 2009).

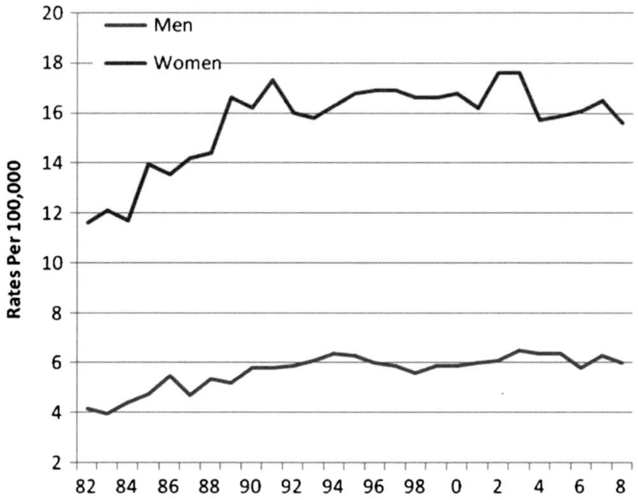

Fig. 5.2 Galbladder cancer mortality Chile 1982–2008 (*Source*: Personal elaboration from Chilean Ministry of Health Statistics)

Most GBC would arise from a metaplasia or dysplasia, while the minority arises from a pre-existing adenoma. These two pathways have different genetic alterations and, possibly, diverse etiologies (Goldin and Roa 2009). The first and most frequent pathway is through GS and chronic inflammation, and the other is associated with abnormality of the pancreatic-bile junction (Wistuba and Gazdar 2004). For the gallstone pathway, early TP53 inactivation in gallstone disease and chronic inflammation has been described, while KRAS mutations are frequent in the second pathway (Wistuba and Gazdar 2004).

5.3.4 Risk Factors of GBC

5.3.4.1 Gallstones

Data about prevalence of GS came from studies of autopsies which showed a wide variation among populations – 3–70% in women and 0–20% in men. In every study, prevalence increased with age, in women, and in Amerindians. In 1976, Brett reviewed autopsies reported worldwide, for the period 1870–1940: gallbladder stones were found in 37.8% of women and 26.2% of men in Vienna; 30% of women and 20% of men in Rochester, Stockholm and Copenhagen; and the lowest rates, 3–5% in women and 3–4% in men, were reported in Japan, Korea and Peking. From 1940 to 1973, reported gallstone prevalence were higher, 52.6% in women and 26.2% in men in Malmo, Sweden, and 40–50% in women and 17–20% in men in

American Indians in USA and Chileans from Santiago. In this same period the lowest prevalence, 3–7% in women and 0.5–6% in men, was reported for Leningrad, Singapore, Thailand, blacks in Johannesburg and blacks in Chicago (Brett and Barker 1976). Diehl (1998) estimated that one third of white American women carried GS, African Americans had half the prevalence of Hispanic-whites, and American Indians had the highest rates, reaching 70% of women. Similarly high rates have been reported in other North American Indians, suggesting a common genetic determinant (Diehl 1998). The prevalence of GS among American women living in USA (33% at age 50–59) (Diehl et al. 1980) was higher than that reported for women of the same age (22%) living in Mexico (Méndez-Sánchez et al. 2004). In South East England, the prevalence of GS in necropsies had little change in the period 1988–1998, varied from 20% to 19% among men and from 30% to 29% among women (Khan et al. 2009). Current studies using ultrasonography also show wide variation in the prevalence of GS worldwide: 15% among Turkish postmenopausal women (Karayalçin et al. 2010), and 10% among women over age 50 in Northern India (Unisa et al. 2011). A study comparing gallstone prevalence in Denmark and Germany confirmed the lower prevalence among Danish (women 13.1%, men 6.0%) than northeast German (women 24.8%, men 10.9%) populations (Friedrich et al. 2009). The highest prevalence (52.7%) was among German women aged 65–74 years. No difference was found in known risk factors between these populations that could explain the excess of risk. The newest report of GS prevalence in the general population came from the US National Health and Nutrition Examination, which gave the figures of 7% for GS and 5% for cholecystectomy (Ruhl and Everhart 2011).

5.3.4.2 Gallstones and GBC

In 1951, Lieber reported the association of GBC with GS in autopsies in Philadelphia, USA, which increased the prevalence of GBC three, five and tenfolds among white women, white men and black men respectively (Lieber 1952). This association has also been reported in the Latino population of USA, where the excess of GS in Mexican American women is accompanied by an excess of GBC death in this group, while other Latino populations in USA, from Cuba and Puerto Rico, do not present excess of lithiasis nor of gallbladder cancer (Diehl 1998). In Chile, prevalence of GBC as a finding in gallbladder surgery is higher in women than in men (3.4% and 1.3%); after age 70, this prevalence increases to 20% among women and 5% among men (Csendes et al. 1991). A recent prospective study of 2,018 subjects with gallstone disease followed for 18 years, found a higher all-cause mortality (Hazard Ratio (HR) 1.3, 95% CI 1.1–1.5), cardiovascular mortality (HR 1.3, 95% CI 1.1–1.5) and cancer mortality (HR 1.3, 95% CI 1.1–1.5); 5 cases of cancer were identified, of which only 1 corresponded to GBC. The risk was similar in subjects with stones than those with cholecystectomy (Ruhl and Everhart 2011). In England, the prevalence ratio of GBC in necropsies among those with GS was 3.4 for women and 4.4 for men (Khan et al. 2009).

Large stones represent a significantly higher risk than small stones (Csendes et al. 2000); compared with stones <1 cm, those ≥3 cm have an odds ratio for cancer of 9 (Diehl 1983) and 10 (Lowenfels et al. 1989). Stones >3 cm were more frequent in GBC (40%) than in similar age controls (12%) (Lowenfels et al. 1989).

Even though GS are always named as a cause of GBC, it is far from being proven (Wistuba and Gazdar 2004). The arguments in favor of stones causing GBC are: the ecological evidence that both diseases coincide in the same population and in the same sex (women); the presence of GS in the vast majority of GBC in high risk areas (85–90% in Chile, 60–70% in India); the positive association of risk of GBC with size of GS (relative risk of 10 for 3 cm vs. 1 cm stones); and the increase in incidence and mortality of GBC observed after a drop in number of gallstone surgeries (Serra et al. 1990; Chianale et al. 1990). Also, cohort studies demonstrated that people with GS have a 6.6-fold risk of dying from cancers, including GBC, than those without GS (Grimaldi et al. 1993). The arguments against its causality are that only 1–3% of gallstone cases will develop GBC and not all GBC are associated with stones (in Japan only 50% of GBC have stones), and that in areas where GBC is low, i.e. low risk areas, the gender preference of GBC tends to disappear, thus suggesting a different pathologic process (Wistuba and Gazdar 2004).

5.3.4.3 Genetic Susceptibility

As mentioned above, GBC is more frequent in populations with a high prevalence of GS, such as the natives of North, (Lemrow et al. 2008) Central and South America (Lazcano-Ponce et al. 2001). GS are strongly related with Amerindian genes (Galman et al. 2004; Méndez-Sánchez et al. 2004) and have been associated with an increase in bile acid synthesis, polymorphisms in the lipid metabolism (Galman et al. 2004), and a high rate of cholesterol lithogenesis (Galman et al. 2004). Recent reports have associated GS and GBC with polymorphisms in the lipid metabolism and insulin sensitivity in Chinese populations (Chang et al. 2008; Andreotti et al. 2008).

5.3.4.4 Lifestyle and Hormonal Factors

In Chile, those residing in rural areas are protected from GBC (Bertran et al. 2010) and this is in agreement with ecological studies suggesting a higher prevalence of environmental risk factors in urban areas, e.g. rate of typhoid infection (Laval and Ferreccio 2007) and hormone use and obesity (Serra et al. 2002), or of protective factors in rural areas, e.g., higher fruit and vegetable intake (Serra et al. 2002) and higher physical activity. The protective effect of rural residence overcomes the fact that the rural population has a higher rate of poverty.

Hsing et al. (2006) reviewed the evidence of the effects of diet on GBC, indicating that dietary components (sugar and fat) play a smaller role compared with obesity, physical activity and energy balance (Hsing et al. 2006).

The review of the evidence of the effects of estrogens in GBC concluded that endogenous estrogens would increase GBC risk by promoting gallstone formation or through a direct effect on the biliary epithelium, but that there was no clear association with exogenous estrogens (Hsing et al. 2006).

5.3.4.5 Environmental Factor

In a review, Pandey concluded that there was not sufficient evidence of an association of GBC with chemical contaminants (Pandey 2006). In Chile, the heavy metal contaminated areas of Northern regions present below average risks of GBC, much lower than Southern areas where chemical contamination is lower (Andia et al. 2008).

A recent analysis of cancer deaths among immigrants in Sweden (Hemminki et al. 2010) found that Indian and Chilean women had an excess of GBC. Interestingly, these women immigrated to Sweden after age 27, thus the environmental agent – biological, chemical or physical – must have acted at earlier ages.

5.3.4.6 Epidemiological Evidence of the Association Between *S. typhi* Infection and GBC

Until recently, the existing methods did not accurately separate infected from non-infected, posing an extra challenge for studies of the association of S. *typhi* infection and GBC. Thus, today the role of *S. typhi* in GBC is highly debated. Here we review the arguments in favor of the theory.

Retrospective Cohort Studies

One of the first epidemiological reports of the association between *S. typhi* infection and GBC came from Welton et al. (1979), who compared the causes of death of 471 chronic carriers of *S. typhi* with 942 controls matched by age-sex, year of death, and place of birth and death, reporting that chronic carriers had a six times higher risk of dying from GBC ($p<0.001$). A similar finding was reported by Mellemgaard and Gaarslev (1988) based on the Danish registry of *S. typhi* carriers, comparing the observed causes of death of 219 chronic carriers with age-time-sex incidence rates for the whole country. A standard mortality ratio (SMR) of 3.85 for hepatobiliary tract cancer was reported. In 1994, Caygill et al. reported the SMR of GBC for chronic carriers of *S typhi*. They included two groups of carriers: 83 cases from the Glasgow carrier register and 387 cases identified during the follow-up of the 1964 Aberdeen outbreak. The same analysis was performed for the remainder subjects of the Aberdeen outbreak who did not become chronic carriers. They found that carriers had a GBC SMR of 167.0 (95% CI 54.1-389), while among the TF cases who did not become carriers, there was no case of GBC. They estimated a lifetime risk of gallbladder cancer of at least 6% for carriers. In 1995, Caygill et al. reanalyzed this data and presented the latency from date of infection to date of death for 20 GBC cases: less than 10 years in 8 cases (40%), 10–20 years in 4 cases (20%), 20–30 years in 3 cases (15%), and more than 30 years in 5 cases (25%). In the follow-up of the Aberdeen cohort, the cumulative incidence of cholecystectomy was 19% among those who became *S. typhi* carriers and 2% among those who cleared the *S. typhi* infection, with a relative risk of 6.1 (95% CI 2.6–14.0). In their cohort, Caygill et al. estimated a 200-fold risk of developing hepatobiliary carcinoma among carriers of *S. typhi* in comparison with those who cleared the infection, whereas carriers of GS only increased their risk 2–15 times (Caygill and Hill 2000).

Case-Control Studies

In 2000, Shukla et al. conducted a case-control study of GBC using as controls subjects with gallstone disease and subjects with non-hepatobiliary diseases, measuring Vi antibodies by indirect hemagglutination. They reported an 8 (95% CI 2–37) times higher prevalence of Vi antibodies in the GBC cases (29.4%) than the non biliary tract disease cases (5%) and three times higher prevalence than the gallstones cases (10.7%). Most interestingly, 40–45% of subjects on either group presented antibodies (somatic O and flagellar H) against other antigens of *S. typhi* not associated with chronic carriage (Shukla et al. 2000), suggesting this is a population highly exposed to *S. typhi*.

In Egypt, el-Zayadi et al. (1991) found a significantly higher rate of carriage of *S. typhi in* stool cultures among cases of bile duct carcinoma (39.1%) and cases of icteric gallstone (34.0%) than in healthy controls (2.0%) (OR 19.5 and 17.0 respectively). Based on Vi antibodies measured by Indirect Hemagglutination assay, Nath et al. (2008) reported a significantly higher risk of GBC among chronic carriers of *S. typhi* (OR 6.1) compared with the healthy controls.

Ecological Studies

Chile represents a unique situation with regards to GBC and TF epidemiology. As already mentioned, Levine estimated a prevalence of *S. typhi* chronic carriers of 6.9 per 1,000 inhabitants for 1980 (Levine et al. 1982, 1984). In 1990, TF experimented a dramatic reduction in incidence (rates/100,000 dropped from 100 to 10), thus practically stopping the input of new carriers into the community. Nevertheless, most carriers that accumulated during the hyperendemic period 1976–1986 were under 40 years old, i.e. they are today in their sixties, the age at which most cases of GBC present. Thus, it is probable that a large part of the excess of GBC cases occurring today in Chile is coming from that reservoir. In that same period (1970–1990), occurred an important fall in gallbladder surgery and some authors demonstrated an inverse association between the rate of gallbladder surgery and GBC mortality (Serra et al. 1990; Chianale et al. 1990). The result was that Chilean men and women began leading GBC incidence and mortality worldwide; GBC became the first cancer killer of women to date.

The association of TF with GBC risk was suggested in an ecological analysis of GBC mortality among the 333 Chilean counties, between 1985 and 2002, with an OR of 3 for TF, after adjusting for most confounders –age-sex composition, poverty, Amerindian ancestry and access to gallbladder surgery- (Andia et al. 2008).

5.3.5 *Interventions to Decrease the Burden of GBC*

Even if TF transmission were controlled immediately, as it happened in Chile in 1990 (Laval and Ferreccio 2007), a significant volume of chronic carriers will still persist in the population, some of whom, if left untreated, may develop GBC as they age. Thus, surveillance, identification and treatment of chronic carriers should be considered, especially among those with additional risk factors for GBC, such as obese women with gallbladder disease and *S. typhi* infection residing in high risk areas of GBC.

Primary prevention measures are related to preventing *S. typhi* infection (discussed above) and lithiasis formation. A diet rich in fat and poor in vegetables and fruits has been associated with higher risk of GS and GBC, thus educating in healthy lifestyles could be considered a primary prevention strategy. As secondary prevention, we could mention gallbladder surgery for GS either symptomatic or silent. Although there is not agreement on this as a general recommendation, it should be considered in areas where the risk of GBC is high. In Chile, an inverse association of gallbladder surgery and GBC mortality has been demonstrated (Chianale et al. 1990; Serra et al. 1990). If a person with GS is also a carrier of *S. typhi*, anti-microbial treatment for *S. typhi* eradication should be considered, since it has been demonstrated that at least 25% of subjects will carry *S. typhi* even after cholecystectomy.

5.4 Conclusions

In conclusion, GBC presents a marked geographical differential, with the coexistence of extremely high risk areas where GBC is the first cancer killer of women (Chile) and areas where it is a rare disease (most western developed countries). The differential may be explained by the distribution of GBC risk factors: propensity to develop GS due to genetic predisposition plus lifestyle factors that facilitate the formation of cholesterol GS, higher prevalence of chronic *S. typhi* infection of the gallbladder, and insufficient access to gallbladder surgery for GS. In the developed world, typhoid was controlled in the 50s, mainly by basic sanitation, before the epidemic of lifestyle factors that are associated with lithogenic bile: poor diet and sedentarism. Contrarily, in Chile and India there was a coexistence of a high typhoid endemic setting with an epidemic of lithogenic lifestyles, aggravated by insufficient access to preventive surgery. GBC is mainly affecting the poorer in each population, among whom, all the above mentioned factors are more prevalent.

The main difference with regard to *S. typhi* in Chile and in India is that, while the first country has controlled TF reaching incidence rates comparable to those of USA and Europe, India still has one of the highest TF incidence rates reported. Thus, it is expected that even without any specific intervention GBC will soon begin a downward trend in Chile, and that in a few years India will take the top position in GBC deaths.

The evidence presented here may seem abundant but there are many more questions that need research. It is not known if *S. typhi* is viable in pre-cancerous or GBC lesions; nor if clearance of *S. typhi* could cause a regression of GB precancerous lesions. Thus it is not possible to predict the effect of *S. typhi* eradication. It is unknown if *S. typhi* interacts with other bacteria, diet, female hormones or bile composition. The very low rate of recovery of *S. typhi* in GBC subjects that present high Vi antibody levels has puzzled investigators. It is unclear if the bacteria are in a niche out of the gallbladder or if they are in non-cultivable form.

GBC is a highly lethal disease that in most cases is diagnosed at advanced stages; it is urgent to understand the biology of the disease to identify early biomarkers that could be incorporated into the medical practice in high risk areas. In comparison with other cancers which affect western countries (like lung or breast cancers), GBC can be considered a neglected disease. If it is proven that *S. typhi* has a causal role, GBC will become a preventable disease.

References

Albores-Saavedra J, Alcántra-Vazquez A, Cruz-Ortiz H et al (1980) The precursor lesions of invasive gallbladder carcinoma. Hyperplasia, atypical hyperplasia and carcinoma in situ. Cancer 45(5):919–927

Ames WR, Robins M (1943) Age and sex as factors in the development of the typhoid carrier state, and a method for estimating carrier prevalence. Am J Public Health Nations Health 33(3): 221–230

Andia ME, Hsing AW, Andreotti G et al (2008) Geographic variation of gallbladder cancer mortality and risk factors in Chile: a population-based ecologic study. Int J Cancer 123(6):1411–1416

Andreotti G, Chen J, Gao YT et al (2008) Polymorphisms of genes in the lipid metabolism pathway and risk of biliary tract cancers and stones: a population-based case-control study in Shanghai, China. Cancer Epidemiol Biomarkers Prev 17(3):525–534

Baker S, Hanage WP, Holt KE (2010) Navigating the future of bacterial molecular epidemiology. Curr Opin Microbiol 13(5):640–645

Batra Y, Pal S, Dutta U et al (2005) Gallbladder cancer in India: a dismal picture. J Gastroenterol Hepatol 20(2):309–314

Bertran E, Heise K, Andia ME et al (2010) Gallbladder cancer: incidence and survival in a high-risk area of Chile. Int J Cancer 127(10):2446–2454

Brett M, Barker DJ (1976) The world distribution of gallstones. Int J Epidemiol 5(4):335–341

Caygill CPJ, Hill MJ (2000) Salmonella typhi/paratyphi and gallbladder cancer. In: Goedart JJ (ed) Infectious causes of cancer: targets for intervention. Humana Press, Totowa

Caygill CPJ, Hill MJ, Braddick M et al (1994) Cancer mortality in chronic typhoid and paratyphoid carriers. Lancet 343(8889):83–84

Caygill CPJ, Braddick M, Knowles RL et al (1995) The association between typhoid carriage, typhoid infection and subsequent cancer at a number of sites. Eur J Cancer Prev 4(2):187–193

Chang SC, Rashid A, Gao YT et al (2008) Polymorphism of genes related to insulin sensitivity and the risk of biliary tract cancer and biliary stone: a population-based case-control study in Shanghai, China. Carcinogenesis 29(5):944–948
Chianale J, del Pino G, Nervi F (1990) Increasing gall-bladder cancer mortality rate during the last decade in Chile, a high-risk area. Int J Cancer 46(6):1131–1133
Crump JA, Mintz ED (2010) Global trends in typhoid and paratyphoid fever. Clin Infect Dis 50(2): 241–246
Crump JA, Luby SP, Mintz ED (2004) The global burden of typhoid fever. Bull World Health Organ 82(5):346–353
Crump JA, Ram PK, Gupta SK et al (2008) Part I. Analysis of data gaps pertaining to Salmonella enterica serotype Typhi infections in low and medium human development index countries, 1984–2005. Epidemiol Infect 136(4):436–448
Csendes A, Becerra M, Smok G et al (1991) Prevalence of gallbladder neoplasms in cholecystectomies. Rev Med Chil 119(8):887–890
Csendes A, Becerra M, Rojas J et al (2000) Number and size of stones in patients with asymptomatic and symptomatic gallstones and gallbladder carcinoma: a prospective study of 592 cases. J Gastrointest Surg 4(5):481–485
Curado MP, Edwards B, Shin HR et al (eds) (2007) Cancer incidence in five continents, vol IX. IARC scientific publication no. 160. Available from http://www.iarc.fr/en/Publications/PDFs-online/Cancer-Epidemiology/IARC-Scientific-Publication-No.-155
Cvjetanovic B, Grab B, Uemura K (1971) Epidemiological model of typhoid fever and its use in the planning and evaluation of antityphoid immunization and sanitation programmes. Bull World Health Organ 45(1):53–75
Diehl AK (1983) Gallstone size and the risk of gallbladder cancer. JAMA 250(17):2323–2326
Diehl AK (1998) Gallstone disease in Mestizo Hispanics. Gastroenterology 115(4):1012–1015
Diehl AK, Stern MP, Ostrower VS et al (1980) Prevalence of clinical gallbladder disease in Mexican-American, Anglo, and black women. South Med J 73(4):438–441
el-Zayadi A, Ghoneim M, Kabil SM et al (1991) Bile duct carcinoma in Egypt: possible etiological factors. Hepatogastroenterology 38(4):337–340
Ferlay J, Bray F, Pisani P et al (2004) GLOBOCAN 2002: Cancer incidence, mortality and prevalence worldwide IARC Cancer Base No. 5 version 2.0. IARC Press, Lyon
Ferreccio C, Levine MM, Manterola A et al (1984) Benign bacteremia caused by Salmonella typhi and paratyphi in children aged younger than 2 years. J Pediatr 104(6):899–901
Friedrich N, Völzke H, Hampe J et al (2009) Known risk factors do not explain disparities in gallstone prevalence between Denmark and northeast Germany. Am J Gastroenterol 104(1):89–95
Gabrielli M, Hugo S, Dominquez A et al (2010) Mortality due to gallbladder cancer: retrospective analysis in three Chilean hospitals. Rev Med Chil 138(11):1357–1364
Galman C, Miquel JF, Pérez RM et al (2004) Bile acid synthesis is increased in Chilean Hispanics with gallstones and in gallstone high-risk Mapuche Indians. Gastroenterology 126(3):741–748
Goldin RD, Roa JC (2009) Gallbladder cancer: a morphological and molecular update. Histopathology 55(2):218–229
Gomez-Roel X, Arrieta O, Leon-Rodriguez E (2007) Prognostic factors in gallbladder and biliary tract cancer. Med Oncol 24(1):77–83
Grimaldi CH, Nelson RG, Pettitt DJ et al (1993) Increased mortality with gallstone disease: results of a 20-year population-based survey in Pima Indians. Ann Intern Med 118(3):185–190
Hemminki K, Mousavi SM, Barndt A et al (2010) Liver and gallbladder cancer in immigrants to Sweden. Eur J Cancer 46(5):926–931
Hofmann E, Chianale J (1993) Blood group antigen secretion and gallstone disease in the Salmonella typhi chronic carrier state. J Infect Dis 167(4):993–994
Hsing AW, Rashid A, Devesa SS et al (2006) Biliary tract cancer. In: Schottenfeld D, Fraumeni JF (eds) Cancer epidemiology and prevention, 3rd edn. Oxford University Press, New York, pp 787–800
International Agency for Research on Cancer (IARC) (2009) Cancer Mondial. http://www-dep.iarc.fr/. Accessed 31 Dec 2009.

Karayalçin R, Genç V, Karaca AS et al (2010) Prevalence of cholelithiasis in a Turkish population sample of postmenopausal women. Turk J Gastroenterol 21(4):416–420
Kayahara M, Nagakawa T, Nakagawara H et al (2008) Prognostic factors for gallbladder cancer in Japan. Ann Surg 248(5):807–814
Khan AA, Shrivastava A (2010) Bacterial infections associated with cancer: possible implication in etiology with special reference to lateral gene transfer. Cancer Metastasis Rev 29(2):331–337
Khan HN, Harrison M, Bassett EE et al (2009) A 10-year follow-up of a longitudinal study of gallstone prevalence at necropsy in South East England. Dig Dis Sci 54(12):2736–2741
Konstantinidis IT, Deshpande V, Genevay M et al (2009) Trends in presentation and survival for gallbladder cancer during a period of more than 4 decades: a single-institution experience. Arch Surg 144(5):441–447
Kothari A, Pruthi A, Chugh TD (2008) The burden of enteric fever. J Infect Dev Ctries 2(4): 253–259
Lai CW, Chan RC, Cheng AF et al (1992) Common bile duct stones: a cause of chronic salmonellosis. Am J Gastroenterol 87(9):1198–1199
Lanata CF, Ristori C, Jimenez L et al (1983) Vi serology in detection of chronic Salmonella typhi carriers in an endemic area. Lancet 322(8347):441–443
Laval E, Ferreccio C (2007) Fiebre Tifoidea: emergencia, cuspide y desaparicion de una enfermedad infecciosa en Chile. Rev Chil Infect 24:435–440
Lazcano-Ponce EC, Miquel JF, Munoz N et al (2001) Epidemiology and molecular pathology of gallbladder cancer. CA Cancer J Clin 51(6):349–364
Lemrow SM, Perdue DG, Stewart SL et al (2008) Gallbladder cancer incidence among American Indians and Alaska Natives, US, 1999-2004. Cancer 113:1266–1273
Levine MM, Black RE, Lanata C (1982) Precise estimation of the numbers of chronic carriers of Salmonella typhi in Santiago, Chile, an endemic area. J Infect Dis 146(6):724–726
Levine MM, Ristori C, Ferreccio C (1984) Reply. J Infect Dis 149(1):125–126
Lieber MM (1952) The incidence of gallstones and their correlation with other diseases. Ann Surg 135(3):394–405
Lin FY, Becke JM, Groves C et al (1988) Restaurant-associated outbreak of typhoid fever in Maryland: identification of carrier facilitated by measurement of serum Vi antibodies. J Clin Microbiol 26(6):1194–1197
Losonsky GA, Ferreccio C, Kotloff KL et al (1987) Development and evaluation of an enzyme-linked immunosorbent assay for serum Vi antibodies for detection of chronic Salmonella typhi carriers. J Clin Microbiol 25(12):2266–2269
Lowenfels AB, Walker AM, Althaus DP et al (1989) Gallstone growth, size, and risk of gallbladder cancer: an interracial study. Int J Epidemiol 18(1):50–54
Lynch MF, Blanton EM, Bulens S et al (2009) Typhoid fever in the United States, 1999-2006. JAMA 302(8):859–865
Mellemgaard A, Gaarslev K (1988) Risk of hepatobiliary cancer in carriers of *Salmonella typhi*. J Natl Cancer Inst 80(4):288
Méndez-Sánchez N, King-Martínez AC, Ramos MH et al (2004) The Amerindian's genes in the Mexican population are associated with development of gallstone disease. Am J Gastroenterol 99(11):2166–2170
Nath G, Singh YK, Kumar K et al (2008) Association of carcinoma of the gallbladder with typhoid carriage in a typhoid endemic area using nested PCR. J Infect Dev Ctries 2(4):302–307
Nath G, Gulati AK, Shukla VK (2010) Role of bacteria in carcinogenesis, with special reference to carcinoma of the gallbladder. World J Gastroenterol 16(43):5395–5404
Nervi F, Raddatz A, Zamorano N (1984) Overestimation of the numbers of chronic carriers of Salmonella typhi in Santiago. J Infect Dis 149(1):124–126
Pandey M (2006) Environmental pollutants in gallbladder carcinogenesis. J Surg Oncol 93(8):640–643
Parry CM, Hien TT, Dougan G et al (2002) Typhoid fever. N Engl J Med 347(22):1770–1782
Prouty AM, Schwesinger WH, Gunn JS (2002) Biofilm formation and interaction with the surfaces of gallstones by Salmonella spp. Infect Immun 70(5):2640–2649

Ristori C, Rodríguez H, Vicent P et al (1982a) Investigation of the Salmonella typhi-paratyphi carrier state in cases of surgical intervention for gallbladder disease. Bull Pan Am Health Organ 16(2):161–171

Ristori C, Rodríguez H, Vicent P et al (1982b) Persistence of the Salmonella typhi-paratyphi carrier state after gallbladder removal. Bull Pan Am Health Organ 16(4):361–366

Roa I, Araya JC, Villaseca M (1996) Preneoplastic lesions and gallbladder cancer: an estimate of the period required for progression. Gastroenterology 111(1):232–236

Roa I, de Aretxabala X, Araya JC et al (2006) Preneoplastic lesions in gallbladder cancer. J Surg Oncol 93(8):615–623

Roa EI, Muñoz NS, Ibacache SG et al (2009) Natural history of gallbladder cancer. Analysis of biopsy specimens. Rev Med Chil 137(7):873–880

Ruhl CE, Everhart JE (2011) Gallstone disease is associated with increased mortality in the United States. Gastroenterology 140(2):508–516

Sears SD, Ferreccio C, Levine MM et al (1984) The use of Moore swabs for isolation of Salmonella typhi from irrigation water in Santiago, Chile. J Infect Dis 149(4):640–642

Sears SD, Ferreccio C, Levine MM (1986) Sensitivity of Moore sewer swabs for isolating *Salmonella typhi*. Appl Environ Microbiol 51(2):425–426

Serra I, Calvo A, Maturana M et al (1990) Changing trends of gall-bladder cancer in Chile, a high-risk area. Int J Cancer 45(2):376–377

Serra I, Yamamoto M, Calvo A et al (2002) Association of chili pepper consumption, low socioeconomic status and longstanding gallstones with gallbladder cancer in a Chilean population. Int J Cancer 102(4):407–411

Shukla VK, Singh H, Pandey M et al (2000) Carcinoma of the gallbladder–is it a sequel of typhoid? Dig Dis Sci 45(5):900–903

Siddiqui FJ, Rabbani F, Hasan R et al (2006) Typhoid fever in children: some epidemiological considerations from Karachi, Pakistan. Int J Infect Dis 10(3):215–222

Tewari M, Mishra RR, Shukla HS (2010) Salmonella typhi and gallbladder cancer: report from an endemic region. Hepatobiliary Pancreat Dis Int 9(5):524–530

Unisa S, Jagannath P, Dhir V et al (2011) Population-based study to estimate prevalence and determine risk factors of gallbladder diseases in the rural Gangetic basin of North India. HPB 13(2): 117–125

Vaishnavi C, Kochhar R, Singh G et al (2005) Epidemiology of typhoid carriers among blood donors and patients with biliary, gastrointestinal and other related diseases. Microbiol Immunol 49(2):107–112

Welton JC, Marr JS, Friedman SM (1979) Association between hepatobiliary cancer and typhoid carrier state. Lancet 313(8120):791–794

Wistuba II, Gazdar AF (2004) Gallbladder cancer: lessons from a rare tumour. Nat Rev Cancer 4(9):695–706

Chapter 6
Ocular Adnexal Lymphoma of MALT-Type and Its Association with *Chlamydophila psittaci* Infection

Andrés J.M. Ferreri, Riccardo Dolcetti, Silvia Govi, and Maurilio Ponzoni

Abstract Ocular adnexal marginal zone lymphoma (OAMZL) presents as an indolent and generally limited-stage disease that involves conjunctiva, eyelid, lachrymal gland and sac, or orbit soft tissue; standard treatments are not well defined, and include "wait and watch" policy, radiotherapy or other topic approach, and also chemotherapy. The development of OAMZL has been linked to *Chlamydophila psittaci (Cp)* infection. *Chlamydiae* are obligate intracellular bacteria that grow in eukaryotic cells and cause a wide spectrum of diseases. They can establish persistent infections, are mitogenic *in vitro*, promote cell proliferation *in vivo* and induce resistance to apoptosis in infected cells. *Cp* determines both local and systemic infections in these patients, and bacterial eradication with antibiotic therapy is often followed by lymphoma regression. Despite the recent advances in the understanding of this bacterium–lymphoma association, several questions remain unanswered. Variations among different geographic areas and related diagnostic and therapeutic implications remain a major investigational issue. We focus on clinical features of OAMZL, standard treatments, *Cp*-infection and its therapeutic implications in lymphomas.

A.J.M. Ferreri (✉)
Unit of Lymphoid Malignancies, Department of Onco-Hematology,
San Raffaele Scientific Institute, Via Olgettina 60, 20132 Milan, Italy
e-mail: andres.ferreri@hsr.it

R. Dolcetti
Department of Oncology, Cancer Bioimmunotherapy Unit,
National Cancer Institute CRO, Aviano, Italy

S. Govi
Unit of Lymphoid Malignancies, Department of Onco-Hematology,
San Raffaele Scientific Institute, Via Olgettina 60, 20132 Milan, Italy

M. Ponzoni
Unit of Lymphoid Malignancies, Department of Onco-Hematology,
San Raffaele Scientific Institute, Via Olgettina 60, 20132 Milan, Italy

Pathology Unit, San Raffaele Scientific Institute, Milan, Italy

A.A. Khan (ed.), *Bacteria and Cancer*, DOI 10.1007/978-94-007-2585-0_6,

Keywords *Chlamydophila psittaci* • Ocular adnexal lymphoma • Marginal zone B-cell lymphoma • Chemotherapy • Radiotherapy • *C. pneumoniae* • *C. trachomatis* • Elementary bodies • Reticulate-bodies • Environmental exposure • Doxycycline

Abbreviations

Cp	*Chlamydophila psittaci*
HCDR	Heavy chain complementarity-determining region
Hp	*Helicobacter pylori*
IGHV-DJ	Immunoglobulin heavy chain variable-diversity-joining regions
MALT	Mucosa-associated lymphoid tissue
OAMZL	Ocular adnexal marginal zone lymphoma
TETR-PCR	Touchdown enzyme time release polymerase chain reaction

6.1 Ocular Adnexal Lymphoma

6.1.1 Epidemiological Data

Non-Hodgkin lymphomas account for about a half of all orbital malignancies, although representing a relatively rare disease (Margo and Mulla 1998; Ferreri et al. 2009). Extranodal lymphomas arising in the ocular adnexae have an incidence of 5–15% and may involve the conjunctiva, lachrymal gland, orbital fat, eyelid, or lachrymal sac (Sasai et al. 2001). The most frequent lymphoma histotypes in these anatomical structures are marginal zone B-cell lymphomas of mucosa-associated lymphoid tissue (MALT)-type (OAMZL), which are indolent tumours, usually presenting as stage I-II disease; followed by follicular lymphomas (15–20%) and diffuse large B-cell lymphomas (7–15%). Mantle-cell (5%) and small lymphocytic (3%) lymphomas rarely originate in the ocular adnexae, but generally involve these structures as part of a disseminated disease (Ferry et al. 2007; Coupland et al. 1998; Sjo 2009). OAMZL constitutes 50–78% of all OAL diagnosed in western countries and 80–90% of OAL diagnosed in Japan and Korea (Mannami et al. 2001); importantly, its incidence is rapidly increasing, with annual rates higher than 6%, and no evidence of peaking (Moslehi et al. 2006). These epidemiologic features seem not correlated to changes in lymphoma classification schemes or improvements in diagnostic procedures and technology (Moslehi et al. 2006).

The pathogenesis of OAMZL is still unclear and requires identification of environmental and genetic risk factors in aetiologic mechanisms, as well as the potential role of infectious agents (Moslehi et al. 2006). Recently, an association between Chlamydial infection and OAMZL was reported (Ferreri et al. 2004) as a new pathogenic insight for the development of innovative targeted therapies. The characterization of genetic alterations may be useful to predict therapeutic response and to identify the candidates for different treatments.

6.1.2 Pathological Features

Whether MALT is present in normal conjunctiva is still a matter of debate. The orbital tissues lack both lymphoid tissue and lymphatic drainage; it was assumed that OAMZL may derive from MALT tissue acquired after chronic inflammatory stimulus or autoimmune disorders (Hara et al. 2001). In analogy with the finding that *Helicobacter pylori* (*Hp*) infection constitutes a chronic antigenic stimulus that would drive the development of gastric MALT lymphoma, a similar pathogenic model of antigen-driven lymphoproliferation may be hypothesized also for OAMZL.

The previously defined 'pseudolymphomas' or 'benign lymphoid hyperplasia' may actually be classified as B-cell lymphomas, containing clonal B cell expansions (Neri et al. 1987). In several studies, different cases were diagnosed as 'lymphoma not further specified', because of the scant diagnostic sample; the majority of these entities, however, entered the 'low-grade lymphoma' category.

OAMZL displays the classical histopathologic and immunophenotypic profile of most MALT lymphomas. The histopathology of MALT lymphomas includes both neoplastic and non-neoplastic cells. Lymphoma cells are heterogeneous, with centrocytic-like cells, monocytoid cells or small-sized lymphocytes co-expressed in the same tissue. A plasma cell differentiation may be observed. Lymphoepithelial lesions are delineated from tumour cells infiltration through glandular and superficial epithelium. Follicular colonization is observed when neoplastic lymphocytes grow within germinal centers, and a few large cells (blasts) are encountered throughout the section. Non-neoplastic cells, including reactive germinal centers, reactive T cells and histiocytes are also present.

MALT lymphoma immunophenotype comprises CD20+, CD79a+, generally IgM + with light chain restriction, PAX5+, bcl-2+, TCL1+, CD11c+/–, CD43+/–, CD21+/–, CD35+/–, and IgD-, CD3-, CD5-, CD10-, CD23-, cyclinD1-, bcl-6-, MUM1- cells.

6.1.3 Molecular Features and Cytogenetics

PCR analysis of immunoglobulin heavy chain gene rearrangement showed a clonal B cell population in 55% of OAMZL (Mannami et al. 2001) and somatic hypermutations in two thirds of these cases (Coupland et al. 1999). The V_H3 family is expressed in half, V_H4 in 23% of cases (Coupland et al. 1999). OAMZL result from clonal expansion of post-germinal-center memory B-cells, where, in two thirds of cases, antigen selection may have occurred (Coupland et al. 1999). Ongoing mutations have been described, especially if follicular dendritic cells are present, supporting a potential role of microenvironmental stimuli (Hara et al. 2001).

Chromosomal translocations are t(11;18)(q21;q21)/API2-MALT1, observed in 0–10% of OAMZL (Remstein et al. 2002; Tanimoto et al. 2006a; Murga Penas et al. 2003; Streubel et al. 2003) and t(14;18)(q32;q21)/IGH-MALT1, noticed in 7–11% (Tanimoto et al. 2006a). These translocations lead to the activation of NF-kB, a critical promotor of lymphocyte proliferation and survival (Isaacson and Du 2004). Several new translocations involving the IGH promoter have been described in ocular adnexal MALT lymphoma, in a recent study of 29 largely extra-gastric MALT lymphomas in which 21 (72%) had balanced translocations (Vinatzer et al. 2008). Ten distinct translocations were identified in lymphomas from nine different sites. The commonest translocation was the t(14;18) (IgH;MALT1), but nine other translocations were identified (FOXP1 on chromosome 3, BCL-6 on chromosome 3, JMJD2C on 9, CCN3 on 1, ODZ2 on 5, and 4 with unknown translocation partners). The cases of ocular adnexal MALT lymphoma showed the new translocations of IgH/ODZ2, IgH-JMJD2C and 2 with unknown partners (Vinatzer et al. 2008). Among 116 OAMZL newly diagnosed patients from Denmark, the presence of translocations involving IGH and/or MALT1 -gene loci was detected in 2 (5%) of 42 investigated specimens (Sjo et al. 2009).

Cytogenetic data on OAMZL are limited and seem to present geographic variability in their incidence (Ruiz et al. 2007; Streubel et al. 2004). Aneuploidy, particularly trisomy 3 and 18, occurs frequently in t(11;18)-negative OAMZL (Remstein et al. 2002; Tanimoto et al. 2006a; Murga Penas et al. 2003; Streubel et al. 2003; Ott et al. 1997). Trisomy 3, trisomy 18 and t(14;18)(q32;q21) deserve to be further investigated as possible predictors of multifocal disease (Raderer et al. 2006). OAMZL with trisomy 18 seems to show definite clinical features: it involves the conjunctiva, occurs in young females, and presents a high recurrence rate (Tanimoto et al. 2006a).

6.1.4 Clinical Aspects

6.1.4.1 Presentation

OAMZL can infiltrate every orbital and ocular adnexal tissue. Conjunctival involvement is observed in 25% of OAMZL; intra-orbital masses are described in 75% of cases; bilateral involvement is present in 10–15% of patients, generally in conjunctival forms (Bhatia et al. 2002; Martinet et al. 2003; Uno et al. 2003; Rosado et al. 2005; Fung et al. 2003). Surgical biopsy is mandatory for histopathological characterization considering that treatment and prognosis remarkably vary among different lymphoma categories; clinical and radiological features do not distinguish between benign hyperplasia and lymphomas or between indolent and aggressive lymphomas.

Median age at presentation is 65 years, with the disease generally arising after the fourth decade of life. Females are predominantly affected with a higher prevalence (Bhatia et al. 2002; Martinet et al. 2003; Uno et al. 2003; Rosado et al. 2005; Fung et al. 2003). The period between clinical onset and diagnosis is variable, with a median of 6–7 months and a range of 1–135 months.

Typical lymphoproliferative lesion usually consists of a gradually enlarging, painless mass, that displaces the normal structures, with minimal or absent pain and inflammation. Conjunctival lymphoma is characterized by a classic "salmon pink patch" appearance with swollen conjunctiva. Intra-orbital lymphoma is associated to exophthalmos (27%), palpable mass (19%), eyelid ptosis (6%), diplopia (2%), eyelid nodule, orbital edema, epiphora, dacryocystitis, gritty sensation, and/or variable ocular motility impairment (Bhatia et al. 2002; Martinet et al. 2003; Uno et al. 2003; Rosado et al. 2005; Fung et al. 2003; Coupland et al. 1998; Knowles et al. 1990; Shields et al. 2001; White et al. 1995; Sullivan et al. 2004; McKelvie et al. 2001).

Tearing is the common feature of lymphoma of the lachrymal sac. Rare presentations include signs of expansive effect of the lesion, causing limitations to excursion of the eye; visual acuity and field defects or choroidal folds are observed only in rare cases of rapidly growing tumour (Sarraf et al. 2005).

6.1.4.2 Staging Procedures

OAMZL at presentation is a limited-stage disease in over 75% of cases (Ann Arbor stage I_E). Regional lymphadenopathies are detected in <5% of cases (stage II_E), and extra-orbital disease, generally in extranodal organs, is observed in 10–15% of cases (stage IV_E); it is exceptional in conjunctival lymphomas (Ferreri et al. 2004; Martinet et al. 2003; Uno et al. 2003; Rosado et al. 2005; Fung et al. 2003). The use of extensive and invasive staging showed that 38% of OAMZL patients have at least one extraorbital site of disease at diagnosis (Raderer et al. 2006). No difference in cause-specific survival between subgroups has been reported (Fung et al. 2003), so the usefulness of extended staging is controversial. However, it remains essential to define therapeutic approach. Gastroscopy and colonoscopy are not mandatory, but they are reserved to patients with specific gastrointestinal symptoms (Raderer et al. 2006).

Neuroimaging techniques are important for staging and definition of clinical responses. At neuroimaging examination, OAMZL presents generally as well-defined lesion, surrounding and displacing extra-ocular muscles, without ocular infiltration. On computed tomography images, OAMZL contrast enhancement is homogeneous, comparable to lachrymal glands and extra-ocular muscles. Magnetic Resonance Imaging (MRI) is useful to differentiate lymphoma from other orbital expansive lesions.

6.1.4.3 Clinical History and Prognosis

OAZML presents a better prognosis than other lymphoma histotypes arising in the ocular adnexae (Fung et al. 2003). Good prognostic factors are limited disease, ECOG-performance status 0–2, absence of systemic symptoms (Rosado et al. 2005; Tanimoto et al. 2006b), while nodal involvement, systemic symptoms, increased lactate dehydrogenase serum levels, and non-conjunctival sites are negative predictors of outcome (Rosado et al. 2005; Martinet et al. 2003). Some cases of spontaneous remission in OAMZL patients have been reported, mostly in Japanese patients with conjunctival localization (Matsuo and Yoshino 2004). However, some of these patients have been equally treated with topical steroids or antibiotics, contributing to tumour regression. Presenting symptoms sometimes require prompt treatment. Local overall responses are different according to applied therapy, with a 5-year relapse-free survival of ~65%. Multiple relapses are frequently observed, both at contralateral orbit and distant extranodal organs. Five-year cause-specific and overall survival of 100% and >90%, respectively (Fung et al. 2003; Uno et al. 2003). High-grade transformation has been reported in 1–3% of cases (Tanimoto et al. 2006b; Ben Simon et al. 2006).

6.1.5 Treatment

If adequately treated, less than 5% of patients with OAMZL die of lymphoma (Ferreri et al. 2008a; Uno et al. 2003; Martinet et al. 2003). Defined guidelines for OAMZL treatment do not exist, because therapeutic knowledge results from a limited number of small, retrospective and variably treated series, which included different lymphoma histotypes (de Cremoux et al. 2006). There is a lack of consensus about the best upfront treatment. The optimal treatment should ideally not be negatively related to side effects. The range of therapeutic options is heterogeneous, from 'watch and wait policy', to systemic chemotherapy. Most patients are usually treated as first-line approach with radiotherapy, with well known late-term toxicities.

6.1.5.1 Surgery

Surgery is an essential diagnostic tool and, in selected cases, a part of therapeutic approach. Complete excision can be performed in conjunctival and lachrymal gland MALT lymphomas but does not influence survival (Tanimoto et al. 2006a). "Wait and watch" strategy in patients with stage I-disease produces similar results to immediate radiotherapy, with a 10-year overall survival of 94% (Tanimoto et al. 2006a).

6.1.5.2 Radiotherapy

Radiotherapy is the most studied treatment for OAMZL. However, only a few series have been focused on OAMZL treated exclusively with radiotherapy (Uno et al. 2003; Fung et al. 2003; Ejima et al. 2006). Target dose of 25–30 Gy in 10–15 fractions (minimal target dose >25 Gy) is indicated (Tsang et al. 2003). Electron beams (4–12 MeV) and 4–9 MV photon beams are used in conjunctival and intra-orbital lymphomas. Radiotherapy is usually well tolerated (Uno et al. 2003; Ejima et al. 2006). The most common toxicities of grade ≥2 are cataract (38% of cases), retinal disorders (17%), xerophthalmia (17%), and glaucoma (2%) (Ejima et al. 2006). Toxicity is more common with doses >36 Gy. After radiation, patients with stage-I OAMZL achieve a slow and gradual response (Uno et al. 2003; Ejima et al. 2006). In-field relapses are rare, generally in cases receiving low radiation doses (Tsang et al. 2003). Relapses involve especially the contralateral orbit (half of relapses) but also distant extranodal organs (Ejima et al. 2006; Tsang et al. 2003; Fung et al. 2003).

6.1.5.3 Chemotherapy and Immunotherapy

Prospective trials addressing chemotherapy have not been conducted in OAMZL; in retrospective series, the largest experience regards chlorambucil, with a 100% overall response rate, a 79% complete remission rate and a 5-year relapse-free survival of 60% (Ben Simon et al. 2006). Relapses after chlorambucil mostly involve extra-orbital tissues, with rare cases (3%) of high-grade transformation (Ben Simon et al. 2006). The efficacy of other drugs is limited to a few prospective trials on unselected MALT lymphomas including a small number of OAMZL patients. The use of anthracycline-based chemotherapy is actually the most common treatment. Rituximab, a chimeric monoclonal antibody directed against the B-lymphocyte antigen CD20 widely used in the treatment of B-cell lymphomas, has displayed activity in MALT lymphomas, and it has been also anecdotally used in patients with OAMZL (Conconi et al. 2003; Nuckel et al. 2004; Ferreri et al. 2005a). When used by intravenous route at conventional dose, rituximab is associated with very high response rate but response duration is usually short-lived (Nuckel et al. 2004; Ferreri et al. 2005a; Blasi et al. 2001). ^{90}Y-ibritumomab tiuxetan (Zevalin®), a radioimmunoconjugate employed in relapsed or refractory follicular lymphomas, was assessed in a small phase II trial as front-line therapy in 12 patients with stage I_E indolent lymphoma of the ocular adnexae (9 with OAMZL) (Esmaeli et al. 2009). A complete remission rate of 83% was achieved without grade 3 or 4 cytopenias, but median follow-up is still too short (Esmaeli et al. 2009).

6.1.5.4 Intralesional and Topic Treatment

Anecdotal cases of conjunctival lymphoma treated with topical cytostatics have been reported, in particular mitomycin instillation, which is used frequently in epithelial and melanocytic tumours of the conjunctiva (Poothullil and Colby 2006), was successfully administered in a single reported case of conjunctival lymphoma (Yu et al. 2008). However, the risk of severe effects on the corneal epithelium requires further investigation (Schallenberg et al. 2008; Ballalai et al. 2009).

Intra-lesional drug delivery is a new, investigational approach to OAMZL. Therapy with interferon-α injection is a simple procedure successfully used for conjunctival lymphomas (Lachapelle et al. 2000; Blasi et al. 2001). Side effects consist of local hemorrhage, chemosis and minor systemic effects (Lachapelle et al. 2000; Blasi et al. 2001). However, follow-up of reported cases is insufficient to establish the efficacy of this approach. Moreover, after an initial interest about this therapeutic modality, additional series treated with this strategy were not reported in the last decade. Recently, we reported a pilot experience with intralesional rituximab in three patients with recurrent conjunctival lymphoma. This approach was well tolerated and was invariably associated with tumour regression. This strategy led to tumour remission even in patients refractory to intravenous rituximab. The limited activity of intravenous rituximab in OAMZL may be the result of the reduced bioavailability of the drug and also its immunologic effectors. The intralesional delivery of rituximab seems to be a valid method to overcome these limitations. Treatment was feasible and well tolerated even in hepatitis B and C virus-positive patient who underwent liver transplantation. Supplementation with autologous serum, introduced in the patient with follicular lymphoma, improved the cytotoxic activity of intralesional rituximab, probably by increasing the local concentration of proteins C3 and C4. Response duration and potential late effects remain to be defined (Ferreri et al. 2011).

6.2 Chlamydiae

6.2.1 Structure and Biology

Chlamydiae are ubiquitous, obligate intracellular bacteria growing in eukaryotic cells and are responsible for a wide spectrum of human and animal diseases. Chlamydial infection is initiated by a dynamic process of attachment of a Chlamydial elementary body (CEB) to the host cell, followed by its entry into the cell. The attachment and internalization mechanism of CEB is a poorly understood process. Within eukaryotic cells, *Chlamydia* undergoes an orderly alternation between a metabolically inactive, highly infective form, the CEB, and a metabolically active, intracellular growth stage form, the Chlamydial reticulate body (CRB) (Byrne and Ojcius 2004). CEB are round (or pear-shaped in the case of *Chlamydophila pneumoniae*), electron-dense structures approximately 0.3 μm in diameter that are metabolically inert outside the host cell. CRB have a diameter ~1 μm, with diffuse and fibrillar nucleic acids and a ribosome-rich cytoplasm.

Intracellular replication of *Chlamydia* occurs within a membrane-bound vesicle or inclusion. *Chlamydiae* communicate with host cells by secreting effector proteins into the host-cell cytoplasm, probably via type III secretion apparatus. Infected cells receive either pro-apoptotic or anti-apoptotic signals according to the growth rate of the Chlamydial pathogen. *Chlamydiae* induced cell apoptosis may be associated with the manifestations of acute disease, whereas inhibition of apoptosis is characteristic of the chronic disease phase. In particular, conditions (for example, interferon-γ-mediated activation of host cells), CRBs cease to divide, do not differentiate to CEBs and maintain a more stable association with the infected host cell in the form of persistent bodies, which may be important in the pathogenesis of chronic Chlamydial infections.

6.2.2 *Chlamydial Infection and Tumours*

The potential oncogenic role of *Chlamydiae* was suggested based on the ability of this pathogen to establish persistent infection, be mitogenic *in vitro*, induce polyclonal cell proliferation *in vivo* and cause resistance to apoptosis in infected cells (Rasanen et al. 1986; Lehtinen et al. 1986; Rajalingam et al. 2001). Infections by *Chlamydia trachomatis* and *C. pneumoniae*, two members of the *Chlamydiaceae* family, are associated with sporadic cases of cervical carcinoma, lung cancer, cutaneous T-cell lymphoma. *Chlamydophila psittaci (Cp)* has a potential pathogenic role in development of OAMZL (Smith et al. 2002; Laurila et al. 1997; Ferreri et al. 2004). In fact, the DNA of *Cp* has been detected in 80% of OAMZL patients and immunohistochemistry data identified cells of the monocyte/macrophage system as likely carriers of the infection (Ferreri et al. 2004) (see paragraph "Worldwide variability of prevalence of *Cp* infection"). The presence of *Cp*, *C. pneumoniae* and *C. trachomatis* in OAMZL was tested in 13 Austrian patients. Gastrointestinal MALT lymphomas and gastritis specimens served as controls. Of 13 OAMZL, seven resulted positive for *Cp* DNA. Only one of 17 gastrointestinal specimens tested was positive. All specimens were negative for *C. trachomatis* and *C. pneumoniae* (Aigelsreiter et al. 2008). In a screening study, we analyzed the presence of *Cp* infection in both nodal and extranodal lymphomas. We reported that 74% of examined OALs were associated with *Cp* infection. Immunohistochemistry and TETR-PCR showed a significantly concordance rate of 70%. *Cp* DNA was equally prevalent in non-OAL, nodal, and extranodal lymphomas: in particular, it was more common in diffuse large B-cell lymphomas of the skin and Waldeyer's ring (Ponzoni et al. 2008).

Recently, we investigated the presence of DNA of *Cp*, *C. pneumoniae* and *C. trachomatis* on diagnostic tissue samples from cutaneous biopsies of 108 cases of primary cutaneous lymphomas by using three different PCR targeting 16S-23S region, outer membrane protein A and heat shock protein 60. *Cp* prevalence in cutaneous lymphomas of diffuse large B-cell lymphoma, MZL, and mycosis fungoides categories was similar to those detected in normal skin samples (controls), while the trend to a high prevalence of *Cp* infection in follicular lymphomas did not reach significant levels. An association between *C. pneumoniae* and mycosis fungoides was not detected in this series. There is not a support to a pathogenic role of *Chlamydiae* in cutaneous lymphomas (Ferreri et al. 2011 Epub ahead of print).

6.2.3 Epidemiology of Chlamydiae

Chlamydiae are responsible for several human and animal diseases. *Cp* is the etiological agent of avian chlamydiosis and psittacosis-ornithosis in humans, a zoonotic disease caused by the exposure to infected animals, mostly birds, but also domestic mammals and pets. The high prevalence of Chlamydial infections in household cats (Sykes 2005) suggests that human *Cp* infections originating from pets other than birds may be under-diagnosed.

Cp has been detected in 467 species of birds (Kaleta and Taday 2003), and in several mammals worldwide (Longbottom and Coulter 2003). All *Cp* genotypes identified so far, can be transmitted to humans (Heddema et al. 2006). Differently to the worldwide distribution of *C. pneumoniae*, with a seroprevalence from 40% to 80% in some geographic areas, *Cp* infection is rare in European individuals: IgG-seropositive rate is only 3% in elderly persons (Koivisto et al. 1999); higher rates have been detected in professionally exposed individuals or people living with infected animals or in rural areas (Bergström et al. 1996). However, serological tests for *Cp* detection have low specificity due to cross-reactivity with other bacteria, like *C. pneumoniae* (Mahmoud et al. 1994; Ni et al. 1996). More reliable results may be obtained by PCR analysis of peripheral blood samples. Blood donors used as control groups have been assessed by PCR and *in vitro* culture techniques, with a *Cp* detection rates of 0–5% (Ferreri et al. 2004, 2008b; Ponzoni et al. 2008). These results confirm that the prevalence of *Cp* infection in the general population is low.

6.3 *C. psittaci* Infection in OAMZL Patients

6.3.1 OAMZL Pathogenesis

As stated above, OAMZL usually arises from acquired MALT as a response to chronic inflammatory or autoimmune disorders. The detection of ongoing immunoglobulin gene mutations in OAMZL is consistent with a chronic antigen-driven stimulation. V_H genes are frequently rearranged in autoantibody production and are often over-expressed in B-cell malignancies, consistently with the occurrence of an antigen selection process during OAMZL development (Pascual and Capra 1992; Coupland et al. 1999; Hara et al. 2001; Adam et al. 2008). The VH3 family is the most common germline VH family used in OAMZL (54% of cases), followed by VH4 (27%) (Coupland et al. 1999). In the largest reported series (Coupland et al. 1999), one third of OAMZL expressed VH genes frequently associated with autoantibodies, DP-8, DP-10, DP-53, DP-63, and DP-49, which have also been detected in Alzheimer's disease, rheumatoid arthritis, idiopathic cold agglutinin disease, and systemic lupus erythematosus (Fang et al. 1995; Isenberg et al. 1993). These features are consistent with the possible role of epitopes from self-antigens in triggering lymphoma cell proliferation, at least in a subset of OAMZL cases. Comparison of heavy chain complementarity-determining region 3 (HCDR3) sequences derived from OAMZL with those of more than 8,000 unique immunoglobulin heavy chain variable-diversity-joining regions (IGHV-D-J) rearrangements from neoplastic and non-neoplastic B cells, including auto-reactive clones, revealed that 10% of immunoglobulin genes expressed in OAMZL display high homology with auto-reactive antibodies detectable in rheumatoid arthritis, Sjögren's syndrome and autoimmune thyroiditis. Analysis of the immunoglobulin gene repertoire suggests that OAMZL may originate from B cells selected for their capability to recognize auto-antigens, which is similar to what is reported for gastric MALT (Adam et al. 2008). Auto-antigen stimulation, as a result of the inflammatory process promoted by Chlamydial infection, may be the driving force of OAMZL pathogenesis, rather than the direct bacterial stimulation of B cells. *Chlamydiae* are linked to chronic infections of the conjunctiva with follicular or "inclusion" conjunctivitis features, which seem to be more common than previously recognized. In OAML patients, *Cp* establishes a systemic infection, as demonstrated by the detection of the DNA of the bacterium in peripheral blood mononuclear cells of 43% of the patients carrying a *Cp*-positive lymphoma (Ferreri et al. 2004). Such a systemic infection persists over time in a high proportion of cases, even more than 5 years, further supporting the possible involvement of *Cp* in sustaining lymphoma cell growth (Ferreri et al. 2004). Microbial persistence is favoured by molecular mimicry, a phenomenon by which antigens derived from micro-organisms are able to induce immune reactions cross-reacting with host self antigens (Oldstone 1998). In fact, the expression of antigenic motifs shared with the host allows the long-lasting persistence of microbial pathogens since the immune system is usually tolerant towards auto-antigens. Similar to *H. pylori*, a pathogenic model consisting of chronic antigenic stimulus that drives lymphoma genesis along a continuum pathway may be also hypothesized for *Cp*-associated OAMZL. *Chlamydiae* may also provide antigens, like heat shock proteins (Lamb et al. 2003), which may act as "molecular mimickers" and trigger autoimmune reactivity. Heat shock proteins produced by *Chlamydiae* may trigger both humoral and cell mediated immune responses that at least partially cross-react against the human protein counterpart (and other related self antigens) (Pockley 2003; Ausiello et al. 2005; Bachmaier and Penninger 2005). This phenomenon may contribute to break local tolerance, leading to a chronic stimulation by antigens that cannot be successfully eliminated by the host and that may ultimately favor the onset of OAML (Ishii et al. 2001; Yamasaki et al. 2004).

The existence of mechanisms regulating lymphocytic homing to the ocular adnexae could offer a possible explanation to the fact that OAMZL is usually a localized disease, followed by local relapses and systemic dissemination only in a few cases, and in a late phase of disease. Available data, however, are preliminary and limited to the absence of $\alpha4\beta7$ integrin expression, which is a regulator of lymphocyte trafficking, and its ligand MAdCAM (Liu et al. 2001), and to the expression of the chemoattractant cytokine CXCL13 on neoplastic lymphocytes (Falkenhagen et al. 2005).

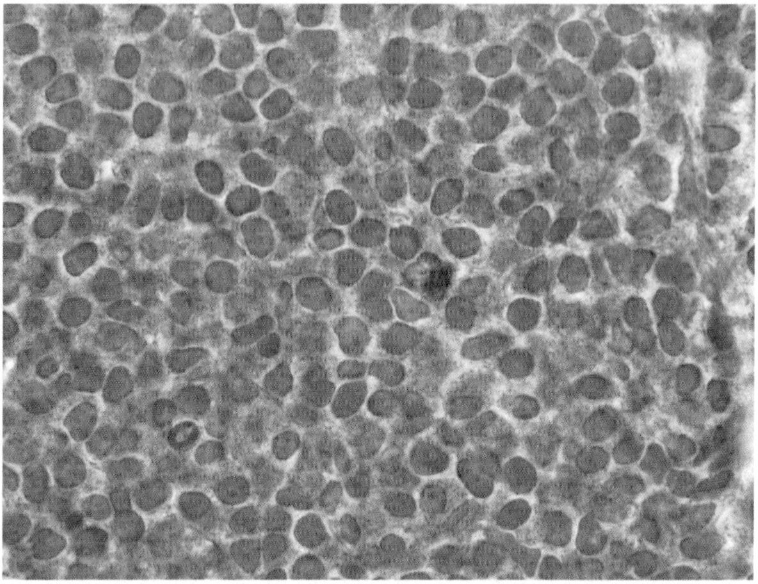

Fig. 6.1 Monoclonal antibody directed against Chlamydial lipopolysaccharide (LPS) showing granular immunoreactivity within the cytoplasm of a macrophage-like cell in a case of OAMZL

6.3.2 *Prevalence of Cp Infection in OAMZL*

Based on the hypothesis that OAMZL could be an antigen-driven lymphoproliferative disorder, we have investigated the possible pathogenic role of some infectious agents (*Cp*, *H. pylori*, hepatitis C virus) in these lymphomas. We have demonstrated that 80% of patients with OAMZL carried *Cp* DNA in the lymphomatous tissue (Ferreri et al. 2004); similar results were observed in a retrospective cohort study (Ponzoni et al. 2008) and a prospective case–control study (Ferreri et al. 2008b) in patients with OAMZL, both reporting *Cp* detection rates of 75% in tumour tissues. Other members of the *Chlamydiaceae* family such as *C. pneumoniae* or *C. trachomatis* were rarely detected in OAMZL. In our experience, patients with OAMZL commonly live in rural areas, and often report a history of chronic conjunctivitis and prolonged contact with household animals (Ferreri et al. 2008c). All these features strongly suggest an increased risk of *Cp* exposure.

Cp DNA and proteins in irregularly scattered monocytes and macrophages infiltrating OAMZL tissues has been demonstrated with different detection methods (Ponzoni et al. 2008). Anti-chlamydial lipopolysaccharide immunoreactivity is usually visualized as cytoplasmic granules by immunohistochemical techniques (Fig. 6.1). Monocytes and macrophages have been isolated by laser capture microdissection and identified as the *Cp* reservoir in lymphomatous tissue (Ponzoni et al. 2008). Electron microscopy confirmed the presence of clusters of intact CEB in the cytoplasm of monocytes and macrophages within OAMZL specimens (Fig. 6.2). Cp is vital and infective both in conjunctival tissue and PBMC (see "*In vitro* growth").

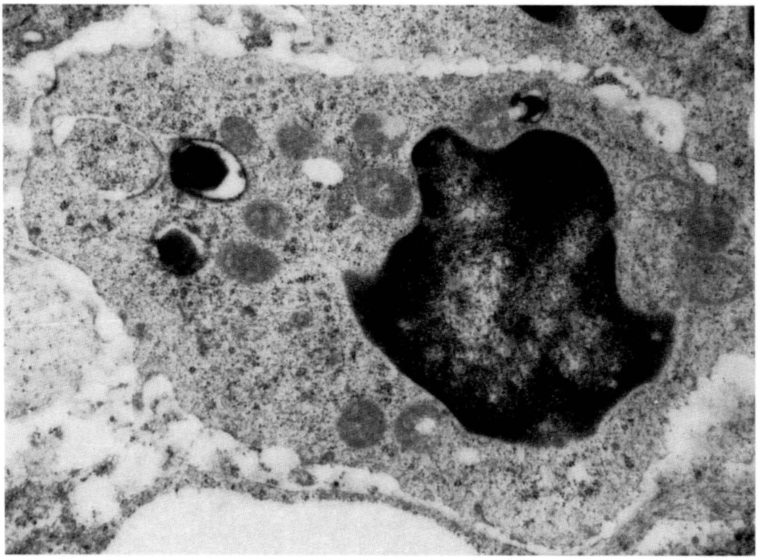

Fig. 6.2 Direct visualization of *Cp* in a macrophage by transmission electron microscopy in a case of OAMZL. Intact Chlamydial elementary bodies (EB), with electron dense 'black' core of nucleic acid condensed onto Chlamydial histone protein, and Chlamydial reticulate bodies (RB) are visible in the cytoplasm. Kindly provided by Luciano Sacchi; Department of Animal Biology, University of Pavia, Pavia, Italy

Cp DNA has been detected in peripheral blood mononuclear cells (PBMC) of 41% of patients with OAMZL (Ferreri et al. 2004), indicating the presence of asymptomatic systemic *Cp* infection at the time of lymphoma diagnosis. Chronic persistence of *Cp* in PBMC could favor lymphoma dissemination and recurrence (Ferreri et al. 2004, 2008b, c). In fact, *Cp* infection in PBMC has been detected in 62% of patients with disseminated OAMZL and in 11% of patients with stage-I disease (Ferreri et al. 2005b). Although a role for *Cp* in lymphoma recurrence can be hypothesized, mechanisms explaining the persistence of OAMZL at the primary site for years and the involvement of contralateral ocular adnexae at the time of relapse remain to be investigated. *Cp* detection in PBMC is a useful tool for monitoring bacterial eradication, persistence or re-infection in patients with OAMZL who were treated with antibiotics (Ferreri et al. 2005b, 2006a).

6.3.3 World-Wide Variability of Prevalence of Cp Infection in OAMZL Patients

A great variability in *Cp* prevalence among countries and among different regions within the same country has been reported (Table 6.1). The prevalence of *Cp* infection in published series of OAMZL ranges from 0% to 87%. The infection was detected in 14% of ocular adnexal lymphomas of other histotype and in 8% of orbit biopsies for non-neoplastic disorders (Decaudin et al. 2008). The highest prevalence rates have been reported in Italy, Korea, Austria, and Germany.

Cp prevalence was tested in 33 Korean patients with OAMZL and compared to 21 non neoplastic ocular lesions. DNA was extracted from paraffin-embedded tissues, and touchdown enzyme time release polymerase chain reaction (TETR-PCR) was performed to identify three *Chlamydia* species. *Cp* DNA was detected in 26/33 (79%) OAL samples compared with 5/21 (23%) non-neoplastic samples ($P<0.001$), suggesting that *Cp* plays a role as a causative antigen in Korean OAL patients. With a median follow-up of 38.5 months (range: 1–105 months), the 5-year progression-free survival and overall survival rates of OAL patients were 72% and 93%, respectively. Clinicopathologic characteristics, recurrence rate and outcome were not associated with *Cp* infection (Yoo et al. 2007).

Seven of 13 OAMZL in Austrian patients resulted *Cp* positive, with a prevalence of 54% (Aigelsreiter et al. 2008). Recently, these authors analyzed the presence of *Chlamydiae* infections in 47 non-gastrointestinal and 14 gastrointestinal MALT lymphomas, 37 nonmalignant control samples, and 27 autoimmune precursor lesions by polymerase chain reaction amplification and direct sequencing. In 47 nongastrointestinal MALT lymphomas, 13 (28%) were positive for *Cp* DNA compared with 4 (11%) of 37 non-malignant control samples ($P=0.09$). *Cp* was detected in MALT lymphomas of several organs: lung, with 100% of prevalence (5/5; $P<0.01$); thyroid gland, in 30% of cases (3/10; $P>0.05$); salivary gland, in 13% of patients (2/15; $P>0.05$); in 15% of OAMZL (2/13); and in 25% of cutaneous lymphomatous lesions (1/4). Of 27 autoimmune precursor lesions, 11 (41%) resulted *Cp* positive. Only one (7%) of 14 gastrointestinal MALT lymphomas was positive for *Cp*. All specimens were negative for *C. trachomatis* and *C. pneumoniae* (Aigelsreiter et al. 2011).

The *Cp*-OAMZL association was not detected in series from some countries, like Japan (Yakushijin et al. 2007), France, China (Zhang et al. 2007), Miami (Matthews et al. 2008), and Kenya (Carugi et al. 2010). Intermediate rates from 10% to 40% were detected in series from Cuba (Gracia et al. 2007), Netherlands, UK, and others (Table 6.1).

Variations in *Cp* prevalence among studies from different countries could be explained by different methodological analysis, by the use of wide-spectrum antibiotics, and also by the presence of other possible microbial agents. Differences have been influenced by tissue samples, experimental conditions, DNA extraction protocols, PCR sensitivity, and amount of DNA template. Most studies used a "multiplex" TETR-PCR (Ferreri et al. 2004), while, sometimes, these primers were used in "monoplex" PCR, and direct sequencing to confirm the specificity of the amplified DNA was not always performed. However, centralized molecular analyses from a multicentre study have confirmed the discrepancies in prevalence among countries and different regions of the same country (Chanudet et al. 2006). Topic or systemic wide-spectrum antibiotics are often employed in patients with conjunctival or orbital lesions firstly considered inflammatory or infectious diseases; this practice in OAMZL could reduce the Chlamydial load, resulting in false negatives at PCR. A large study involving an adequate number of cases from several countries deserves to be conducted with the same methodological diagnostic approach.

Table 6.1 Prevalence of *C. psittaci* (*Cp*) infection in patients with OAMZL

Geographical area	Number of OAMZL pts	Number of *Cp*+pts	% of *Cp*+cases (95% CI)	Reference
Austria	13	7	54 (27–81)*	(Aigelsreiter et al. 2008)
Cuba	19	2	10 (1–33)	(Gracia et al. 2007)
France	6	0	0 (0–46)	(De Cremoux et al. 2006)
Germany	19	9	47 (24–71)	(Chanudet et al. 2006)
Germany	23	0	0 (0–12)	(Goebel et al. 2007)
Italy	24	21	87 (68–97)	(Ferreri et al. 2006a)
Italy	15	2	13 (2–40)	(Chanudet et al. 2006)
Italy	20	15	75 (56–94)	(Ferreri et al. 2004)
Italy	35	26	74 (60–88)	(Ponzoni et al. 2008)
Japan	18	0	0 (0–19)	(Daibata et al. 2006)
Japan	12	0	0 (0–26)	(Liu et al. 2006)
South Korea	33	23	77 (58–90)	(Yoo et al. 2007)
Southern China	37	4	11 (3–25)	(Chanudet et al. 2006)
The Netherlands	19	0	0 (0–18)	(Mulder et al. 2006)
The Netherlands	21	6	29 (11–52)	(Chanudet et al. 2006)
United Kingdom	33	4	12 (3–28)	(Chanudet et al. 2006)
USA, East Coast	17	6	35 (14–62)	(Chanudet et al. 2006)
USA, Florida	46	0	0 (0–8)	(Rosado et al. 2005)
USA, North-East	7	0	0 (0–41)	(Vargas et al. 2006)
USA, North-East	31	0	0 (0–10)	(Ruiz et al. 2007)
USA, North-East	28	0	0 (0–11)	(Zhang et al. 2007)

6.3.4 *In Vitro Growth*

In a recent prospective case-control study (Ferreri et al. 2008b), conjunctival swabs and peripheral blood samples from patients with OAMZL and healthy blood donors were collected and analyzed by *in vitro* cultures and TETR-PCR. Viability and infectivity of *Cp* were demonstrated by *in vitro* isolation and growth assays in 25% of patients with OAMZL, but not in the healthy blood donors used as controls (Ferreri et al. 2008b). TETR-PCR displayed a higher sensitivity in *Chlamydia* detection compared with *in vitro* cultures (Ferreri et al. 2008b). This is likely due to the fact that in some micro-environmental conditions, *Chlamydiae* can enter a persistent phase in infected cells, with very limited, or absent, production of CEB (Byrne and Ojcius 2004). CEBs do not grow in cultures but can be detected by PCR-based techniques.

The isolation in *in vitro* cultures (Ferreri et al. 2008b; Ponzoni et al. 2008), the observation of intact CEB by electron microscopy (Ponzoni et al. 2008) and *Cp* eradication after doxycycline treatment (Ferreri et al. 2006a) confirm that *Cp* is viable and infectious in biological samples obtained from patients with OAMZL. These findings fulfilled the second Koch's postulate – the organism can be isolated and grown *in vitro* from disease lesions. Koch's postulates are still useful to evaluate a cause-effect relationship between an infectious agent and a disease. To date,

H. pylori and *Cp* are the only bacteria with lymphomagenic potential isolated from patients with lymphoma and that have demonstrated growth potential in *in vitro* cultures (Ponzoni et al. 2008).

6.3.5 Other Infections in OAMZL

C. pneumoniae DNA has been detected in a few cases of OAMZL (Chanudet et al. 2006; Chan et al. 2006). *C. trachomatis*, herpes simplex virus 1 and 2 and adenovirus 8 and 19 are commonly related to chronic eye diseases (Ferreri et al. 2004). *Chlamydophila pneumoniae* DNA and mRNA transcripts were investigated by PCR and RT-PCR in fresh CSF and PBMC specimens collected from 14 patients with definite multiple sclerosis (MS) and 19 patients with other inflammatory and non-inflammatory neurological controls. A positivity for *C. pneumoniae* DNA and mRNA was detected in CSF and PBMCs of nine MS patients (64.2%) with evidence of disease activity, whereas only three controls were positive for Chlamydial DNA. These findings could suggest that *C. pneumoniae* may occur in a persistent and metabolically active state at both peripheral and intrathecal levels in MS, but not in other neurological controls (Contini et al. 2008).

Gastric infection by *H. pylori* has been detected in one third of Italian patients with OAMZL, but with no obvious correlation with particular disease characteristics (Ferreri et al. 2006b). *H. pylori*-eradicating antibiotic therapy was not associated with lymphoma regression (Ferreri et al. 2006b). Worldwide data on the presence of *H. pylori* in the conjunctiva of healthy individuals are lacking.

Scattered reports throughout the literature, however, have also linked infection with *Hp* to MALT lymphoma arising in extragastric organs, and regressions of extragastric manifestations following *Hp* eradication have anecdotally been reported. To date, no systematic prospective investigation on the role of *Hp* infection in patients with primary extragastric MALT lymphomas has been performed. A total of 77 patients with extragastric MALT lymphoma were prospectively studied. The presence of *Hp* was tested by histology, breath test, and serology. Evidence of *Hp* infection was present in 35 of 77 patients (45%). All patients with *Hp*-infection underwent eradication, 16 before initiation of further therapy. Only a patient with lymphoma involving parotid and colon, obtained regression with eradication as exclusive treatment, the remaining 15 showed no regression of the disease at a median follow-up of 14 months (range, 8–48+ months). No correlation between *Hp*-status, localization, stage, autoimmune diseases, and genetic findings was seen (Grunberger et al. 2006).

Some sporadic cases of extra-gastrointestinal MZL associated to *Hp*-gastric lymphoma had been described: the presence of *Hp* in other sites of disease was not reported. Eradicating treatment resulted in lymphoma regression in all involved organs (Alkan et al. 1996; Berrebi et al. 1998; Arima and Tsudo 2003).

Hepatitis C virus (HCV) has been detected in 13–36% of OAMZL patients, and it is related to more aggressive lymphomas (Ferreri et al. 2006c; Arcaini et al. 2007).

In line with the well-known activity reported in other indolent lymphomas, a few HCV+patients with OAMZL were successfully treated with anti-viral treatment with interferon and ribavirine.

Thirteen patients with histologically proven low-grade B-NHL characterized by an indolent course and positive for HCV infection underwent antiviral treatment alone with pegylated interferon and ribavirin. Seven of them (58%) achieved complete response and two (16%) partial hematologic response at 14.1 +/– 9.7 months (range, 2–24 months, median follow-up, 14 months). Two patients presented stable disease while only one patient had progressive disease. Overall response rate was 75% and was significantly associated to clearance or decrease in serum HCV viral load following treatment (P=0.005). Treatment-related toxicity caused discontinuation of therapy in only two patients (Vallisa et al. 2005).

6.3.6 Cp Infection in Other Lymphomas

In an archival series of 205 non-consecutive lymphomas (Ponzoni et al. 2008), *Cp* infection was found in diffuse large B-cell lymphomas of the skin and Waldeyer's ring, two extranodal organs considered as first barriers to air-transported antigens. *Cp* transmission usually occurs by inhalation of aerosol or direct contact (Harkinezhad et al. 2009), however, *Cp* infection is rarely found in lymphomas of the lungs and the gastrointestinal tract (Chanudet et al. 2007; Ponzoni et al. 2008; Aigelsreiter et al. 2008). Recently, the presence of *Chlamydiae* infections was investigated in 47 nongastrointestinal MALT lymphomas and 37 nonmalignant control samples (Aigelsreiter et al. 2011). In 47 nongastrointestinal MALT lymphomas, 13 (28%) were positive for *Cp* DNA compared with 4 (11%) of 37 nonmalignant control samples (P=0.09). *Cp* was detected in 100% of lung lymphomas (5/5; P<0.01); 30% of thyroid gland (3/10; P>0.05); 13% of salivary gland (2/15; P>0.05); and 25% of the skin, (1/4). Conversely, our study suggests that Chlamydial infections are not related to development and growth of the most common forms of cutaneous lymphomas (Ferreri et al. 2011 Epub ahead of print).

6.3.7 Cp Detection Methods

In a number of lymphoma studies conducted in Italy (Ferreri et al. 2004, 2005b, 2006a, 2007, 2008b; Ponzoni et al. 2008), *Cp* DNA detection has been performed by PCR amplifications of three different genomic regions: a portion of the *16S rRNA* gene and of the *16S–23S* spacer region (Madico et al. 2000; Ferreri et al. 2004), the major outer membrane protein gene (*ompA*) (Tong and Sillis 1993) and the heat shock protein 60 (*hsp 60*) gene (Hill et al. 2005). Although results obtained with these assays showed high concordance rates, the multiplex touchdown enzyme time-release PCR assay (TETR-PCR) should be considered as the recommended protocol for *Cp* detection (Madico et al. 2000; Ferreri et al. 2004). TETR-PCR simultaneously detects *C. trachomatis, C. pneumoniae,* and *Cp* DNA at bacterial loads lower than one Inclusion Forming Unit. The DNA fragment amplified with this protocol includes the end of the *16S rRNA* gene and the beginning of the *16S–23S* spacer region. PCR products are analyzed by electrophoresis and DNA fragment size is quantified by image analysis. We confirmed the specificity of the amplified DNA strand by sequencing; false-positivity may be a critical issue, since there is a high degree of homology between several genomic regions among the various Chlamydial species.

Several other real-time PCR assays have been recently proposed (Pantchev et al. 2008; Sachse et al. 2008), but these have not been used for the analysis of lymphomatous tissue. PCR-based assays for evaluating Chlamydial prevalence in lymphoma series should include: (1) sequencing analysis of the amplicons to confirm specificity; (2) analysis of multiple genetic targets to exclude the presence of mixed infections with *Chlamydiaceae* other than *Cp*; (3) consideration and reporting of any potential interfering factors, such as geographical area and year of diagnosis; and (4) a detailed report of experimental conditions. The quality and amount of DNA template and PCR protocol sensitivity levels are crucial parameters for assessment of the prevalence of *Cp* in biopsies. Recently, a large study confirmed that immunohistochemistry for Chlamydial lipopolysaccharide could be a fast, easy and cost-effective surrogate to PCR for the detection of *Chlamydiae*, with a 76% concordance rate between these techniques (Ponzoni et al. 2008). Indirect tests for the detection of antibodies to *Cp* are serological tests that include the complement fixation test (CFT), the micro-immunofluorescence (MIF) assay and several enzyme-linked immunosorbent assays (ELISA). Serological tests used in the screening of large series of patients and controls in lymphoma-*Chlamydia* studies have a low specificity and have therefore produced confusing results (Ferreri et al. 2004). In this context, efforts should be made to use direct detection methods, preferably nucleic acid amplification methods, wherever possible.

6.4 *Chlamydophila psittaci* Eradicating Antibiotic Therapy

6.4.1 Therapeutic Modality and Results

Eradicating treatment with doxycycline (100 mg administered *per os,* twice a day, for 3 weeks) has been proposed as a valid therapeutic strategy in patients with OAMZL. In a multicentre phase II trial (Ferreri et al. 2006a), 11 patients with *Cp*-positive and 16 with *Cp*-negative OAMZL have been treated with doxycycline, obtaining an overall response rate of 48% at a median follow-up of 14 months. Regressions were slow and gradual, even in patients with multiple relapses, occurring both in *Cp*-positive (65%) and *Cp*-negative (38%) patients. After a median follow-up of 28 months, the 3-year progression-free survival was 68%, and the median time to progression was >31 months, which is in line with prolonged remissions reported by others (Abramson et al. 2005).

To date, with the single exception of one case of tumour remission in a patient with *Cp*-associated diffuse large B-cell lymphoma of the bronchus after doxycycline treatment (Ferreri et al. 2007), all patients treated with this strategy had OAMZL, at diagnosis or relapse, with limited or advanced disease. Response varied amongst the different tumour lesions in patients with advanced disease, which might be attributed to the heterogeneous distribution of tumour clones that are dependent on antigenic stimulation (Ferreri et al. 2008c). Moreover, response to doxycycline could be influenced by re-infection due to prolonged contact with infected pets (Ferreri et al. 2007) or the effect of other concomitant infections (for example, hepatitis C virus).

6.4.2 Risk of Cp Re-infection

Up to 85% of patients with OAMZL have a history of chronic conjunctivitis and/or prolonged contact with pets (Ferreri et al. 2008b). This is an important issue in OAMZL eligible for doxycycline treatment considering that sub-clinical Chlamydial infections are common in household animals (Greco et al. 2005). Recently, we reported a case of prolonged exposure to a sub-clinically infected animal, which resulted in repeated *Cp* re-infections, asymptomatic *Cp* persistence and to the development of two metachronus independent *Cp*-associated lymphomas, clonally unrelated (an OAMZL and a DLBCL of the bronchus) (Ferreri et al. 2007). Sustained contact with potentially infected animals should be investigated in *Cp*-related lymphomas. The treatment or removal of the animal from the household may prevent *Cp* re-infection and eliminate the risk of lymphoma development (Ferreri et al. 2007).

6.5 Conclusion

Cp is associated with the development of OAMZL in some geographical areas. This is suggested by molecular features of this lymphoma and some biological properties of *Cp*. Evidence supporting this pathogenic association includes the presence of *Cp* within macrophages and monocytes infiltrating the lymphomatous tissue (Ferreri et al. 2004; Ponzoni et al. 2008), *Cp* isolation in *in vitro* cultures from conjunctival swabs and from the peripheral blood of patients with OAMZL (Ferreri et al. 2008b) and the regression of lymphoma after *Cp*-eradication with doxycycline (Ferreri et al. 2005b, 2006a). This therapeutic approach in OAMZL appears a cheap, fast, save, and active therapy in *Cp*-related OAMZL. Presently, the use of antibiotics in OAMZL is still an experimental therapy within clinical trials. We need a longer follow-up to define antibiotics response duration, and further investigations to reliably distinguish between *Cp* persistence and re-infection in unresponsive patients.

Several questions on this association remain unanswered. The prevalence of *Cp* infection in different geographical areas and the identification of additional infectious agents associated with OAMZL are two of the most important investigational issues, since we observed that also *Cp*-negative patients obtained clinical responses to doxycycline: so we supposed that other bacteria responsive to doxycycline may be present in OAMZL (Ferreri et al. 2007). Antibiotic treatment appears a promising alternative source with a significant rate of success in OAMZL. In the case of *Cp* persistence, a different schedule of antibiotic therapy or a new antibiotic should be considered. Re-infection from an external source may frequently occur. The presence of an external reservoir for *Cp* in the domestic or professional environment should be always investigated. These concerns represent the essential endpoints of a recently conducted multicentre prospective phase II trial (IELSG #27 trial) with centralized molecular analysis, under the sponsorship of the International Extranodal Lymphoma Study Group. The role of antibiotic therapy in OAMZL and the correlation between therapeutic activity and genetic abnormalities were additional endpoints in this trial. The development of *in vitro* tests for patients with OAMZL that can assess lymphocyte proliferation in response to Chlamydial antigen stimulation could improve our understanding of the pathogenesis of these lymphomas. In order to establish a causative role of *Cp* in OAMZL, experimental animal models that confirm the lymphomagenic potential of *Cp* need to be developed. Finally, studies that characterize the mechanisms of antibiotic resistance, *Cp* re-infection and reactivation might improve the efficacy of antibiotic therapy in patients with OAMZL.

References

Abramson DH, Rollins I, Coleman M (2005) Periocular mucosa-associated lymphoid/low grade lymphomas: treatment with antibiotics. Am J Ophthalmol 140(4):729–730

Adam P, Haralambieva E, Hartmann M et al (2008) Rare occurrence of IgVH gene translocations and restricted IgVH gene repertoire in ocular MALT-type lymphoma. Haematologica 93(2):319–320

Aigelsreiter A, Leitner E, Deutsch AJ et al (2008) *Chlamydia psittaci* in MALT lymphomas of ocular adnexals: the Austrian experience. Leuk Res 32(8):1292–1294

Aigelsreiter A, Gerlza T, Deutsch AJ et al (2011) *Chlamydia psittaci* infection in nongastrointestinal extranodal MALT lymphomas and their precursor lesions. Am J Clin Pathol 135(1):70–75

Alkan S, Karcher DS, Newman MA et al (1996) Regression of salivary gland MALT lymphoma after treatment for *Helicobacter pylori*. Lancet 348(9022):268–269

Arcaini L, Burcheri S, Rossi A et al (2007) Prevalence of HCV infection in nongastric marginal zone B-cell lymphoma of MALT. Ann Oncol 18(2):346–350

Arima N, Tsudo M (2003) Extragastric mucosa-associated lymphoid tissue lymphoma showing the regression by *Helicobacter pylori* eradication therapy. Br J Haematol 120(5):790–792

Ausiello CM, Palazzo R, Spensieri F et al (2005) 60-kDa heat shock protein of *Chlamydia pneumoniae* is a target of T-cell immune response. J Biol Regul Homeost Agents 19(3–4):136–140

Bachmaier K, Penninger JM (2005) *Chlamydia* and antigenic mimicry. Curr Top Microbiol Immunol 296:153–163

Ballalai PL, Erwenne CM, Martins MC et al (2009) Long-term results of topical mitomycin C 0.02% for primary and recurrent conjunctival-corneal intraepithelial neoplasia. Ophthal Plast Reconstr Surg 25(4):296–299

Ben Simon GJ, Cheung N, McKelvie P et al (2006) Oral chlorambucil for extranodal, marginal zone, B-cell lymphoma of mucosa-associated lymphoid tissue of the orbit. Ophthalmology 113(7):1209–1213

Bergström K, Domeika M, Vaitkiene D et al (1996) Prevalence of *Chlamydia trachomatis, Chlamydia psittaci* and *Chlamydia pneumoniae* antibodies in blood donors and attendees of STD clinics. Clin Microbiol Infect 1(4):253–260

Berrebi D, Lescoeur B, Faye A et al (1998) MALT lymphoma of labial minor salivary gland in an immunocompetent child with a gastric *Helicobacter pylori* infection. J Pediatr 133(2):290–292

Bhatia S, Paulino AC, Buatti JM et al (2002) Curative radiotherapy for primary orbital lymphoma. Int J Radiat Oncol Biol Phys 54(3):818–823

Blasi MA, Gherlinzoni F, Calvisi G et al (2001) Local chemotherapy with interferon-alpha for conjunctival mucosa-associated lymphoid tissue lymphoma: a preliminary report. Ophthalmology 108(3):559–562

Byrne GI, Ojcius DM (2004) *Chlamydia* and apoptosis: life and death decisions of an intracellular pathogen. Nat Rev Microbiol 2(10):802–808

Carugi A, Onnis A, Antonicelli G et al (2010) Geographic variation and environmental conditions as cofactors in *Chlamydia psittaci* association with ocular adnexal lymphomas: a comparison between Italian and African samples. Hematol Oncol 28(1):20–26

Chan CC, Shen D, Mochizuki M et al (2006) Detection of *Helicobacter pylori* and *Chlamydia pneumoniae* genes in primary orbital lymphoma. Trans Am Ophthalmol Soc 104:62–70

Chanudet E, Zhou Y, Bacon CM et al (2006) *Chlamydia psittaci* is variably associated with ocular adnexal MALT lymphoma in different geographical regions. J Pathol 209(3):344–351

Chanudet E, Adam P, Nicholson AG et al (2007) *Chlamydiae* and *Mycoplasma* infections in pulmonary MALT lymphoma. Br J Cancer 97(7):949–951

Conconi A, Martinelli G, Thieblemont C et al (2003) Clinical activity of rituximab in extranodal marginal zone B-cell lymphoma of MALT type. Blood 102(8):2741–2745

Contini C, Seraceni S, Castellazzi M et al (2008) *Chlamydophila pneumoniae* DNA and mRNA transcript levels in peripheral blood mononuclear cells and cerebrospinal fluid of patients with multiple sclerosis. Neurosci Res 62(1):58–61

Coupland SE, Krause L, Delecluse HJ et al (1998) Lymphoproliferative lesions of the ocular adnexa. Analysis of 112 cases. Ophthalmology 105(8):1430–1441

Coupland SE, Foss HD, Anagnostopoulos I et al (1999) Immunoglobulin VH gene expression among extranodal marginal zone B-cell lymphomas of the ocular adnexa. Invest Ophthalmol Vis Sci 40(3):555–562

Daibata M, Nemoto Y, Togitani K et al (2006) Absence of *Chlamydia psittaci* in ocular adnexal lymphoma from Japanese patients. Br J Haematol 132(5):651–652

De Cremoux P, Subtil A, Ferreri AJ et al (2006) Evidence for an association between *Chlamydia psittaci* and ocular adnexal lymphomas. J Natl Cancer Inst 98(5):365–366

Decaudin D, Dolcetti R, de Cremoux P et al (2008) Variable association between *Chlamydophila psittaci* infection and ocular adnexal lymphomas: methodological biases or true geographical variations? Anti-Cancer Drugs 19(8):761–765

Ejima Y, Sasaki R, Okamoto Y et al (2006) Ocular adnexal mucosa-associated lymphoid tissue lymphoma treated with radiotherapy. Radiother Oncol 78(1):6–9

Esmaeli B, McLaughlin P, Pro B et al (2009) Prospective trial of targeted radioimmunotherapy with Y-90 ibritumomab tiuxetan (zevalin) for front-line treatment of early-stage extranodal indolent ocular adnexal lymphoma. Ann Oncol 20(4):709–714

Falkenhagen KM, Braziel RM, Fraunfelder FW et al (2005) B-cells in ocular adnexal lymphoproliferative lesions express B-cell attracting chemokine 1 (CXCL13). Am J Ophthalmol 140(2):335–337

Fang Q, Kannapell CC, Fu SM et al (1995) VH and VL gene usage by anti-beta-amyloid autoantibodies in alzheimer's disease: detection of highly mutated V regions in both heavy and light chains. Clin Immunol Immunopathol 75(2):159–167

Ferreri AJ, Guidoboni M, Ponzoni M et al (2004) Evidence for an association between *Chlamydia psittaci* and ocular adnexal lymphomas. J Natl Cancer Inst 96(8):586–594

Ferreri AJ, Ponzoni M, Martinelli G et al (2005a) Rituximab in patients with mucosal associated lymphoid tissue-type lymphoma of the ocular adnexa. Haematologica 90(11):1578–1579

Ferreri AJ, Ponzoni M, Guidoboni M et al (2005b) Regression of ocular adnexal lymphoma after *Chlamydia psittaci*-eradicating antibiotic therapy. J Clin Oncol 23(22):5067–5073

Ferreri AJ, Ponzoni M, Guidoboni M et al (2006a) Bacteria-eradicating therapy with doxycycline in ocular adnexal MALT lymphoma: a multicenter prospective trial. J Natl Cancer Inst 98(19):1375–1382

Ferreri AJ, Ponzoni M, Viale E et al (2006b) Association between *Helicobacter pylori* infection and MALT-type lymphoma of the ocular adnexa: clinical and therapeutic implications. Hematol Oncol 24(1):33–37

Ferreri AJ, Viale E, Guidoboni M et al (2006c) Clinical implications of hepatitis C virus infection in MALT-type lymphoma of the ocular adnexa. Ann Oncol 17(5):769–772

Ferreri AJ, Dolcetti R, Magnino S et al (2007) A woman and her canary: a tale of *Chlamydiae* and lymphomas. J Natl Cancer Inst 99(18):1418–1419

Ferreri AJ, Dolcetti R, Du MQ et al (2008a) Ocular adnexal MALT lymphoma: an intriguing model for antigen-driven lymphomagenesis and microbial-targeted therapy. Ann Oncol 19(5):835–846

Ferreri AJ, Dolcetti R, Dognini GP et al (2008b) *Chlamydophila psittaci* is viable and infectious in the conjunctiva and peripheral blood of patients with ocular adnexal lymphoma: results of a single-center prospective case-control study. Int J Cancer 123:1089–1093

Ferreri AJ, Dognini GP, Ponzoni M et al (2008c) *Chlamydia psittaci*-eradicating antibiotic therapy in patients with advanced-stage ocular adnexal MALT lymphoma. Ann Oncol 19(1):194–195

Ferreri AJ, Dolcetti R, Magnino S et al (2009) Chlamydial infection: the link with ocular adnexal lymphomas. Nat Rev Clin Oncol 6(11):658–669

Ferreri AJ, Govi S, Colucci A et al (2011) Intralesional rituximab: a new therapeutic approach for patients with conjunctival lymphomas. Ophthalmology 118(1):24–28

Ferreri AJ, Pnzoni M, Govi S, et al (2011) Prevalence of Chlamydial infection in a series of 108 primary cutaneous lymphomas. Br J Dermatol (Epub ahead of print)

Ferry JA, Fung CY, Zukerberg L et al (2007) Lymphoma of the ocular adnexa: a study of 353 cases. Am J Surg Pathol 31(2):170–184

Fung CY, Tarbell NJ, Lucarelli MJ et al (2003) Ocular adnexal lymphoma: clinical behavior of distinct world health organization classification subtypes. Int J Radiat Oncol Biol Phys 57(5):1382–1391

Goebel N, Serr A, Mittelviefhaus H et al (2007) *Chlamydia psittaci*, *Helicobacter pylori* and ocular adnexal lymphoma-is there an association? The German experience. Leuk Res 31(10):1450–1452

Gracia E, Froesch P, Mazzucchelli L et al (2007) Low prevalence of *Chlamydia psittaci* in ocular adnexal lymphomas from Cuban patients. Leuk Lymphoma 48(1):104–108

Greco G, Corrente M, Martella V (2005) Detection of *Chlamydophila psittaci* in asymptomatic animals. J Clin Microbiol 43(10):5410–5411, author reply 5410–1

Grunberger B, Wohrer S, Streubel B et al (2006) Antibiotic treatment is not effective in patients infected with *Helicobacter pylori* suffering from extragastric MALT lymphoma. J Clin Oncol 24(9):1370–1375

Hara Y, Nakamura N, Kuze T et al (2001) Immunoglobulin heavy chain gene analysis of ocular adnexal extranodal marginal zone B-cell lymphoma. Invest Ophthalmol Vis Sci 42(11):2450–2457

Harkinezhad T, Geens T, Vanrompay D (2009) *Chlamydophila psittaci* infections in birds: a review with emphasis on zoonotic consequences. Vet Microbiol 135(1–2):68–77

Heddema ER, van Hannen EJ, Duim B et al (2006) Genotyping of *Chlamydophila psittaci* in human samples. Emerg Infect Dis 12:1989–1990

Hill JE, Goh SH, Money DM et al (2005) Characterization of vaginal microflora of healthy, nonpregnant women by chaperonin-60 sequence-based methods. Am J Obstet Gynecol 193(3 Pt 1):682–692

Isaacson PG, Du MQ (2004) MALT lymphoma: from morphology to molecules. Nat Rev Cancer 4(8):644–653

Isenberg D, Spellerberg M, Williams W et al (1993) Identification of the 9 G4 idiotope in systemic lupus erythematosus. Br J Rheumatol 32(10):876–882

Ishii E, Yokota K, Sugiyama T et al (2001) Immunoglobulin G1 antibody response to *Helicobacter pylori* heat shock protein 60 is closely associated with low-grade gastric mucosa-associated lymphoid tissue lymphoma. Clin Diagn Lab Immunol 8(6):1056–1059

Kaleta EF, Taday EM (2003) Avian host range of *Chlamydophila* spp. based on isolation, antigen detection and serology. Avian Pathol 32(5):435–461

Knowles DM, Jakobiec FA, McNally L et al (1990) Lymphoid hyperplasia and malignant lymphoma occurring in the ocular adnexa (orbit, conjunctiva, and eyelids): a prospective multiparametric analysis of 108 cases during 1977 to 1987. Hum Pathol 21(9):959–973

Koivisto AL, Isoaho R, Von Hertzen L et al (1999) Chlamydial antibodies in an elderly finnish population. Scand J Infect Dis 31(2):135–139

Lachapelle KR, Rathee R, Kratky V et al (2000) Treatment of conjunctival mucosa-associated lymphoid tissue lymphoma with intralesional injection of interferon alfa-2b. Arch Ophthalmol 118(2):284–285

Lamb DJ, El-Sankary W, Ferns GA (2003) Molecular mimicry in atherosclerosis: a role for heat shock proteins in immunisation. Atherosclerosis 167(2):177–185

Laurila AL, Anttila T, Laara E et al (1997) Serological evidence of an association between *Chlamydia pneumoniae* infection and lung cancer. Int J Cancer 74(1):31–34

Lehtinen M, Rantala I, Aine R et al (1986) B cell response in *Chlamydia trachomatis* endometritis. Eur J Clin Microbiol 5(5):596–598

Liu YX, Yoshino T, Ohara N et al (2001) Loss of expression of alpha4beta7 integrin and L-selectin is associated with high-grade progression of low-grade MALT lymphoma. Mod Pathol 14(8):798–805

Liu YC, Ohyashiki JH, Ito Y et al (2006) *Chlamydia psittaci* in ocular adnexal lymphoma: Japanese experience. Leuk Res 30(12):1587–1589

Longbottom D, Coulter LJ (2003) Animal chlamydioses and zoonotic implications. J Comp Pathol 128(4):217–244

Madico G, Quinn TC, Boman J et al (2000) Touchdown enzyme time release-PCR for detection and identification of *Chlamydia trachomatis, C. pneumoniae*, and *C. psittaci* using the 16S and 16S-23S spacer rRNA genes. J Clin Microbiol 38(3):1085–1093

Mahmoud E, Elshibly S, Mardh PA (1994) Seroepidemiologic study of *Chlamydia pneumoniae* and other chlamydial species in a hyperendemic area for trachoma in the Sudan. Am J Trop Med Hyg 51(4):489–494

Mannami T, Yoshino T, Oshima K et al (2001) Clinical, histopathological, and immunogenetic analysis of ocular adnexal lymphoproliferative disorders: characterization of malt lymphoma and reactive lymphoid hyperplasia. Mod Pathol 14(7):641–649

Margo CE, Mulla ZD (1998) Malignant tumors of the orbit. Analysis of the florida cancer registry. Ophthalmology 105(1):185–190

Martinet S, Ozsahin M, Belkacemi Y et al (2003) Outcome and prognostic factors in orbital lymphoma: a rare cancer network study on 90 consecutive patients treated with radiotherapy. Int J Radiat Oncol Biol Phys 55(4):892–898

Matsuo T, Yoshino T (2004) Long-term follow-up results of observation or radiation for conjunctival malignant lymphoma. Ophthalmology 111(6):1233–1237

Matthews JM, Moreno LI, Dennis J et al (2008) Ocular adnexal lymphoma: no evidence for bacterial DNA associated with lymphoma pathogenesis. Br J Haematol 142(2):246–249

McKelvie PA, McNab A, Francis IC et al (2001) Ocular adnexal lymphoproliferative disease: a series of 73 cases. Clin Experiment Ophthalmol 29(6):387–393

Moslehi R, Devesa SS, Schairer C et al (2006) Rapidly increasing incidence of ocular non-hodgkin lymphoma. J Natl Cancer Inst 98(13):936–939

Mulder MM, Heddema ER, Pannekoek Y (2006) No evidence for an association of ocular adnexal lymphoma with *Chlamydia psittaci* in a cohort of patients from the Netherlands. Leuk Res 30(10):1305–1307

Murga Penas EM, Hinz K, Roser K et al (2003) Translocations t(11;18)(q21;q21) and t(14;18)(q32;q21) are the main chromosomal abnormalities involving MLT/MALT1 in MALT lymphomas. Leukemia 17(11):2225–2229

Neri A, Jakobiec FA, Pelicci PG et al (1987) Immunoglobulin and T cell receptor beta chain gene rearrangement analysis of ocular adnexal lymphoid neoplasms: clinical and biologic implications. Blood 70(5):1519–1529

Ni AP, Lin GY, Yang L et al (1996) A seroepidemiologic study of *Chlamydia pneumoniae*, *Chlamydia trachomatis* and *Chlamydia psittaci* in different populations on the mainland of China. Scand J Infect Dis 28(6):553–557

Nuckel H, Meller D, Steuhl KP et al (2004) Anti-CD20 monoclonal antibody therapy in relapsed MALT lymphoma of the conjunctiva. Eur J Haematol 73(4):258–262

Oldstone MB (1998) Molecular mimicry and immune-mediated diseases. FASEB J 12(13):1255–1265

Ott G, Katzenberger T, Greiner A et al (1997) The t(11;18)(q21;q21) chromosome translocation is a frequent and specific aberration in low-grade but not high-grade malignant non-hodgkin's lymphomas of the mucosa-associated lymphoid tissue (MALT-) type. Cancer Res 57(18):3944–3948

Pantchev A, Sting R, Bauerfeind R et al (2008) New real-time PCR tests for species-specific detection of *Chlamydophila psittaci* and *Chlamydophila abortus* from tissue samples. Vet J 181(2):145–150

Pascual V, Capra JD (1992) VH4-21, a human VH gene segment overrepresented in the autoimmune repertoire. Arthritis Rheum 35(1):11–18

Pockley AG (2003) Heat shock proteins as regulators of the immune response. Lancet 362(9382):469–476

Ponzoni M, Ferreri AJ, Guidoboni M et al (2008) Chlamydia infection and lymphomas: association beyond ocular adnexal lymphomas highlighted by multiple detection methods. Clin Cancer Res 14(18):5794–5800

Poothullil AM, Colby KA (2006) Topical medical therapies for ocular surface tumors. Semin Ophthalmol 21(3):161–169

Raderer M, Wohrer S, Streubel B et al (2006) Assessment of disease dissemination in gastric compared with extragastric mucosa-associated lymphoid tissue lymphoma using extensive staging: a single-center experience. J Clin Oncol 24(19):3136–3141

Rajalingam K, Al-Younes H, Muller A et al (2001) Epithelial cells infected with *Chlamydophila pneumoniae* (*Chlamydia pneumoniae*) are resistant to apoptosis. Infect Immun 69(12):7880–7888

Rasanen L, Lehto M, Jokinen I et al (1986) Polyclonal antibody formation of human lymphocytes to bacterial components. Immunology 58(4):577–581

Remstein ED, Kurtin PJ, James CD et al (2002) Mucosa-associated lymphoid tissue lymphomas with t(11;18)(q21;q21) and mucosa-associated lymphoid tissue lymphomas with aneuploidy develop along different pathogenetic pathways. Am J Pathol 161(1):63–71

Rosado MF, Byrne JG, Ding F et al (2005) Ocular adnexal lymphoma: a clinicopathological study of a large cohort of patients with no evidence for an association with *Chlamydia psittaci*. Blood 107(2):467–472

Ruiz A, Reischl U, Swerdlow SH et al (2007) Extranodal marginal zone B-cell lymphomas of the ocular adnexa: multiparameter analysis of 34 cases including interphase molecular cytogenetics and PCR for *Chlamydia psittaci*. Am J Surg Pathol 31(5):792–802

Sachse K, Laroucau K, Hotzel H et al (2008) Genotyping of *Chlamydophila psittaci* using a new DNA microarray assay based on sequence analysis of ompA genes. BMC Microbiol 8:63

Sarraf D, Jain A, Dubovy S et al (2005) Mucosa-associated lymphoid tissue lymphoma with intraocular involvement. Retina 25(1):94–98

Sasai K, Yamabe H, Dodo Y et al (2001) Non-hodgkin's lymphoma of the ocular adnexa. Acta Oncol 40(4):485–490

Schallenberg M, Niederdraing N, Steuhl KP et al (2008) Topical mitomycin C as a therapy of conjunctival tumours. Ophthalmologe 105(8):777–784

Shields CL, Shields JA, Carvalho C et al (2001) Conjunctival lymphoid tumors: clinical analysis of 117 cases and relationship to systemic lymphoma. Ophthalmology 108(5):979–984

Sjo LD (2009) Ophthalmic lymphoma: epidemiology and pathogenesis. Acta Ophthalmol 87; Thesis 1: 1–20

Sjo LD, Heegaard S, Prause JU et al (2009) Extranodal marginal zone lymphoma in the ocular region: clinical, immunophenotypical, and cytogenetical characteristics. Invest Ophthalmol Vis Sci 50(2):516–522

Smith JS, Munoz N, Herrero R et al (2002) Evidence for *Chlamydia trachomatis* as a human papillomavirus cofactor in the etiology of invasive cervical cancer in Brazil and the Philippines. J Infect Dis 185(3):324–331

Streubel B, Lamprecht A, Dierlamm J et al (2003) T(14;18)(q32;q21) involving IGH and MALT1 is a frequent chromosomal aberration in MALT lymphoma. Blood 101(6):2335–2339

Streubel B, Simonitsch-Klupp I, Mullauer L et al (2004) Variable frequencies of MALT lymphoma-associated genetic aberrations in MALT lymphomas of different sites. Leukemia 18(10):1722–1726

Sullivan TJ, Grimes D, Bunce I (2004) Monoclonal antibody treatment of orbital lymphoma. Ophthal Plast Reconstr Surg 20(2):103–106

Sykes JE (2005) Feline chlamydiosis. Clin Tech Small Anim Pract 20(2):129–134

Tanimoto K, Sekiguchi N, Yokota Y et al (2006a) Fluorescence *in situ* hybridization (FISH) analysis of primary ocular adnexal MALT lymphoma. BMC Cancer 6:249

Tanimoto K, Kaneko A, Suzuki S et al (2006b) Long-term follow-up results of no initial therapy for ocular adnexal MALT lymphoma. Ann Oncol 17(1):135–140

Tong CY, Sillis M (1993) Detection of *Chlamydia pneumoniae* and *Chlamydia psittaci* in sputum samples by PCR. J Clin Pathol 46(4):313–317

Tsang RW, Gospodarowicz MK, Pintilie M et al (2003) Localized mucosa-associated lymphoid tissue lymphoma treated with radiation therapy has excellent clinical outcome. J Clin Oncol 21(22):4157–4164

Uno T, Isobe K, Shikama N et al (2003) Radiotherapy for extranodal, marginal zone, B-cell lymphoma of mucosa-associated lymphoid tissue originating in the ocular adnexa: a multiinstitutional, retrospective review of 50 patients. Cancer 98(4):865–871

Vallisa D, Bernuzzi P, Arcaini L et al (2005) Role of anti-hepatitis C virus (HCV) treatment in HCV-related, low-grade, B-cell, non-Hodgkin's lymphoma: a multicenter Italian experience. J Clin Oncol 23(3):468–473

Vargas RL, Fallone E, Felgar RE et al (2006) Is there an association between ocular adnexal lymphoma and infection with *Chlamydia psittaci*? The University of Rochester experience. Leuk Res 30(5):547–551

Vinatzer U, Gollinger M, Mullauer L et al (2008) Mucosa-associated lymphoid tissue lymphoma: novel translocations including rearrangements of ODZ2, JMJD2C, and CNN3. Clin Cancer Res 14(20):6426–6431

White WL, Ferry JA, Harris NL et al (1995) Ocular adnexal lymphoma. A clinicopathologic study with identification of lymphomas of mucosa-associated lymphoid tissue type. Ophthalmology 102(12):1994–2006

Yakushijin Y, Kodama T, Takaoka I et al (2007) Absence of chlamydial infection in Japanese patients with ocular adnexal lymphoma of mucosa-associated lymphoid tissue. Int J Hematol 85(3):223–230

Yamasaki R, Yokota K, Okada H et al (2004) Immune response in *Helicobacter pylori*-induced low-grade gastric-mucosa-associated lymphoid tissue (MALT) lymphoma. J Med Microbiol 53(Pt 1):21–29

Yoo C, Ryu M, Huh J et al (2007) *Chlamydia psittaci* infection and clinicopathologic analysis of ocular adnexal lymphomas in Korea. Am J Hematol 82(9):821–823

Yu CS, Chiu SI, Ng CS et al (2008) Localized conjunctival mucosa-associated lymphoid tissue (MALT) lymphoma is amenable to local chemotherapy. Int Ophthalmol 28(1):51–54

Zhang GS, Winter JN, Variakojis D et al (2007) Lack of an association between *Chlamydia psittaci* and ocular adnexal lymphoma. Leuk Lymphoma 48(3):577–583

Chapter 7
Possible Strategies of Bacterial Involvement in Cancer Development

Puneet, Gopal Nath, and V.K. Shukla

Abstract Infections have been implicated in around 18% of all malignancy in humans. The common infections include viruses, bacteria and Schistosomes. The role of bacteria in carcinogenesis is now quite evident particularly in *H. pylori* induced gastric cancer and mucosa associated lymphnoid tissue (MALT) lymphoma and *Salmonella typhi* causing carcinoma gallbladder. The chronic inflammation is the important mechanism involved in the majority of bacteria induced malignancies. The chronic inflammation is mediated by various pro and anti – inflammatory cytokines including IL-1, IL-6, IL-17, TNF-α and IL-10. The key factor in the inflammatory process is the activation of NF-kB. Various toxins are produced by different bacteria which cause direct damage to the host cells by DNA damage or affecting DNA repair mechanism resulting in alteration in the enzyme transcription or translation. It has also been postulated that bacterial infection activate inflammatory/immune cells to generate reactive oxygen species (ROS) and reactive nitrogen species (RNS) which causes DNA damage leading to cancer. The free radicals generated also acts at different levels which affects the cellular homeostasis.

The epigenetic alteration in form of DNA methylation and histone modification has also been reported in bacteria induced carcinogenesis. The other alternative molecules has been evolved by the bacteria that involved in adhesions of bacteria to cell surfaces, modulation of cytoskeleton or junctional activities and affecting specific signaling pathways.

The studies have shown that bacterial infection also causes immune modulation and result in persistence of the infection. It is also seen that not only the bacterial

Puneet • V.K. Shukla (✉)
Department of Surgery, Institute of Medical Sciences,
Banaras Hindu University, Varanasi 221005, India
e-mail: vkshuklabhu@gmail.com

G. Nath
Department of Microbiology, Institute of Medical Sciences,
Banaras Hindu University, Varanasi 221005, India

A.A. Khan (ed.), *Bacteria and Cancer*, DOI 10.1007/978-94-007-2585-0_7,

pathogen but the host genetic factors also determine the susceptibility and persistence of infection. The certain genotypes of inflammatory mediators are associated with increased risk of cancer. Thus, the bacteria induced carcinogenesis is a multifaceted, complex process.

Keywords Bacteria • Cancer • *H. pylori* • Gastric cancer • Lung Cancer • Gallbladder cancer • Carcinogenesis • Bacterial toxin • Chronic inflammation • Mucosa associated lymphoid tissue (MALT) lymphoma

Abbreviations

4HNE	4-hydroxynonenal
AP-1	Activator Protein-1
BMP	Bone Morphogenetic Protein
Cag A	Cytotoxin Associated Gene Antigen
Cag PAI	Cag A pathogenicity Associated Island
CDT	Cytolethal Distending Toxin
Cif	Cycle Inhibiting Factor
CNF	Cytotoxic Necrotizing Factor
COX-2	Cyclooxygenase-2
CpG	—C—phosphate—G—
dG	deoxyguanosine
EDIN	Epidermal Differentiation Inhibiting Factor
ERK	Extracellular Regulated Tyrosine Kinase
FAK	Focal Adhesion Kinase
H. pylori	*Helicobacter pylori*
HAT	Histone Acetyl Transferases
HDACs	Histone Deacetylase
HlpA	Histone Like Protein
ICAM-1	Intercellular Cell Adhesion Molecules
IKK	Inhibitory Kappa B Kinase
IL	Interleukin
INF	Interferon
iNOS	inducible Nitric Oxide Synthase
Kb	Kilobase
KDa	Kilo Dalton
LPS	Lipopolysaccharide
LTA	Lipoteichoic acid
MALT	Mucosa Associated Lymphoid Tissue
MAPK	Mitogen Activated Protein Kinase
MIP	Macrophage inflammatory protein
NF-κB	Nuclear Factor Kappa B
NG	Nitroguanine
PMT	*Pasturella multocida* toxin

RN	Recepter Antgonist
RNS	Reactive Nitrogen Species
ROS	Reactive Oxygen Species
RR	Relative Risk
Shh	Sonic Hedgehog Homolog
SRF	Serum Response Factor
T4SS	Type IV Secretion Apparatus
TH1	Type 1Helper T-cell
TLR4	Toll like Receptor 4
TNF	Tumour Necrosis factor
VacA	Vacuolating Cytotoxin
VEGF	Vascular Endothelial Growth Factor

7.1 Introduction

The infection is one of the most important causes of cancer (de Martel and Franceschi 2009; Pisani et al. 1997). In 2002, it was estimated that 18% of all malignancies were attributed to infectious agents (de Martel and Franceschi 2009; Ferlay et al. 2004). The infections involving viruses, bacteria and Schistosomes have been linked with higher risk of malignancy (de Martel and Franceschi 2009). Carcinogenesis is a protracted, multistage and complex process and may take decades to reach its culmination. The time gap between the initiation and development of cancer is too long to lid the role of bacteria in cacinogenesis. Therefore, a direct link between bacterial infection and cancer is many times difficult to establish. The involvement of bacteria in carcinogenesis remain controversial partly because of absence of agreement on its molecular mechanism which leads to development of cancer (Nath et al. 2010), and also the presence of bacteria at the site of tumour itself does not implicate its causation. But the evidence of bacterial infection causing cancer is quite evident in *H. pylori* infections causing gastric cancer and mucosa associated lymphoid tissue (MALT) lymphoma, *Salmonella typhi* resulting in gallbladder cancer. The other bacterial infections causing cancer includes *Streptococcus bovis* with colon cancer and *Chlamydia pneumoniae* with lung cancer (de Martel and Franceschi 2009; Nath et al. 2010; Vogelmann and Amieva 2007).

William Russel in 1890 first gave the possibility of bacteria induced carcinogenesis. In 1926, Thomas Glover stated that specific bacteria could be isolated consistently from neoplastic tissues. In 1931, Hodgkin's Lymphoma was found to be associated with the infection of acid fast bacteria. Few years later in 1941, George Mazet reported that both leukemia and Hodgkin's disease were associated with infection of bacteria. Various authors like WM Crofton, V Livingston, EJ Villesquez, IC Diller between 1936 and 1955, have isolated and reported the presence of microbes in cancer tissue. In 1969, a group of scientists from NCI, USA suggested positive association between bacteria and cancer, although they contradicted their previous hypothesis. In 1994, the International Agency for Research on cancer classified *H. pylori* as a group I carcinogen (Nath et al. 2010).

7.2 Possible Strategies for Bacterial Involvement in Carcinogenesis

7.2.1 Chronic Inflammation

Chronic inflammation represents the major pathologic basis for majority of the bacteria induced malignancies. The role of inflammation in carcinogenesis has been first proposed by Rudolf Virchow in 1863, when he noticed the presence of leukocytes in the neoplastic tissue (Balkwill and Mantovani 2001). Since then, various studies linked inflammation and cancer and data started supporting that tumours can originate at the sites of infection or chronic inflammation. Approximately 25% of all cancers are associated with chronic infection and inflammation (Kundu and Surh 2008; Hussain and Harris 2007). The inflammation act as adaptive host defense against infection but inadequate resolution or persistent inflammatory response may lead to chronic ailments like cancer (Kundu and Surh 2008). Chronic infection from bacteria may increase the risk of cancer through various mechanisms like direct DNA damage, alteration in the cell cycle regulation, inhibition of apoptosis, subversion of immunity and stimulation of angiogenesis. The chronic inflammation also affects proliferation, adhesions and cellular transformation (Nath et al. 2010; Kundu and Surh 2008; McNamara and El-Omar 2008).

H. pylori induced chronic inflammation is mediated by an array of pro and anti-inflammatory cytokines. The proinflammatory cytokines including IL-1, IL-6, and IL-17 participates in the inflammation associated carcinogenesis (Kundu and Surh 2008). TNF-α plays a dual role in bacteria induced neoplasm. IL-10 is an anti-inflammatory cytokines. Chemokines are usually produced by proinflammatory cytokines. The central role of chemokines is to recruit leukocytes at the site of inflammation. Central to the inflammatory process induced by *H. pylori* is the activation of NF-κB in the gastric epithelial cells. NF-κB has a key role as a mediator in mucosal inflammation. *H. pylori* activation of NF-κB through nuclear translocation stimulates increased production of IL-8, IL-1, TNF-α, and various other chemokines (Kundu and Surh 2008; McNamara and El-Omar 2008). The activation of nuclear factor kappa B (NF-κB) is an important link between inflammation and cancer (Nath et al. 2010; Karin and Greten 2005). The NF-κB pathway is activated by bacterial infection and cytokines (TNF-α, and IL-1), resulting in IKK complex formation (Hacker and Karin 2006) and degradation of NF-κB inhibitors (Nath et al. 2010). NF-κB enters into the nucleus and results in the transcription of target genes. It up regulate the genes involved in cell cycle control (cyclin D1, CDK2 kinase) and down regulates the genes responsible of decreased apoptosis (p21, p53 and pRb).The proinflammatory and proangiogenic cytokines such as IL-1B, IL-6, VEGF are upregulated, however TNF is down regulated by NF-κB. Genes responsible for invasion of tumour cell and metastasis are also upregulated by NF-κB (Nath et al. 2010). NF-κB is required for induction of tissue repair genes and results in increased expression of both inducible nitric oxide synthase (iNOS) and intercellular cell adhesion molecules including ICAM-1. *H. pylori* stimulate the adaptive immune system by activation of NF-κB through a multistep process triggered by lipopolysaccharide (LPS) binding with a specific transmembrane pattern recognition receptor i.e. toll like receptor 4 (TLR4) (McNamara and El-Omar 2008; Pasare and Medzhitov 2004). *H. pylori* infection also up regulates signaling molecules like activation of the extracellular regulated tyrosine kinase (ERK) signaling pathway results in an increase in important transcription factors such as activation protein-1 (AP-1) and serum response factor (SRF) which might be responsible for up regulation of pro-inflammatory cytokines seen in *H. pylori* infection (Lax and Thomas 2002; Mitsuno et al. 2001).

7.2.2 Bacterial Toxins

The several bacterial toxins have been implicated in the carcinogenesis. They act through the different pathway and can kill cells or can modify the cellular process that control DNA repair, proliferation, apoptosis and differentiation. These bacterial toxins can cause direct damage to the host cells by DNA damage or affecting the DNA repair mechanism and alteration in the enzyme transcription or translation. It may cause carcinogenesis by inflicting chronic inflammatory response and production of free radicals (Nath et al. 2010; Nougayrède et al. 2005) (Table 7.1).

The cytolethal distending toxin (CDT) is a cell cycle inhibitor produced by several gram negative bacteria like *Campylobacter jejuni* and *Salmonella typhi*. The cdtB unit of CDT is a DNAse that causes double stranded DNA breaks causing cell cycle arrest particularly at the G_2 checkpoint (Mager 2006; Haghjoo and Galan 2004). The CNF (cytotoxic necrotizing factor) is a toxin found in many uropathogenic *Escherichia coli* and also *Campylobacter jejuni, Salmonella typhi, Bartonella* spp. CNF induces elevation in the expression of COX-2 and activation of Rho family protein which act as a key to molecular switches that integrate signals from several different signal transduction pathways. The action of CNF on Rho proteins also leads to perturbation of the cell cycle by inducing DNA synthesis and affect apoptosis. CNF itself can suppress apoptosis by affecting the transcription levels of Bcl-2 family proteins (Lax and Thomas 2002; Fiorentini et al. 1998).

Pasteurella multocida toxin (PMT) is a potent mitogen produced by *Pasteurella multocida.* Its target molecule is unknown but it acts intracellulary to stimulate several signaling cascade mediated by proto-oncogenes. The subsequent activation of ERK-1 and 2 MAPKs (mitogen activated protein kinase) stimulates the DNA synthesis and cellular proliferation. *In vitro* studies has shown that PMT promote Rho A mediated signal transduction result in activation of focal adhesion kinase (FAK) and Src family kinases which are often increased in many cancers (Lax and Thomas 2002; Lax and Grigoriadis 2001; Thomas et al. 2001).

Epidermal differentiation inhibiting factor (EDIN) is expressed by some strains of *Staphylococcus aureus* and related toxins from *Bacillus cereus* and *Clostridium*

Table 7.1 Bacterial toxins and their possible mechanism of action in carcinogenesis

Bacterial toxins		Mechanism of action
1.	Cytolethal distending toxin (CDT) (Mager 2006; Haghjoo and Galan 2004)	Cell cycle inhibitor (usually at G2 check point) (cdtB subunit: chromatin fragmentation)
2.	Cytotoxic necrotizing factor (CNF) (Lax and Thomas 2002; Fiorentini et al. 1998)	Modifies Rho family protein (proliferation) Bcl-2 family protein (suppress apoptosis)
3.	Pasturella multocida toxin (PMT) (Lax and Grigoriadis 2001; Thomas et al. 2001).	Activate mitogen activated protein kinase (MAPK) Promote Rho A
4.	Epidermal differentiation inhibiting factor (EDIN) (Lax and Thomas 2002; Sugai et al. 1992)	Modifies Rho family protein
5.	Cycle inhibiting factor (Cif) (Marches et al. 2003)	Arrest cell cycle in G2M phase
6.	Vascular endothelial growth factor (VEGF) (Lax and Thomas 2002; Kempf et al. 2001)	Proliferation and angiogenesis
7.	Cytotoxin associated gene-A (Cag-A) (Kusters et al. 2006; Blaser and Atherton 2004).	Binding with SH-2 domain
8.	Vacuolating cytotoxin- A (Vac-A) (Kusters et al. 2006; Blaser and Atherton 2004).	Membrane binding and pore formations

spp. modified Rho proteins, and induces hyperplasia (Lax and Thomas 2002; Sugai et al. 1992). Cycle inhibiting factor (Cif) is a cell cycle inhibitor found in *E. coli*. Cif arrest the cell at the G2M phase (Marches et al. 2003) and causes unique alterations in the host cell that result in the attachment of the cytoskeleton to the host cell membrane. These changes inhibit mitosis and causes cellular and nuclear enlargement. The DNA synthesis is initiated but the nuclear division does not occur and endoreduplication result in the cellular DNA content of 8–16n (Nougayrede et al. 2005). *Bartonella* spp. infection promotes the production of vascular endothelial growth factor (VEGF). VEGF is a potent proteins mitogen and stimulate tumour angiogenesis (Lax and Thomas 2002; Kempf et al. 2001). The cytotoxin associated gene antigen (Cag A) and Vacuolating cytotoxin (Vac A) are the important virulence factor for *H. pylori* induced gastric carcinoma and are described below (Kusters et al. 2006; Blaser and Atherton 2004).

7.2.3 Generation of Reactive Oxygen and Nitrogen Intermediate

The oxidative and nitrosative stress are the crucial mechanism that can results in neoplastic transformation. In response to bacterial infection, activated inflammatory/immune cells generate reactive oxygen species (ROS) and reactive nitrogen species

(RNS), which act as chemical effector in inflammation induced carcinogenesis. The generation of ROS and RNS causes DNA damage leading to cancer (Kundu and Surh 2008). *H. pylori* infection causes gastric mucosal inflammation and infiltration of macrophages and polymorphonuclear neutrophils containing inducible nitric oxide synthase (iNOS) (Blaser and Atherton 2004; Bereswill et al. 2000, 1998). This enzyme lead to formation of large amount of nitric oxide (NO), a high reactive molecule which plays an important role in inflammation associated carcinogenesis by direct modification of DNA and inactivation of DNA repair enzymes. *H. pylori* induced oxidative DNA damage is caused by formation of 8-oxo 7, 8 dihydro-2 deoxyguanosine (8-oxo-dG). Reactive nitrogen species, peroxynitrite (product formed by a reaction between NO radical and superoxide anion) causes DNA damage by forming 8-nitroguanine (8-NG) (Kundu and Surh 2008; Ohshima et al. 2006; Yermilov et al. 1995). Thus, oxidative and nitrosative DNA damage products (8-oxo-dG and 8-NG) have been implicated in the initiation of the inflammation induced carcinogenesis. ROS and RNS can also induce lipid peroxidation to generate other reactive species like malondialdehyde and 4-hydroxynonenal (4HNE) which are capable of causing DNA damage (Kundu and Surh 2008; Bartsch and Nair 2005). These free radicals also causes alteration in the cellular protein function involved in DNA cross linkage and cellular homeostasis. As the precancerous process advances, iNOS is seen in the cytoplasm of dysplastic and neoplastic epithelial cells. Nitrosative stress also plays a critical role in inflammation associated carcinogenesis by activating activator protein-1 (AP-1), which is involved in cell transformation and proliferation (Kundu and Surh 2008).

7.2.4 Epigenetic Alterations

The epigenetic alteration induced by bacterial infection can influence gene expression by DNA methylation and histone modification and also contribute to inflammation induced cancer development. The CpG hypermethylation of E-cadherin gene in intestinal metaplasia in *H. pylori* infected patients suggest DNA hypermethylation as an important and early event in the development of gastric carcinoma (Kundu and Surh 2008). *H. pylori* infection also causes DNA hypermethylation of tumour suppressor p16 gene suggesting epigenetic alterations in bacterial infection induced carcinogenesis. The inappropriate activation or inactivation of HDACs (histone deacetylase) and HAT (histone acetyl transferases) has been seen in response to inflammation induced carcinogenesis. The transcriptional activation of NF-κB and IL-8 induced by LPS and TNF-α (proinflammatory stimuli) was dependent on p38 mitogen activated protein kinase (MAPK) and inhibitory kappa B kinase (IKK)-α mediated phosphorylation of histone-3 (Saccani et al. 2002; Yamamoto et al. 2003). Similarly, transcriptional activation of COX-2 (cyclooxygenase-2) is promoted by the acetylation of histone-4 through the HDAC-1 degradation. Thus the bacterial infections causing in epigenetic alterations, result in carcinogenesis (Kundu and Surh 2008).

7.2.5 *Direct Injection Effect*

Bacteria have evolved several sophisticated molecules that interact specifically with host eukaryotic cells. These molecules are involved in adhesion of bacteria to cell surfaces, activating and inhibiting entry into host cells, modulation of cytoskeleton or junctional activities and activating specific signaling pathways. Some bacteria have shown to directly affect the cellular process resulting in carcinogenesis (Nath et al. 2010; Vogelmann and Amieva 2007; Kundu and Surh 2008).

H. pylori also possess a T4SS (a type IV secretion apparatus), encoded in the Cag A pathogenecity island (Cag PAI). The effectors protein Cag A is injected into the host cell cytoplasm through this molecular needle. On reaching into the host cytoplasm, Cag A gets phosphorylated by Src family kinases and activates certain activities like activation of growth factor like signaling pathway, loss of cellular polarity, activated epithelial cell proliferation and NF-κB (Vogelmann and Amieva 2007; Hussain and Harris 2007). Similarly *Bartonella* acts through the injection of effector proteins (BEPs) directly into the endothelial cell stimulating cell proliferation, cytoskeletal rearrangements, activation of NF-κB and inhibition of apoptosis (Vogelmann and Amieva 2007).

7.2.6 *Immune Modulation and Resistance to Phagocytosis*

The various studies have shown that humoral immunity to *H. pylori* is only marginal. This antibody production does not lead to eradication of infection but may contribute to tissue damage. *H. pylori* attachment to the cell leads to the production of auto-antibodies against lewis carbohydrate epitopes on the surface of acid producing parietal cells (Kusters et al. 2006). This results in loss of the parietal cells and subsequent hyperproliferation of gastric stem cells producing adenomatous lesions. *H. pylori* infection results in induction of Th-1 polarized response. The Th-2 cell stimulate B cells in response to extracellular pathogens, Th-1 cells are induced mostly in response to intracellular pathogens. Since *H pylori* is a noninvasive bacteria, so Th-2 cell response would be expected but paradoxically it is Th-1 (Harris et al. 2000). This Th-1 orientation may be due to increased antral production of interleukin-18 in response to *H. pylori* infection (Suerbaum and Michetti 2002).

In experimental studies, it is shown that transfer of unfractionated splenocytes from *H. pylori* infected mice induces gastritis, delayed-type hypersensitivity and metaplasia in mice. This study supports the hypothesis that *H. pylori* induced pathology is being mediated by T-cell. This adaptive immune response is initiated and maintained by the monocytes and Th1 lymphocytes rather than epithelial cells. The presence of *H. pylori* in the gastric mucosa is associated with strong IL-12 production and the presence of large numbers of Th-1 cells. This mechanism of cytokines production by monocytes rather gastric epithelial cell is different way of creating protective immunity (Kusters et al. 2006).

H. pylori infection up regulates the MIP-3α gene expression in gastric epithelial cells resulting in an influx of monocytic cells into the lamina propria. But these cells are functionally not able to perform phagocytosis of *H. pylori* because of reduced anti-*H. pylori* activity of macrophages and also altered processing of *H. pylori* antigen by activated macrophages (dendritic cells)(Kusters et al. 2006).

7.2.7 Host Genetic Factor

It is not only the bacterial pathogen but also the host genetics play an important role in determining the susceptibility and severity of the infections. Genetic polymorphism directly influences the expression of gene product and ultimately determines the clinical outcome (McNamara and El-Omar 2008). There are many genetic polymorphisms that affect the expression levels of inflammatory mediators in *H. pylori* infection. The IL-1 cytokine is encoded by a cluster that contains the polymorphic IL-1B (encoding the IL-1β cytokine) and IL-RN (encoding the IL-1 receptor antagonist) genes. IL-1β is a potent pro-inflammatory cytokine and the most potent known inhibitor of acid secretion (Kusters et al. 2006; Calam 1999). The IL-1 gene cluster contains several polymorphisms such as IL-1B*-31C, IL-B*-511T, and IL-1RN*2*2, which lead to high level expression of IL-1β (Kusters et al. 2006). These genotype are at increased risk of developing hypochlorhydria, gastric atrophy and gastric cancer. The risk of malignancy is increased by two to three-folds in these genotypes (El-Omar et al. 2000, 2003). The genetic polymorphism of IL-1 has increased risk of both intestinal and diffuse types of noncardia gastric cancer. Similarly, polymorphism is observed in other inflammation associated genes like tumour necrosis factor alpha (TNF-α) and IL-10. The TNF A*308 A genotype is associated with increased TNF-α production which along with the IL-1 influence the acid secretion by parietal cells (Kusters et al. 2006).

7.2.8 Affecting Stem Cell Homeostasis

The gastrointestinal epithelium is regulated by the specific stem cell. In order to secure vast supply of differentiated cells, stem cell homeostasis is highly regulated to avoid uncontrolled proliferation of cell and cancer development. The stem cell niche regulates stem cell homeostasis through growth factors secreted by supporting cells and cell-cell adhesion of stem cells and their surrounding cells. The Cag A injection into the epithelial cells results in activation of growth factor pathway, loss of epithelial cell-cell adhesion and degradation of basement membrane (Vogelmann and Amieva 2007).

The stem cells (gastrointestinal epithelium) are controlled by signaling pathways such as Wnt /β-catenin, Hedgehog, TNF β/ BMP (transforming growth factor β/ bone morphogenetic protein) and Notch. In the presence of bacterial infection, these signaling pathways deliver signals to the epithelium and have shown to increase β-catenin transcriptional activity in Cag A positive strains in both *in vivo* and *in vitro* (Vogelmann and Amieva 2007). In animal models (Mongolian gerbils), *H. pylori* decreases Sonic hedgehog homolog (Shh) expression and causes increased epithelial cell proliferation (Suzuki et al. 2005). It has been shown by various studies that Cag A positive *H. pylori* directly affect the gastric stem cell homeostasis.

7.3 Some Representative Examples of Bacteria Associated Cancers

7.3.1 H. pylori Causing Gastric Cancer and Lymphoma

Chronic *H. pylori* infection induced inflammation leads to loss of the normal gastric mucosal architecture and results in the destruction of gastric gland and eventually fibrosis and intestinal type epithelium. This atrophic gastritis and intestinal metaplasia is common where more severe inflammation is present. Thus, the risks for atrophic gastritis depends on the distribution and pattern of chronic active inflammation and extend multifocally with time (Kusters et al. 2006; Blaser and Atherton 2004). The risk of development of gastric cancer depends on the extent and severity of gastric atrophy and varies from 5 to 90-fold (Sipponen et al. 1985). *H. pylori* infection increases the risk of gastric cancer with the process of atrophy and metaplasia, is now well proven. It is also supported by the geographical association seen between the prevalence of *H. pylori* infection and incidence of gastric cancer. *H. pylori* designated as class-I carcinogen by WHO and it is established that *H. pylori* colonization increases the risk of gastric cancer by approximately ten folds (IARC 1994). *H. pylori* is a spiral shaped gram negative bacterium which colonizes on gastric epithelium for a life time in the absence of specific antimicrobial therapy (Kusters et al. 2006). *H. pylori* has been associated with non cardia gastric adenocarcinoma and gastric lymphoma. The epidemiological studies including several case-control studies nested in large population base cohorts have established *H. pylori* as a strong risk factor for non cardia gastric adenocarcinoma with relative risk (RR) of around 3 (Helicobacter and Cancer Collaborative Group 2001). *H. pylori* infection ability to foster a chronic inflammatory response is the best understood explanation of the bacterium's carcinogenic potential. The certain genotypes of *H. pylori* that cause more inflammation are more closely related to malignancy. Several studies now have well described virulence factor including cytotoxin associated gene antigen (Cag A), Vacuolating cytotoxin A (Vac A) (Kusters et al. 2006; Blaser and Atherton 2004). *H. pylori* specifically adapted to colonize and survive in the hostile acidic gastric environment. There are several factors including urease enzyme and urea transport protein maintains a neutral pH microenvironment essential for the survival of the organism.

The certain *H. pylori* strains are more virulent and cause morphological changes, vacuolization and successive degeneration of cell during *in vitro* studies. This activity is due to presence of Cag A protein of molecular mass of approximately 140 kDa. The Cag A protein is highly immunogenic and encoded by Cag A gene (Kusters et al. 2006; Covacci et al. 1993). It is marker for the presence of genomic PAI of 40 Kb and encodes between 27 and 31 proteins. Cag A gene is present in approximately 50–70% *H. pylori* strains and has higher inflammatory response and more risk of developing peptic ulcer and gastric cancer in Western populations. Although, this association is not seen in Asian population (Kusters et al. 2006).

Cag PAI encoded protein also serve as building blocks of type IV secretion apparatus, which forms a syringe like structure. This helps in translocation of Cag A, peptidoglycan and other bacterial factors into the gastric epithelial cell cytoplasm. Cag A induces epithelial cell proliferation and division after tyrosine phosphorylation and binding with the Src homology 2 (SH-2) domain- containing tyrosine phosphatase 2, by activation of nuclear factor kappa B (NF-κB) and secretion of interleukin 8 (IL-8) (Vogelmann and Amieva 2007; McNamara and El-Omar 2008; Kusters et al. 2006). The amount of Cag A tyrosine phosphorylation is directly related with the number of repeats. *H pylori* strains having Cag A with larger number of these repeats produces more pronounced morphological changes in the epithelial cells in culture studies and also associated with risk of gastric carcinoma. The other effect of Cag PAI encoded proteins include rearrangement of the actin cytoskeleton, inhibition of apoptosis, epithelial cell DNA damage, activation of transcription factor AP-1, increased expression of proto-oncogenes *c-fos* and c-*jun* and increased expression of prostaglandin E2 (Keates et al. 1999; Lee et al. 2005). The two type of Cag A are described i.e. Western and East Asian. The increased binding affinity and tyrosine phosphorylation is responsible for more severe gastritis, gastric atrophy and ultimately to gastric carcinoma. The Cag A also affects the immune response of the host by its ability to induce apoptosis of T-cells (Nath et al. 2010).

Vacuolating cytotoxin A is a highly immunogenic protein encoded by VacA gene and induces massive vacuolization in the epithelial cells *in vitro*, but does not seem to appear *in vivo*. Vac A protein can cause membrane channel formation, disruption of endosomal and lysosomal activity, effects on integrin receptor induced signalling, induction of apoptosis and immune modulation. The sequence heterogeneity within the VacA gene is present at the signal region (s) and middle region (m). The signal (s) region occurs as either s1 or s2 whereas (middle) m region exists as m1 or m2 type. Strains with VacA m1 alleles are more toxigenic than those with m2 alleles which are mildly or non-toxigenic and within the m1 group, strains with s1a/m1 alleles are more toxigenic than those with s1b/m1 alleles. The vacuolating activity of *H. pylori* is high in s1m1 genotypes, intermediate in s1/m2 genotype and absent in s2/m2 genotypes. At the same time, VacA s1/m1 genotypes are more frequently associated with gastric carcinoma (Kusters et al. 2006). VacA protein mediated effects are directly or indirectly occurs from membrane binding and pore formations. VacA also enters the cytosol and accumulates in the mitochondrial inner membrane and causes apoptosis. This pro-apoptotic effect of VacA is cell type dependent and is limited to the gastric epithelial cells e.g. parietal cells. It results in reduced acid secretion in stomach and predisposes for the development of gastric carcinoma. The interaction of VacA with immune cells causes inhibition of antigen presenting cells and T cell proliferation. It is also caused by the active inhibition of lymphocyte activation. The virulence factors are the important determinants of outcome of *H. pylori* infection. A better understanding of their modes of action, their interactions and genetic variations can help to identify the risk group for gastric cancer.

Gastric lymphoma is the most common extra-nodal lymphoma but is a rare malignancy since it represents 6% of gastric cancer. The stomach mucosa do not contains any lymphoid tissue. *H. pylori* colonization induces MALT and almost all MALT lymphoma patients are *H. pylori* positive. The true incidence of MALT lymphoma in *H. pylori* positive individuals is not known. However, it occurs in less than 1% *H. pylori* positive cases. MALT arises in an area of chronic inflammation and derives from marginal zone lymphocytes which surround B-cell follicles in gastric lymphoid tissue. The near universal presence of *H. pylori* in gastric biopsies of patients with MALT lymphoma has been demonstrated. The serological nested case control study in two different populations showed a RR for developing gastric lymphoma of 6.3 (de Martel and Franceschi 2009; Parsonnet et al. 1994). The eradication of *H. pylori* could achieve complete regression of stomach lymphoma in 60–100% patients with low grade MALT lymphoma and 60% in early stage high grade lymphoma (Chen et al. 2005).

7.3.2 Salmonella typhi and Gallbladder Cancer

An association of chronic typhoid carriage and carcinoma of the gallbladder was first reported by Axelrod et al. (1971). The various researchers have reported the increased incidence of cancer of hepatobiliary system in typhoid carriers. A case control study which compared individual who experienced acute infection with those who subsequently became chronic carriers following 1922 typhoid out break in New York. In this study, the carriers were six times more likely to die of hepatobiliary carcinoma than matched controls (Mager 2006; Welton et al. 1979). In individuals infected with *S. typhi* in Aberdeen typhoid outbreak in 1964, the lifetime risk of developing gallbladder cancer was 6% (Caygill et al. 1994). Studies from Mexico reported 12 time increased risk of gallbladder cancer in individual with history of typhoid fever (Nath et al. 2010). North Indian studies have shown 7.9-fold increased risk of gallbladder cancer in patients with chronic typhoid carrier diagnosed by Vi-serology (Shukla et al. 2000). In a large cohort of 113,394 from Japan in an area with extremely low prevalence of typhoid fever, the relative risk of developing carcinoma gall bladder was 2.1 (Yagyu et al. 2008). The study on chronic typhoid carriers in carcinoma gallbladder patients using nested PCR techniques showed that 67.3% of carcinoma gallbladder patients were typhoid carriers as compared to 8.3% of the healthy population (Nath et al. 2010, 2008). It has been hypothesized that metabolites and toxins produced by the bacteria are concentrated 10 times in the gallbladder. The various carcinogens produced by *S. typhi* have been suggested including bacterial glucuronidase, it is a bacterial enzyme which act on primary bile acid and produce secondary bile acids at very high concentrations and also nitroso coumpounds from nitrates (Nath et al. 2010). *S. typhi* also produce cytolethal distending toxin (CDT), which is a tripartite complex. CDT has active unit, CdtB which has structural and functional homolog with mammalian DNAase-I and the other two units CdtA and CdtC mediate the binding of toxin to plasma membrane. *S. typhi* produced unique CdtB dependent CDT that required bacterial internalization into host cells (Haghjoo and Galan 2004). CdtB after being delivered to the cytosol it reaches the nucleus of the target cell where it causes DNA damage. The CdtB subunit also causes fragmentation of chromatin in cell culture studies. The persistence of *S. typhi* infection may be facilitated by CDT as this toxin is also known for its immunomodulated activity (Shenker et al. 1999).

The researchers have shown that patients with *Salmonella* infection lack IL-12 Rβ1 chain expression (Mager 2006; de Jong et al. 1998). Interleukin-12 (IL-12) is a cytokine that promotes cell mediated immunity to intracellular pathogens such as *S. typhi* by inducing type1 helper T cell (TH1) response and interferon-γ (INF- γ) production. IL-12 binds to the high affinity β1/β2 heterodimeric IL-12 receptor (IL-12R) complexes on T- cell and natural killer cells. The IL-12 Rβ1 sequence analysis revealed genetic mutations that resulted in premature stop codons in the extracellular domain. The lack of IL-12 R signaling and INF- γ production represent immune deficiency with extreme susceptibility to *Salmonella* and *Mycobacteria* infection but not to any other viral, bacterial or fungal pathogens (Mager 2006). It is also proposed that *S. typhi* internalization causes subsequent deviation from usual endocytic pathway that leads to lysosomes and reaches to unusual membrane bound compartment, where it can survive and replicate. This unique feature of *S. typhi* leads to its long, persistent infection in humans (Nath et al. 2010).

7.3.3 *Streptococcus bovis* and Colon Cancer

Streptococcus bovis is a transient flora of human gastrointestinal tract (Waisberg et al. 2002). The normal human colon is a significant reservoir of *S. bovis* in 2.5–15% of individuals (Waisberg et al. 2002; Burns et al. 1985). It is a non-enterococcal *Streptococcus* in Lancefield's group D and cause bacteremia, endocarditis and urinary infection (Waisberg et al. 2002; Bayliss et al. 1983). The micro-organism may penetrate into the blood stream through epithelial, oropharyngeal, dermal, respiratory, gastrointestinal or urogenital lesions (Waisberg et al. 2002). *S. bovis* causes infectious endocarditis mostly in patients above 60 year of age (Teitz et al. 1995). It is found in 7–14% of subacute infectious endocarditis cases (Bisno 1991).

In 1951, McCoy and Mason first suggested the relationship between colonic carcinoma and infectious endocarditis but the association of *S. bovis* and colorectal carcinoma was recognized in 1974 (zur Hausen 2006; McCoy and Mason 1951; Keusch 1974). The incidence of colonic neoplasia associated with *S. bovis* is detected in 18–62% (Waisberg et al. 2002; Leport et al. 1987; Murray and Roberts 1978). In colon cancer patients, faecal carriage of *S. bovis* is increased by about fivefolds (Tjalsma et al. 2006). Colonic neoplasm may arise many years after the development of bacteremia or infectious endocarditis (Friedrich et al. 1982). Bacterial endocarditis from other *Streptococci* like *S. faecalis, S. equinus, S. sanguis* and *S. salivarius* are also related to colonic neoplasm (Waisberg et al. 2002).

S. bovis is competed out by the normal gut flora in the bowel of the healthy individuals. *S. bovis* has ability to attach to polyps and early colon cancer sites by the bacterial surface proteins that have affinity for specific surface proteins expressed by tumour cells (Tjalsma et al. 2006). *S. bovis* finds a niche for the survival in the bowel and can cause local tumour associated infection. This infection with *S. bovis* might promote tumour development by stimulating the cycloxygenase-2 pathway including hyperproliferation, invasion and angiogenesis where as apoptosis is inhibited (Tjalsma et al. 2006; Biarc et al. 2004; Wendum et al. 2004). *S. bovis* enters the portal circulation through bacterial translocation and causes hepatic dysfunction which modified the hepatic secretion of bile salts and production of immunoglobulins. The studies have shown presence of hepatic cirrhosis in 11–19% patients (Waisberg et al. 2002; Bisno 1991).

The surface attached histone like protein (HIpA) has been identified from *S. bovis* that elicit an immune response in several colon cancer patients. *In vitro* studies have shown that extra cellular release of HIpA from *Streptococcus pyogenes* was present in complexed form with soluble lipoteichoic acid (LTA), the main component of cell walls from gram positive bacteria. It is also reported by *in vitro* studies that these HIpA-LTA complexes could bind to heparin sulphate proteoglycan on the surfaces of human epithelial cells (Stinson et al. 1998). It is also suggested that attachment and internalization of *S. bovis* to colon adenocarcinoma cell lines is mediated by interactions between epithelial hepavan sulfate proteoglycans and bacterial heparin binding proteins. Thus, it has been speculated that surface (LTA) attached HIpA from *S. bovis* has affinity for surface component expressed by colon tumour cells (Tjalsma et al. 2006).

The active role of *S. bovis* in the promotion of colonic carcinogenesis is further strengthen by studies on animal models. When adult rats treated with azoxymethane for 2 weeks and subsequently received either injection of *S. bovis* bacteria or wall extracted antigens twice weekly. The researchers observed progression of preneoplastic lesions, enhanced expression of proliferative markers and increased production of interleukin-8 in the colonic mucosa (Ellmerich et al. 2000). The same group of researchers used a partially purified *S. bovis* S 300 fraction representing 12 different proteins and triggered the synthesis of proinflammatory proteins (IL-8 and prostaglandin E2) in human colon carcinoma cells (Caco-2) and in rat colonic mucosa and correlated with *in vitro* over expression of COX-2 (Biarc et al. 2004). The IL-8 and other cytokines lead to the formation of nitric oxide and free radicals such as superoxide, peroxynitrites, hydroxyl radicals as well as alkyl peroxy radicals which are mutagenic and contribute to the neoplastic process by the cellular DNA modification. The mechanism of development of colonic carcinoma is suspected similar to gastric carcinoma after persistent *Helicobacter pylori* infections (Fox et al. 2006). The high intake of red meat has also been defined as risk factor for the colon cancer. *S. bovis* is one of the contaminants frequently detected in commercially available meat (Knudtson and Hartman 1993).

S. bovis endocarditis is considered as a clinical indication by several authors for colonoscopy since majority of these patients present with either colon carcinoma or polyps. In the presence of carcinoma of colon, the relative risk of developing *S. bovis* linked endocarditis is in range of 3–6% (Vogelmann and Amieva 2007). Whereas, 65–70% of patients with *S. bovis* endocarditis simultaneously revealed malignancy of gastrointestinal tract was not previously diagnosed (Grinberg et al. 1990). In addition, patients with history of *S. bovis* endocarditis with no colonoscopic abnormalities at that time may develop colon tumours 2–4 year later (Waisberg et al. 2002; Bisno 1991). *S. bovis* septicemia and / or endocarditis is also related to the presence of villous or tubular villous adenoma in the large intestine (Waisberg et al. 2002; Cutait et al. 1988). In a case control study comparing patients with *S. bovis* endocarditis who underwent colonoscopy with sex and age matched unaffected patients showed colonic adenomatous polyps in twice as many cases as control (15 of 32 vs 15 of 64) and colorectal cancer approximately 3 times (3 of 32 vs 2 of 64) (Hoen et al. 1994). In spite of these findings, colon evaluation is not performed routinely in most centers because of co-morbidities, advanced age or infection outside the gastrointestinal tract (Gold et al. 2004). *S. bovis* group of bacteria has been recently reclassified based on DNA-DNA hybridization and phylogenetic analyses of 16S rRNA sequences (Schlegel et al. 2003). The biotypes I and II.2 in this classification were renamed as *S. gallolyticus*. The *S. gallolyticus* subsp. *gallolyticus* is the new name of *S. bovis* biotype-I has been more commonly associated with occult cancer (Ruoff et al. 1989). Thus the need for endoscopic screening is strongly recommended in this group of patients.

7.3.4 Chlamydia pneumoniae and Lung Cancer

Lung cancer is the most common cancer world wide with 1.35 million incident cases annually (Chaturvedi et al. 2010; Alberg et al. 2005). Cigarette smoking is the major risk factor for lung cancer but the chronic pulmonary infection from *Chlamydia pneumoniae* is also implicated in lung carcinogenesis (Chaturvedi et al. 2010; Alberg et al. 2005). The association of *C. pneumoniae* infection with lung cancer risk has been variable with relative risk estimates ranges from 0.7 to 9.0 among seropositive individuals (Chaturvedi et al. 2010).

Chlamydophila (formerly *Chlamydia*) *pneumoniae* is a gram negative bacillus and intracellular parasite that causes respiratory infection in more than 50% adults (Mager 2006). The route of transmission is usually aerosol and in most of these infections is usually mild. This bacteria cause infections like pneumonia, bronchitis, sinusitis, rhinitis and chronic obstructive pulmonary disease (Hahn et al. 2002).

The respiratory infections form *C. pneumoniae* vary in different countries and populations, as it is endemic in United State and epidemic in Scandinavian countries (Kocazeybeck 2003). After acute infection with *C. pneumoniae*, metabolically inert atypical "persistent" inclusions are developed. These inclusions contain increased quantities of *Chlamydia* heat shock protein 60, which is a highly immunogenic protein implicated in the pathogenesis of chronic *Chlamydia* infections (Chaturvedi et al. 2010). The various researchers have suggested that persistent *C. pneumoniae* inflammation correlates with increased risk of lung cancer (Chaturvedi et al. 2010; Mager 2006).

The prospective and retrospective studies have reported 50–100% increased risks of lung cancer among those with elevated IgA antibody titers to *C. pneumoniae* (Mager 2006; Kocazeybeck 2003). Chronic inflammation and its sequelae may be part of the causal pathway (Litman et al. 2004). However, the exact mechanism is not known. It is accepted that agents that cause chronic inflammation may cause prolonged irritation, resulting in cell death and increased mitotic activity (Litman et al 2004). The subsequent cell division that occurs during repair of the damaged tissue may increase the risk of cancer at the affected site. The infection with *C. pneumoniae* stimulates the release of inflammatory mediators such as TNF-α, IL-β and IL-8 which may cause genetic damage (Gaydos 2000). Interleukin-8 also acts as a promoter of tumour growth for human non small cell lung carcinoma through its angiogenic properties. *C. pneumoniae* can impair or even block apoptosis of infected cells by induction of interleukin-10, resulting in chronic infection and increased risk of malignant transformation (Fan et al. 1998; Geng et al. 2000).

The infection with *C. pneumoniae* also acts synergistically with cigarette smoking to increase the risk of lung cancer (Litman et al. 2005). Smoking impairs lung immunity and increases secretion of interleukin-4 (IL-4). IL-4 production is associated with predominantly humoral (Th-2) T-helper cell response, which is ineffective at clearing infection. This predisposes for *C. pneumoniae* to localize more easily in the lung of smokers. The activated monocyte produces and secretes superoxide oxygen radicals, tumour necrosis factor, IL-1β and IL-8. These mediator of inflammation causes lung tissue and DNA damage results in carcinogenesis (Koyi et al. 1999). Chaturvedi et al. (2010) showed that Chlamydial heat shock protein-60 (CHSP-60) is consistently expressed by *C. pneumoniae* during chronic infection. They also found that lung cancer risk increased with elevated titers suggest that chronic inflammation from *C. pneumoniae* infection acts to promote lung cancer development.

Kocazeybeck (2003) examined chronic *C. pneumoniae* infection in 123 patients of lung carcinoma in which 70 had small cell, 28 squamous cell, 7 large cell carcinoma and 18 had adenocarcinoma. The elevations in antibody titers were found in total of 62 (50.4%) cases, 54% of the male patients and 36% of female patients. Chronic *C. pneumoniae* infection were seen statically more often in male patients with carcinoma who were age 55 years or younger than control ($P<0.001$).Although no difference was reported between male patients with lung carcinoma over age 55 and controls in blood titre between female patients and controls. The epidemiological study by Litman et al. (2004) also showed that persons with serologic evidence of past infection with *C. pneumoniae* had higher risk of lung cancer than controls.

7.4 Conclusion

The infection is implicated as an important cause of malignancy. The recent research on *H. pylori* and gastric cancer, *Salmonella typhi* and carcinoma gallbladder explained possible strategies of bacteria in carcinogenesis. The chronic inflammation, bacterial toxin and free radical are the key components. The other mechanisms like epigenetic alteration, immune modulation are also been widely implicated. The role of host genetic factor has also been evolved since last decade in determining the susceptibility and persistence of the bacterial infection. The exact mechanism of bacteria induced carcinogenesis is complex and involves interplay between the bacterial products, chronic inflammation affecting cell signaling, cell biology and stem cell homeostasis. The future research may uncover the other possible interactions between bacteria and cellular mechanism associated with cancer.

References

Alberg AJ, Brock MV, Samet JM (2005) Epidemiology of lung cancer: looking to the future. J Clin Oncol 23(14):3175–3185

Axelrod L, Munster AM, O'Brien TF (1971) Typhoid cholecystitis and gallbladder carcinoma after interval of 67 years. JAMA 217(1):83

Balkwill F, Mantovani A (2001) Inflammation and cancer: back to Virchow? Lancet 357(9255): 539–545

Bartsch H, Nair J (2005) Accumulation of lipid peroxidation-derived DNA lesions: potential lead markers for chemoprevention of inflammation-driven malignancies. Mutat Res 591:34–44

Bayliss R, Clarke C, Oakeley CM et al (1983) The microbiology and pathogenesis of infective endocarditis. Br Heart J 50(6):513–519

Bereswill S, Waidner U, Odenbreit S et al (1998) Structural, functional and mutational analysis of the pfr gene encoding a ferritin from *Helicobacter pylori*. Microbiology 144:2505–2516

Bereswill S, Greiner S, van Vliet AHM et al (2000) Regulation of ferritin-mediated cytoplasmic iron storage by the ferric uptake regulator homolog (Fur) of *Helicobacter pylori*. J Bacteriol 182(21):5948–5953

Biarc J, Nguyen IS, Pini A et al (2004) Carcinogenic properties of proteins with pro-inflammatory activity from *Streptococcus infantarius* (formerly *S. bovis*). Carcinogenesis 25(8):1477–1484

Bisno AL (1991) Streptococcal infection. In: Harrison TR, Wilson JD, Isselbacher KJ et al (eds) Harrison's principles of internal medicine, 12th edn. McGraw-Hill, New York, pp 563–569

Blaser MJ, Atherton JC (2004) *Helicobacter pylori* persistence: biology and disease. J Clin Invest 113(3):321–333

Burns CA, McCaughey R, Lauter CB (1985) The association of *Streptococcus bovis* fecal carriage and colon neoplasia; possible relationship with polyps and their premalignant potential. Am J Gastroenterol 80(1):42–44

Calam J (1999) *Helicobacter pylori* modulation of gastric acid. Yale J Biol Med 72:195–202

Caygill CP, Hill MJ, Braddick M et al (1994) Cancer mortality in chronic typhoid and paratyphoid carriers. Lancet 343(8889):83–84

Chaturvedi AK, Gaydos CA, Agreda P et al (2010) *Chlamydia pneumoniae* infection and risk for lung cancer. Cancer Epidemiol Biomarkers Prev 19(6):1498–1505

Chen LT, Lin JT, Tai JJ et al (2005) Long-term results of anti-*Helicobacter pylori* therapy in early-stage gastric high-grade transformed MALT lymphoma. J Natl Cancer Inst 97(18):1345–1353

Covacci A, Censini S, Bugnoli M et al (1993) Molecular characterization of the 128-kDa immunodominant antigen of *Helicobacter pylori* associated with cytotoxicity and duodenal ulcer. Proc Natl Acad Sci USA 90(12):5791–5795

Cutait R, Mansur A, Habr-Gama A (1988) Endocardite por streptococcus bovis e polipos de colon. Rev Bras Coloproctol 8:109–110

de Jong R, Altare F, Haagen IA et al (1998) Severe mycobacterial and salmonella infections in interleukin-12 receptor-deficient patients. Science 280(5368):1435–1438

de Martel C, Franceschi S (2009) Infections and cancer: established associations and new hypotheses. Crit Rev Oncol Hematol 70(3):183–194

Ellmerich S, Scholler M, Duranton B et al (2000) Promotion of intestinal carcinogenesis by *Streptococcus bovis*. Carcinogenesis 21(4):753–756

El-Omar EM, Carrington M, Chow WH et al (2000) Interleukin-1 polymorphisms associated with increased risk of gastric cancer. Nature 404(6776):398–402

El-Omar EM, Rabkin CS, Gammon MD et al (2003) Increased risk of noncardia gastric cancer associated with proinflammatory cytokine gene polymorphisms. Gastroenterology 124(5): 1193–1201

Fan T, Lu H, Hu H et al (1998) Inhibition of apoptosis in Chlamydia- infected cells: blockade of mitochondrial cytochrome C release and caspase activation. J Exp Med 187(4):487–496

Ferlay J, Bray F, Pisani P et al (2004) GLOBOCAN 2002: cancer incidence, mortality and prevalence worldwide, IARC CancerBase No.5, version 2.0. IARC Press, Lyon

Fiorentini C, Matarrese P, Straface E et al (1998) Toxin-induced activation of Rho GTP-binding protein increases Bcl-2 expression and influences mitochondrial homeostasis. Exp Cell Res 242(1):341–350

Fox J, Wang TC, Parsonnet J (2006) Helicobacter, chronic infection and cancer. In: Zur Hausen H (ed) Infectious causes of human cancer. Wiley- VCH, Weinheim, pp 386–466

Friedrich IA, Wormser GP, Gottfried EB (1982) The association of remote *Streptococcus bovis* bacteremia with colonic neoplasia. Am J Gastroenterol 77(2):82–84

Gaydos CA (2000) Growth in vascular cells and cytokine production by *Chlamydia pneumoniae*. J Infect Dis 181(Suppl 3):S473–S478

Geng Y, Shane RB, Berencsi K et al (2000) *Chlamydia pneumoniae* inhibits apoptosis in human peripheral blood mononuclear cells through induction of IL-10. J Immunol 164(10): 5522–5529

Gold JS, Bayar S, Salem RR (2004) Association of *Streptcoccus bovis* bacteremia with colonic neoplasia and extra-colonic malignancy. Arch Surg 139(7):760–765

Grinberg M, Mansur AJ, Ferreira DO et al (1990) Endocardite Por *Streptococcus bovis* e neoplasias de colon e reto. Arq Bras Cardiol 54(4):265–269

Häcker H, Karin M (2006) Regulation and function of IKK and IKK-related kinases. Sci STKE 357:re13

Haghjoo E, Galán JE (2004) *Salmonella typhi* encodes a functional cytolethal distending toxin that is delivered into host cells by a bacterial-internalization pathway. Proc Natl Acad Sci USA 101(13):4614–4619

Hahn DL, Azenabor AA, Beatty WL et al (2002) *Chlamydia pneumoniae* as a respiratory pathogen. Front Biosci 7:66–76

Harris PR, Smythies LE, Smith PD et al (2000) Inflammatory cytokines mRNA expression during early and persistent *Helicobacter pylori* infection in nonhuman primates. J Infect Dis 181(2):783–786

Helicobacter and Cancer Collaborative Group (2001) Gastric cancer and *Helicobacter pylori*: a combined analysis of 12 case control studies nested within prospective cohorts. Gut 49(3):347–353

Hoen B, Briancon S, Delahaye F et al (1994) Tumors of the colon increase the risk of developing *Streptococcus bovis* endocarditis: case-control study. Clin Infect Dis 19(2):361–362

Hussain SP, Harris CC (2007) Inflammation and cancer: an ancient link with novel potentials. Int J Cancer 121(11):2373–2380

International Agency for Research on Cancer (1994) IARC monographs on the evaluation of carcinogenic risks to humans, vol 61, Schistosomes, liver flukes and Helicobacter pylori. International Agency for Research on Cancer, Lyon

Karin M, Greten FR (2005) NF-kappaB: linking inflammation and immunity to cancer development and progression. Nat Rev Immunol 5(10):749–759
Keates S, Keates AC, Warny M et al (1999) Differential activation of mitogen-activated protein kinases in AGS gastric epithelial cells by cag+and cag−*Helicobacter pylori*. J Immunol 163(10):5552–5559
Kempf AJ, Volkmann B, Schaller M et al (2001) Evidence of a leading role for VEGF in Bartonella henselae-induced endothelial cell proliferations. Cell Microbiol 3(9):623–632
Keusch GT (1974) Opportunistic infection in colon carcinoma. Am J Clin Nutr 27(12):1481–1485
Knudtson LM, Hartman PA (1993) Comparison of fluorescent gentamicin thallous- carbonate and KF streptococcal agars to enumerate enterococci and fecal streptococci in meats. Appl Environ Microbiol 59(3):936–938
Kocazeybeck B (2003) Chronic *Chlamydophila pneumoniae* infection in lung cancer, a risk factor: a case-control study. J Med Microbiol 52:721–726
Koyi H, Branden E, Gnarpe J et al (1999) *Chlamydia pneumonia* may be associated with lung cancer. Preliminary report on a seroepidemiological study. APMIS 107(9):828–832
Kundu JK, Surh YJ (2008) Inflammation: gearing the journey to cancer. Mutat Res 659(1–2): 15–30
Kusters JG, van Vliet AHM, Kuipers EJ (2006) Pathogenesis of *Helicobacter pylori* Infection. Clin Microbiol Rev 19(3):449–490
Lax AJ, Grigoriadis AE (2001) *Pasteurella multocida* toxin: the mitogenic toxin that stimulates signalling cascades to regulate growth and differentiation. Int J Med Microbiol 291(4):261–268
Lax AJ, Thomas W (2002) How bacteria could cause cancer: one step at a time. Trends Microbiol 10(6):293–299
Lee WP, Tai DI, Lan KH et al (2005) The -251T allele of the interleukin-8 promoter is associated with increased risk of gastric carcinoma featuring diffuse-type histopathology in Chinese population. Clin Cancer Res 11(18):6431–6441
Leport C, Bure A, Leport J et al (1987) Incidence of colonic lesions in *Streptococcus bovis* and enterococcal endocarditis. Lancet 1(8535):8748
Litman AJ, White E, Jackson LA et al (2004) *Chlamydia pneumonia* infection and risk for lung cancer. Cancer Epidemiol Biomarkers Prev 13(10):1624–1630
Litman AJ, Jackson LA, Vaughan TL (2005) Chlamydia pneumoniae and lung cancer. Epidemiologic Evidence 14(4):773–778
Mager DL (2006) Bacteria and cancer: cause, coincidence or cure? a review. J Transl Med 4:14
Marches O, Ledger TN, Boury M et al (2003) Enteropathogenic and enterohaemorrhagic *Escherichia coli* deliver a novel effector called Cif, which blocks cell cycle G2/M transition. Mol Microbiol 50(5):1553–1567
McCoy WC, Mason JM (1951) Enterococcal endocarditis associated with carcinoma of the sigmoid: report of a case. J Med Assoc State Ala 21(6):162–166
McNamara D, El-Omar E (2008) *Helicobacter pylori* infection and the pathogenesis of gastric cancer: A paradigm for host–bacterial interactions. Dig Liver Dis 40:504–509
Mitsuno Y, Yoshida H, Maeda S et al (2001) *Helicobacter pylori* induces transactivation of SRE and AP-1 through the ERK signalling pathway in gastric cancer cells. Gut 49(1):18–22
Murray HW, Roberts RB (1978) *Streptococcus bovis* bacteremia and underlying gastrointestinal disease. Arch Intern Med 138(7):1097–1099
Nath G, Singh YK, Kumar K et al (2008) Association of carcinoma of the gallbladder with typhoid carriage in a typhoid endemic area using nested PCR. J Infect Dev Ctries 2(4):302–307
Nath G, Gulati AK, Shukla VK (2010) Role of bacteria in carcinogenesis, with special reference to carcinoma of the gallbladder. World J Gastroenterol 16(43):5395–5404
Nougayrède JP, Taieb F, De Rycke J et al (2005) Cyclomodulins: bacterial effectors that modulate the eukaryotic cell cycle. Trends Microbiol 13(3):103–110
Ohshima H, Sawa T, Akaike T (2006) 8-Nitroguanine, a product of nitrative DNA damage caused by reactive nitrogen species: formation, occurrence, and implications in inflammation and carcinogenesis. Antioxid Redox Signal 8:1033–1045
Parsonnet J, Hansen S, Rodriguez L et al (1994) *Helicobacter pylori* infection and gastric lymphoma. N Engl J Med 330(18):1267–1271

Pasare C, Medzhitov R (2004) Toll-like receptors: linking innate and adaptive immunity. Microbes Infect 6(15):1382–1387

Pisani P, Parkin DM, Munoz N et al (1997) Cancer and infection: estimates of the attributable fraction in 1990. Cancer Epidemiol Biomarkers Prev 6(6):387–400

Ruoff KL, Miller SI, Garner CV et al (1989) Bacteremia with *Streptococcus bovis* and *Streptococcus salivarius*: clinical correlates of more accurate identification of isolates. J Clin Microbiol 27(2):305–308

Saccani S, Pantano S, Natoli G (2002) p38-Dependent marking of inflammatory genes for increased NF-kB recruitment. Nat Immunol 3(1):69–75

Schlegel L, Grimont F, Ageron E et al (2003) Reappraisal of the taxonomy of the *Streptococcus bovis/Streptococcus equinus* complex related species: description of *Streptococcus gallolyticus* subsp. *gallolyticus subsp. nov., S. gallolyticus subsp. macedonius* subsp. nov. and *Streptococcus gallolyticus* subsp. Pasteurianus subsp.nov. Int J Syst Evol Microbiol 53:631–645

Shenker BJ, McKay T, Datar S (1999) *Actinobacillus actinomycetemcomitans* immunosuppressive protein is a member of the family of cytolethal distending toxins capable of causing a G2 arrest in human T cells. J Immunol 162(8):4773–4780

Shukla VK, Singh H, Pandey M et al (2000) Carcinoma of the gallbladder-is it a sequel of typhoid? Dig Dis Sci 45(5):900–903

Sipponen P, Kekki M, Haapakoski J et al (1985) Gastric cancer risk in chronic atrophic gastritis: statistical calculations of cross-sectional data. Int J Cancer 35(2):173–177

Stinson MW, McLaughlin R, Choi SH et al (1998) Streptococcal histone-like protein: primary structure of hlpA and protein binding to lipoteichoic acid and epithelial cells. Infect Immun 66(1):259–265

Suerbaum S, Michetti P (2002) *Helicobacter pylori* infection. N Engl J Med 347(15):1175–1186

Sugai M, Hashimoto K, Kikuchi A et al (1992) Epidermal cell differentiation inhibitor ADP-ribosylates small GTP-binding proteins and induces hyperplasia of epidermis. J Biol Chem 267(4):2600–2604

Suzuki H, Minegishi Y, Nomoto Y et al (2005) Down-regulation of a morphogen (sonic hedgehog) gradient in the gastric epithelium of *Helicobacter pylori*-infected Mongolian gerbils. J Pathol 206(2):186–197

Teitz S, Guidetti-Sharon A, Manor H et al (1995) Pyogenic liver absence: warning indication of silent colonic cancer. Report of a case and review of the literature. Dis Colon Rectum 38(11): 1220–1223

Thomas W, Pullinger GD, Lax AJ et al (2001) *Escherichia coli* cytotoxic necrotizing factor and *Pasteurella multocida* toxin induce focal adhesion kinase autophosphorylation and Src association. Infect Immun 69(9):5931–5935

Tjalsma H, Guinard MS, Lasonder E et al (2006) Profiling the humoral immune response in colon cancer patients: diagnostic antigens from *Streptococcus bovis*. Int J Cancer 119(9):2127–2135

Vogelmann R, Amieva MR (2007) The role of bacterial pathogens in cancer. Curr Opin Microbiol 10(1):76–81

Waisberg J, de O Matheus C, Pimenta J (2002) Infectious endocarditis from *Streptococcus bovis* associated with colonic carcinoma: case report and literature review. Arq Gastroenterol 39(3): 177–180

Welton JC, Marr JS, Friedman SM (1979) Association between hepatobiliary cancer and typhoid carrier state. Lancet 1(8120):791–794

Wendum D, Masliah J, Trugnan G et al (2004) Cyclooxygenase-2 and its role in colorectal cancer development. Virchows Arch 445(4):327–333

Yagyu K, Kikuchi S, Obata Y (2008) Cigarette smoking, alcohol drinking and the risk of gallbladder cancer death: a prospective cohort study in Japan. Int J Cancer 122(4):924–929

Yamamoto Y, Verma UN, Prajapati S et al (2003) Histone H3 phosphorylation by IKK-alpha is critical for cytokine-induced gene expression. Nature 423(6940):655–659

Yermilov V, Rubio J, Becchi M et al (1995) Formation of 8-nitroguanine by the reaction of guanine with peroxynitrite *in vitro*. Carcinogenesis 16(9):2045–2050

Zur Hausen H (2006) *Streptococcus bovis*: casual or incidental involvement in cancer of the colon? Int J Cancer 119(9):xi–xii

Chapter 8
Bacteria as a Therapeutic Approach in Cancer Therapy

Sazal Patyar, Ajay Prakash, and Bikash Medhi

Abstract In the wake of growing global burden of cancer, newer cancer prevention and control modalities are being explored. One such novel experimental strategy is the implication of natural and genetically modified non-pathogenic bacterial species as potential antitumor agents. This therapy is based on the fact that live, attenuated or genetically modified non-pathogenic bacterial species are capable of multiplying selectively in tumours and inhibiting their growth. Moreover due to their selectivity for tumour tissues, these bacteria and their spores also serve as ideal vectors for delivering therapeutic proteins to tumours. Bacterial toxins too have been explored for their anti-cancer potential. Although the oncolytic potential of bacteria was recognized several hundred years back yet the bacterial therapy failed to establish because of certain drawbacks associated with it like toxicity, lack of specificity and inconvenient administration of bacteria. However the emergence of gene therapy and recombinant DNA technology has revived the interest in bacterial therapy and a variety of applications employing bacteria have been investigated. Out of these, the most potential and promising strategies are bacteria based gene-directed enzyme prodrug therapy, anaerobic bacteria vector-mediated cancer therapy and immunotherapy. These therapies have demonstrated significant efficacy in preclinical studies and some are currently under clinical investigation.

Keywords Bacteria • Bacterial therapy • *Bifidobacterium* • Bacterial gene-directed enzyme prodrug therapy • β-glucuronidase • *Clostridium* • Cancer • Cytosine deaminase • Diptheria • 5-fluorocytosine • 5-fluorouracil • Gene therapy • Genetically engineered bacteria • Hypoxia • Microorganisms • Nitroreductase • *Pseudomonas* • *Salmonella* • Spores • Tumour targeting • Toxin • Vector

---

S. Patyar • A. Prakash • B. Medhi (✉)
Department of Pharmacology, Postgraduate Institute of Medical Education & Research, Research Block B, 4th Floor, Room no 4043, Chandigarh 160012, India
e-mail: drbikashus@yahoo.com

A.A. Khan (ed.), *Bacteria and Cancer*, DOI 10.1007/978-94-007-2585-0_8,

Abbreviations

5FC	5-fluorocytosine
5FU	5-fluorouracil
ALL	Acute lymphoblastic leukaemia
AMP	Adenosine monophosphate
BCG	Bacillus Calmette-Guerin
BoNT	Botulinum neurotoxin
BR96 sFv-PE40	Single-chain immunotoxin. SGN-10 is composed of the fused gene products encoding the translocating and ADP-ribosylating domains of Pseudomonas exotoxin (PE40) and the variable heavy (V(H)) and variable light (V(L)) regions of BR96 monoclonal antibody
CD	Cytosine deaminase
CDTs	Cytolethal distending toxins
Cif	Cycle inhibiting factor
CMV	Cytomegalovirus
CNF	Cytotoxic necrotizing factor
COBALT	Combination bacteriolytic therapy
CPE	*Clostridium perfringens* enterotoxin
CPE-R	CPE receptor
CPG2	Carboxypeptidase G2
DCA	Dichloroacetate
DNA	Deoxyribonucleic acid
DT	Diphtheria toxin
EF2	Elongation factor-2
EGF	Epidermal growth factor
EHEC	Enterohaemorrhagic *E. coli*
EPEC	Enteropathogenic *E. coli*
erb-38	A recombinant immunotoxin that targets the erbB2 receptor
G-CSF	Granulocyte colony stimulating factor
HA22	(CAT-8015) is an immunotoxin composed of an anti-CD22 variable fragment linked to a 38 kDa truncated protein derived from Pseudomonas exotoxin A.
HAMLET	Human alpha-lactalbumin made lethal to tumour cells
HB-EGF	Heparin-binding epidermal growth factor like growth factor
hGM-CSF	Human granulocyte-macrophage-colony stimulating factor
hIL-12	Human interleukin-12
HSVTK	Herpes simplex virus thymidine kinase
IL13-PE38QQR	Cintredekin besudotox
IL-3	Interleukin-3
IL-4	Interleukin-4
IL-4-PE38KDEL	A chimeric protein composed of circularly permuted IL-4 and a truncated form of Pseudomonas exotoxin (PE), into recurrent malignant high-grade gliomas

IL4-P	Aeruginosa exotoxin (IL4-PE NBI-3001), tumour growth factor (TGF) alpha-*P. aeruginosa* exotoxin (TP-38), IL13-P. aeruginosa exotoxin (IL13-PE38), and transferrin-*C. diphtheriae* toxin (TransMID(trade mark), Tf-CRM107
IL4-PE	Interleukin-4-Pseudomonas exotoxin
LMB-1	Is composed of monoclonal antibody B3 chemically linked to PE38 a genetically engineered form of Pseudomonas exotoxin
LMB-2	CD25-directed immunotoxin
LMB-2	Recombinant immunotoxin anti-Tac(Fv)-PE38
Mab	Monoclonal antibody
mGM-CSF	granulocyte-macrophage-colony stimulating factor
mIL-12	Murine interleukin 12
NBI-3001	IL-4 Pseudomonas exotoxin protein
NR	Nitroreductase
OVB3-PE	An immunotoxin composed of a murine monoclonal antibody reactive with human ovarian cancer and conjugated to Pseudomonas exotoxin (PE)
PE38	Is a 38-kDa derivative of the 66-kDa Pseudomonas exotoxin (PE)
TAPET	Tumour Amplified Protein Expression Therapy
TAPET-CD, VNP20029	VNP20029genetically modified *Salmonella typhimurium* expressing cytosine deaminase
Tf	Transferrin
Tf-CRM 107	Transferrin-DT conjugate
TGF-α	Transforming growth factor α
TNF-α	Tumour necrosis factor α
TP-38	A recombinant chimeric targeted toxin composed of the EGFR binding ligand TGF-alpha and a genetically engineered form of the Pseudomonas exotoxin PE-38
TP40	Transforming growth factor alpha-Pseudomonas exotoxin-40
VEGF	Vascular endothelial growth factor

8.1 Introduction

Cancer is a disease characterized by unrestricted and incursive multiplication of cells which may metastasize to other parts of the body. Despite the relentless incorporation of newer drugs and therapies into the oncological armamentarium, cancer remains a leading cause of morbidity and mortality throughout the world. Currently a wide range of conventional anticancer therapies are available which include surgical resection, radiotherapy and chemotherapy. But their inefficacy for about half of the cancer suffering population has ushered the development of alternative approaches.

Moreover, systemic toxicity arising from the non-specificity of most of anticancer therapies ushers the need to focus on alternative approaches with targeted therapeutics specific to cancer cells. Certain new medical therapies intending or claiming to treat cancer by improving, supplementing or replacing conventional methods are classified under the term *Experimental cancer treatment* which includes photodynamic therapy, HAMLET (human alpha-lactalbumin made lethal to tumour cells), gene therapy, telomerase therapy, hyperthermia therapy, dichloroacetate (DCA), noninvasive RF cancer treatment, complementary and alternative therapy, diet therapy, insulin potentiating therapy and bacterial treatment (Jain 2001). The role of bacteria in cancer is quite uncertain. Several scientific findings have indicated an association of certain species of bacteria with carcinogenesis. However many other studies have reported that some bacterial species display selective replication in tumour cells or preferential accumulation in the tumour microenvironment thus offering a great potential for cancer therapy. Furthermore certain characteristics of bacteria like motility, capacity to both carry and express multiple therapeutic proteins as well as eradication by antibiotics project them as a novel and promising strategy for cancer therapy (Nauts et al. 1953).

8.2 Historical Background

Bacteria were recognized as tumour regressor agents almost 100 years back. Busch and Fehleisen, two German physicians had independently observed that accidental erysipelas (*Streptococcus pyogenes*) infections occurring in hospitalized patients regressed certain types of cancers (Nauts 1980). An American physician, William Coley pioneered the use of bacteria and their toxins to treat end stage cancers after observing that a patient suffering from neck cancer began to recover following an infection with erysipelas. He developed a safer vaccine composed of two killed bacterial species, *Streptococcus pyogenes* and *Serratia marcescens* to simulate an infection with the accompanying fever without the risk of an actual infection (Richardson et al. 1999; Zacharski and Sukhatme 2005). This vaccine was widely used to successfully treat sarcomas, carcinomas, lymphomas, melanomas and myelomas (Hoption Cann et al. 2003). *Coley's toxins* composed of toxic bacterial derivatives too exhibited potential anticancer activity during initial stages (Nauts and McLaren 1990).

8.3 Bacteria as Cancer Therapy

Later on other scientific discoveries reported that certain species of anaerobic bacteria, such as those belonging to the genus *Clostridium* can only survive and consume oxygen poor cancerous tissue while they die within the tumour's oxygenated sides, indicating their safety and specificity (Malmgren and Flanigan

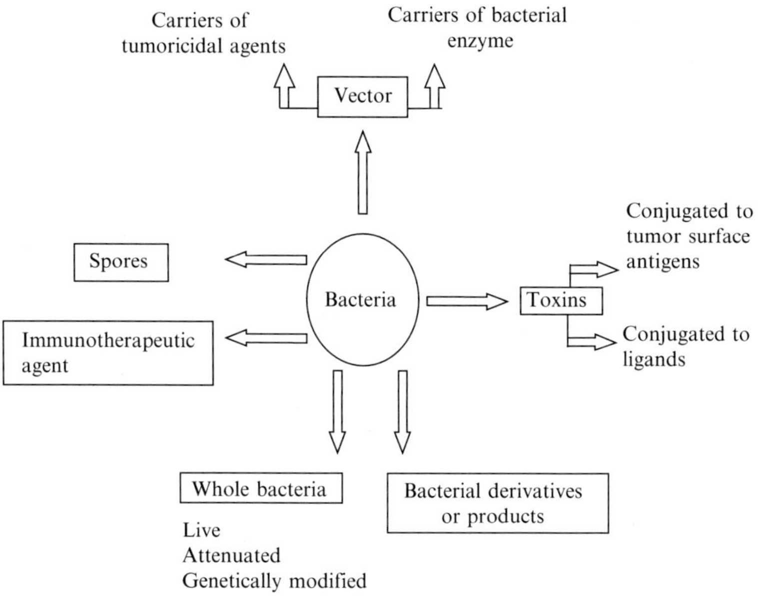

Fig. 8.1 Schematic representation of different functions of bacteria in cancer therapy

1955). All these observations provided the grounds for exploiting the bacteria as oncolytic agents. But as the bacteria alone do not completely consume the malignant tissue, they were tested in combination with chemotherapeutic treatments. Therefore bacteria can be implied as sensitizing agents for chemotherapy. Bacterial products like endotoxins (lipopolysaccharides) too have been tested for cancer treatment. Similarly bacterial toxins were employed as cancer vaccines (Carswell et al. 1975). Bacteria have been employed as delivery agents and vectors for anticancer drugs and gene therapy respectively. Spores of anaerobic bacteria can be used instead of bacteria because only spores that reach an oxygen starved area of a tumour will germinate, multiply and become active. Currently genetically modified bacteria and bacterial gene-directed enzyme prodrug therapy are considered promising because of their selective destruction of tumours. The overall bacterial approach is summarized in Fig. 8.1 and hypothesized mechanism in cancer therapy in Fig. 8.2.

8.3.1 Whole Bacteria

Live, attenuated or genetically modified, non-pathogenic bacteria have been employed as potential antitumour agents, either as direct tumoricidal agents or as carriers of tumoricidal molecules. Live bacteria in the form of probiotics have

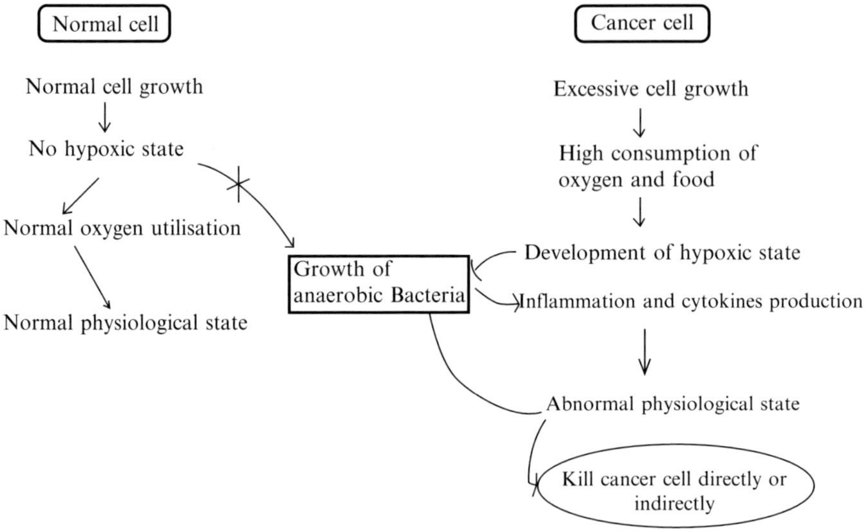

Fig. 8.2 Hypothesized mechanism of anaerobic bacteria in cancer therapy

been studied for the anticancer effect in the field of colorectal, breast and bladder cancer. The mounting evidence has suggested the ability of probiotic strains to prevent colorectal cancer. Although the anti-cancer effects of probiotics are quite evident from animal studies yet there is a lack of evidence from human studies. These probiotics usually refer to highly selected lactic acid bacteria; for example, *Lactobacillus sp.*, *Bifidobacterium sp.* and *Streptococcus sp. Lactobacillus sp.* which are considerably safe have been reported to inhibit colon cancer. Data collected from some epidemiological studies have indicated that consumption of large quantities of fermented milk products containing *Lactobacillus* or *Bifidobacteria* are associated with a lower incidence of colon cancer while others have reported negligible or no effect at all (Liong 2008; Rafter 2004). Studies have indicated that oral consumption of *Lactobacillus casei strain Shirota* can reduce bladder cancer recurrence. Recently a study has reported that *L. rhamnosus GG* (LGG), a probitoic originally isolated from the human gut, induces tumour regression in mice bearing orthotopic bladder tumours and has the potential to replace BCG immunotherapy for the treatment of bladder cancer (Seow et al. 2010). Furthermore, a clinical trial has concluded that synbiotic intervention composed of the prebiotic-oligofructose-enriched inulin (SYN1) and probiotics-*Bifidobacterium lactis* Bb12 (BB12) reduced cancer risk factors in polypectomized and colon cancer patients (Roller et al. 2007).

Similarly some live anaerobic bacteria have been employed in the cancer therapy based on their ability to proliferate preferentially within necrotic (anaerobic) regions of tumours as compared to normal tissues. As evident from experimental studies, certain pathogenic species of the anaerobic *Clostridia* resulted in tumour regression but was accompanied by acute toxicity which caused illness or death in most of the animals (Malmgren and Flanigan 1955; Minton 2003). Therefore, a nonpathogenic strain of *Clostridium* such as 'M55' was tested which despite colonizing the anaerobic parts of the tumour failed to cause significant tumour regression (Carey et al. 1967). Several other anaerobic bacteria-species like *Bifidobacteria*, *Lactobacilli* and pathogenic *Clostridia* have been screened for their ability to accumulate in experimental tumours in animals. *Clostridium novyi* demonstrated significant antitumour effects, but again culminated in death.

C. novyi-NT, an attenuated strain obtained after deletion of a lethal toxin encoding gene has exhibited good results and produced toxicity too. Thus, *C. novyi-NT* spores have been administered in combination with conventional chemotherapeutic agents like dolastatin-10, mitomycin C, vinorelbine and docetaxel. This approach is known as *combination bacteriolytic therapy* (COBALT). But despite significant antitumor efficacy, this strategy was not devoid of animal deaths (Dang et al. 2001). *C. novyi* has also been investigated in conjunction with radiotherapy, radioimmunotherapy, and chemotherapy in experimental tumour models (Bettegowda et al. 2003; Wei et al. 2007). The results have demonstrated the potential of combined multi-modality approaches as developmental future cancer therapies. *C. novyi-NT* has been exploited to enhance the release of liposome encapsulated drugs within tumours because of its evident membrane-disrupting potential. The bacterial factor responsible for the enhanced drug release has been identified as liposomase. Remarkable eradication of the tumours in mice bearing large, established tumours by employing *C. novyi-NT* plus a single dose of liposomal doxorubicin has led to further studies in the field (Cheong et al. 2006). To enhance the accessibility of the drugs in the poorly vascularized regions of tumours, *C. novyi-NT* was used in combination with anti-microtubule agents. It has demonstrated that the microtubule destabilizers such as HTI-286 and vinorelbine, but not the microtubule stabilizers such as the taxanes, docetaxel and MAC-321, radically reduced blood flow to tumours thereby enlarging the hypoxic region favourable for spores germination (Dang et al. 2004).

VNP20009, a derivative strain of *Salmonella typhimurium,* was obtained through deletion of two of its genes – *msbB* and *purI* – which resulted in its complete attenuation (by preventing toxic shock in animal hosts) and dependence on external sources of purine for survival. Due to this dependence bacteria can selectively grow in tumorous tissue where purine is abundant and are unable to replicate in normal tissue such as the liver or spleen. This approach has showed long-lasting efficacy against a broad range of experimental tumours and was even able to target metastatic lesions (Low et al. 1999; Luo et al. 1999). The advantage of using *Salmonella* over *Clostridium* or *Bifidobacterium* is its ability to grow in both aerobic and anaerobic conditions, indicating its usefulness against small tumours. VNP20009 has been investigated successfully in Phase I clinical trials in cancer patients. Other live, attenuated bacteria, such as *Clostridia* and *Bifidobacterium*, may be evaluated in human clinical trials in the future. Furthermore *Salmonella choleraesuis*, *Vibrio cholerae*, *Listeria monocytogenes* and even *Escherichia coli* are being screened for anticancer potential (Bermudes et al. 2002).

8.3.2 Bacterial Derivatives/Products

As discussed above, live or attenuated bacteria have narrow spectrum of applications because they are often associated with toxicity or side-effects and in many cases invoke immune response. This shifts the focus on the use of bacterial derivatives or products instead of whole bacteria. Various antibiotics obtained from bacteria have shown prominent antitumor activity. They intercalate between DNA strands and interfere with its template function. Amongst these the important ones belong to anthracyclins, bleomycin, actinomycin, mitomycin and aureolic acid families. Anthracyclines (anthracycline antibiotics) are derived from *Streptomyces* bacteria (more specifically, *Streptomyces peucetius var. caesius*) and are used to treat a wide range of cancers, including leukemias, lymphomas, breast, uterine, ovarian and lung cancers. Bleomycin, a glycopeptide antibiotic produced by the bacterium *Streptomyces verticillus* is used as an anti-cancer agent. It is used in the treatment of Hodgkin lymphoma, squamous cell carcinomas, and testicular cancer (Takimoto and Calvo 2008). Similarly, the actinomycins are a class of polypeptide antibiotics isolated from soil bacteria of the genus *Streptomyces*, of which the most significant is actinomycin D. It was the first antibiotic isolated by *Selman Waksman*. Actinomycin D, marketed under the trade name Dactinomycin, is one of the older chemotherapy drugs, and has been used in therapy for many years. It is commonly used in treatment of a variety of cancers including gestational trophoblastic neoplasia, Wilms' tumour and rhabdomyosarcoma (Turan et al. 2006; Abd El-Aal et al. 2005; Khatua et al. 2004). The mitomycins are a family of aziridine-containing natural products isolated from *Streptomyces caespitosus* or *Streptomyces lavendulae.* Mitomycin C, is used as a chemotherapeutic agent due to its antitumor antibiotic activity. It is employed to treat upper gastro-intestinal (e.g. esophageal carcinoma), anal cancers, and breast cancers, as well as by bladder instillation for superficial bladder tumours (Renault et al. 1981). Kinamycins, a group of bacterial polyketide secondary metabolites containing a diazo group are known for their cytotoxicity and are considered of interest for potential use in anti-cancer therapies (Ballard and Melander 2008).

Several other bacterial products like lipopolysaccharides (LPS), lipotechoic acid, peptidoglycans and epothilones have been reported to possess anti-cancer activity. Some detoxified bacterial LPS preparations with or without additional components have shown varying anti-tumour efficacy (Ribi et al. 1975). The anti-tumour effect of bacterial lipid A has been demonstrated in both animals and clinical trials (Reisser et al. 2002). Similarly lipotechoic acid obtained from *Streptococcus pyogenes* and *Enterococcus faecalis* has shown tumour protective effect both alone as well as in combination with peptidoglycan of *Moraxella catarrhalis* and LPS from *E. coli* (Keller et al. 1995). A new class of non-taxane cancer drugs is epothilones, which are also known as *microtubule depolymerization inhibitors* as they prevent cancer cells from dividing by interfering with tubulin. Basically these epothilones are secondary metabolites produced by the myxobacterium *Sorangium cellulosum.* Till now, epothilones A to F have been identified and characterised (Julien and Shah

2002). Early studies in cancer cell lines and in human cancer patients have suggested superior efficacy to the taxanes. Although both have similar mechanism of action yet because of their simpler chemical structure, better water solubility and better ability to inhibit the growth of tumour cells over-expressing the P-glycoprotein (a factor responsible for drug resistance in many tumours), epothilones are preferable than taxol. Several epothilone analogs are currently undergoing clinical development for treatment of various cancers. One analog, ixabepilone, was approved in October 2007 by the United States Food and Drug Administration for use in the treatment of aggressive metastatic or locally advanced breast cancer no longer responding to currently available chemotherapies. Ixabepilone and patupilone, have shown significant activity in men with castration-resistant prostate cancer (CRPC) in clinical studies. A randomized multicenter phase II trial of patupilone (at two different doses, 8 mg/m^2 or 10 mg/m^2 given every 3 weeks) vs. docetaxel in CRPC is currently underway (Prakash and Winston 2010). Azurin, a copper-containing oxido-reductase obtained from *Pseudomonas aeruginosa* has shown anti-cancer activity both *in vitro* and *in vivo* in a nude mouse model. Recently it has been reported that azurin induces apoptosis in the human breast cancer cell line MCF-7 (Punj et al. 2004).

8.3.3 Bacterial Toxins

Bacterial toxins have already been tested as cancer therapy. Bacterial toxins directly kill cells or alter cellular processes which control proliferation, apoptosis and differentiation. Therefore alterations in any of these processes may either stimulate cellular aberrations or inhibit normal cell controls thereby leading to carcinogenesis. Cell cycle inhibitors, such as cytolethal distending toxins (CDTs) and the cycle inhibiting factor (Cif), block mitosis and are thought to compromise the immune system by inhibiting clonal expansion of lymphocytes. In contrast, cell cycle stimulators such as the cytotoxic necrotizing factor (CNF) promote cellular proliferation and interfere with cell differentiation (Nougayrede et al. 2005). Bacterial toxins that subvert the host eukaryotic cell cycle have been classified as cyclomodulins. For example, CNF is a cell cycle stimulator released by certain bacteria, such as *E. coli*. CNF triggers G_1-S transition and induces DNA replication. The number of cells does not increase, however. The cells become multinucleated instead, perhaps by the toxin's ability to inhibit cell differentiation and apoptosis (Oswald et al. 1994; Fiorentini et al. 1998). CDTs are found in several species of gram-negative bacteria, including *Campylobacter jejuni* and *S. typhi* while Cif is found in enteropathogenic (EPEC) and enterohaemorrhagic (EHEC) *E. coli*. The anti-tumour effect of toxins is probably with reduced side effects compared to traditional tumour treatment. Bacterial toxins *per se* or when combined with anti-cancer drugs or irradiation could therefore possibly increase the efficacy of cancer treatment (Carswell et al. 1975).

8.3.3.1 Bacterial Toxins Binding to Tumour Surface Antigens

Certain bacterial toxins act by binding to antigens present on tumour surface. Diphtheria toxin (DT) binds to the surface of cells expressing the heparin-binding epidermal growth factor like growth factor (HB-EGF) precursor. DT-HB-EGF complex is internalized after endocytosis via clathrin-vesicles. Subsequently DT undergoes several post-translational modifications resulting in a catalytically active toxin, called DT fragment A. This catalytically ribosylates elongation factor-2 (EF-2) leading to inhibition of protein synthesis with subsequent cell lysis and/or induction of apoptosis (Louie et al. 1997; Frankel et al. 2002; Lanzerin et al. 1996; Falnes et al. 2000). Like DT, *Pseudomonas* exotoxin A is also known to catalytically ribosylate EF-2 and thus leading to inhibition of protein synthesis. Extremely high cytotoxicity of this toxin with a lethal dose of 0.3 μg after i.v., injection in mice makes it a potential candidate for targeted cancer therapy (Pastan 1997). *Clostridium perfringens* type A strain, the causative agent of gastroenteritis, produces *Clostridium perfringens* enterotoxin (CPE). The C-terminal domain of CPE is responsible for high affinity binding to the CPE receptor (CPE-R) and the N-terminal is assumed to be essential for cytotoxicity (Kokai Kun and Mcclane 1997; Kokai Kun et al. 1999). Studies have shown that purified CPE exerts an acute cytotoxic effect on pancreatic cancer cells and led to tumour necrosis and inhibition of tumour growth *in vivo*. It is being investigated for colon, breast and gastric cancers. Moreover, before evaluating CPE for systemic cancer therapy, its long term efficiency and lack of toxicity *in vivo* need to be demonstrated (Michl et al. 2001; Hough et al. 2000; Kominsky et al. 2004). A recent study has demonstrated for the first time that botulinum neurotoxin (BoNT) briefly opens tumour vessels, allowing more effective destruction of cancer cells by radiotherapy and chemotherapy. It has been proposed that BoNTs act by an effect on the tumour microenvironment rather than by a direct cytotoxic effect on tumour cells (Ansiaux and Gallez 2007). Some bacterial toxins (alfa-toxin from *Staphylococcus aureus*, AC-toxin from *Bordetella pertussis*, shiga like toxins, and cholera toxin) are presently being studied on two cell lines, mesothelioma cells (P31) and small lung cancer cells (U-1690). Preliminary results with AC-toxin showed increasing cytotoxicity with increasing dose of AC-toxin in both cell lines and the toxin markedly increased apoptosis. However, cholera toxin did not induce apoptosis (Nougayrede et al. 2005).

8.3.3.2 Bacterial Toxins Conjugated to Ligands

Pseudomonas exotoxin (PE), diphtheria toxin (DT), and ricin are among the most potent cell killing agents. To bypass their lethality and maximize the therapeutic efficacy specific targeting of these toxins on the surface of cancer cells is highly desirable. This is achieved by conjugating the toxins to cell binding proteins such as monoclonal antibodies or growth factors thereby avoiding their association to toxin receptors. These ligand-conjugated toxins bind and kill cancer cells selectively thus sparing normal cells, which do not bind the conjugates. A wide variety of DT ligands such as IL-3, IL-4, granulocyte colony stimulating factor (G-CSF), transferrin (Tf), EGF and vascular endothelial growth factor (VEGF) have been studied for targeted tumours (Frankel et al. 2002). The transferring-DT conjugate (Tf-CRM 107) and DT-EGF have reached the stage of clinical trials in patients of brain tumour and metastatic carcinomas respectively (Hagihara et al. 2000). Similarly a large variety of antibodies and ligands to surface antigens over expressed in different tumours have been conjugated to PE. Important ones tested in clinical trials are IL-4, IL-13, monoclonal antibody recognizing a carbohydrate antigen Lewis Y, reacting with metastatic adenocarcinoma cells (Mab B3) and transforming growth factor (TGF-α) (Fan et al. 2002).

Alternatively genetically modified or recombinant toxins have been produced by deleting the DNA coding for the toxin binding region and replacing it with various complementary DNA encoding other cell binding proteins. These chimeric toxins may be useful in designing future toxin-based anticancer therapies. For targeted DT therapy, deletions within the DT-receptor binding domain (amino acid residues 390–535) or targeted mutations of the critical HB-EGF precursor binding loop (amino acid residues 510–530) have been used (Frankel et al. 2002; Greenfield et al. 1987). Recently, a recombinant interleukin-4-*Pseudomonas* exotoxin (IL4-PE) for therapy of glioblastoma has been developed. *In vivo* experiments with nude mice have demonstrated that IL4-PE has significant antitumor activity against human glioblastoma tumour model. Intratumor administration of IL4-PE is being investigated for the treatment of malignant astrocytoma in a phase I clinical trial (Puri 1999).

8.3.4 Bacterial Spores

Spores act as a resting or dormant stage in the bacterial life cycle which preserves the bacterium through periods of unfavorable conditions. Of all the anaerobic bacteria discussed above, almost all of them form highly resistant spores which allow them to survive in oxygen rich conditions. Under favorable conditions, such as the dead areas inside tumours, the spores germinate and the bacteria thrive, thereby making them ideal to target cancers. Spores of *C. novyi-NT* have shown targeted action without any systemic side-effects. Intratumoral injection of *C. histolyticum* spores as well as intravenous administration of *C. sporogenes* spores resulted in marked lysis of tumour tissues in mice. Pharmacologic and toxicological evaluation of *C. novyi-NT* spores reported that spores were rapidly cleared from the circulation by the reticuloendothelial system. Moreover it was observed that *Clostridium* was detected only in tumours and not in normal tissues of mice receiving an intravenous injection of bacteria (Thiele et al. 1963). Even after the administration of large doses, no clinical toxicity was observed in healthy mice or rabbits. However like in case of any bacterial infection, in tumour bearing mice, toxicity appeared related to tumour size and spore dose (Diaz et al. 2005). Furthermore, bacterial spores are being tested as delivery agents for anticancer agents, cytotoxic peptides, therapeutic proteins, and as vectors for gene therapy.

8.3.5 Bacteria as Vector for Gene Therapy

Gene therapy is being explored for cancer treatment but the basic obstacle in cancer gene therapy is the specific targeting of therapy directly to a solid tumour. On the other hand, the major problem with using bacteria as anti-cancer agents is toxicity (at the dose required for therapeutic efficacy) and reducing the dose results in diminished efficacy. Therefore an alternative approach to overcome these limitations is the use of genetically engineered bacteria which can express a specific therapeutic gene. These bacterial vectors can provide a powerful adjuvant therapy to various cancer treatments as the protein of interest will be produced specifically in the tumour micro-environment. Thus, bacteria serve as vectors or vehicles for delivering anti-cancer agents, cytotoxic peptides, therapeutic proteins or prodrug converting enzymes to solid tumours.

8.3.5.1 Bacteria as Carriers of Tumoricidal Agents

Attenuated form of *S. typhimurium* is preferred to deliver cytokines locally to liver, with an effect on hepatic metastases as it naturally colonizes in liver. A genetically engineered strain of *S. typhimurium*, ×4550 has been developed which is cya/crp mutant (genes encoding proteins involved in the regulation of cyclic AMP levels) and expresses interleukin-2 for the treatment of liver cancer in preclinical models (Saltzman et al. 1996; 1997). hIL-12, hGM-CSF, mIL-12 and mGM-CSF have been cloned under the control of a cytomegalovirus (CMV) promoter, into SL3261, an auxotrophic *S. typhimurium*. It has been reported that oral administration of *Salmonella* expressing mGM-CSF or mGM-CSF plus mIL-12 caused tumour regression in mice bearing Lewis lung carcinomas (Yuhua et al. 2001). Other therapeutic proteins like TNF-α and platelet factor 4 fragment too have been cloned and expressed in VNP20009 (Lin et al. 1999; Karsten et al. 2001). Similarly, functional TNF-α has been cloned and expressed in *C. acetobutylicum*. Recently *Bifidobacterium adolescentis* has been used as a delivery system for the anti-angiogenic protein endostatin. Studies have reported a strong inhibition of angiogenesis and reduction in tumour growth after systemic administration of its spores via tail vein of tumour bearing mice (Li et al. 2003).

8.3.5.2 Bacterially Directed Enzyme Prodrug Therapy

Another lucrative approach to overcome the unacceptable side effects of bacterial therapy is the tumour selective activation of prodrugs. This strategy uses anaerobic bacteria that have been transformed with an enzyme that can convert a non-toxic prodrug into a toxic drug. In addition, as the bacteria proliferate in the necrotic and hypoxic areas of the tumour, the enzyme is expressed selectively in the tumour. Therefore the systemically administered prodrug gets metabolized to the toxic drug only in the tumour (Mengesha and Dubois 2009). For significant efficacy, both the prodrug and the activated drug must be able to cross biological membranes, because the prodrug will be activated within bacterial cells and the active drug will then need to enter the tumour cells. Several enzyme/prodrug systems are available. Cytosine deaminase (CD), which converts 5-fluorocytosine (5FC) to 5-fluorouracil (5FU), and nitroreductase (NR), which converts the prodrug CB1954 to a DNA cross-linking agent, have been tested with *Clostridium sporogenes*. Although these combinations can kill tumour cells *in vitro* and deliver high concentrations of enzymes to model tumours, to date, the results *in vivo* have been disappointing. Similarly, CD expressed in *Clostridium acetobutylicum* has demonstrated a selective delivery of the active exogenous enzyme into tumours (Theys et al. 2001; Liu et al. 2002). Recently, it was demonstrated that CD can be successfully cloned and expressed in the same strain of *Clostridium* and CD expression was enhanced significantly by the vascular targeting agent combretastatin A-4 phosphate. The enhancement may be due to the enlargement of the necrotic area in tumours (Theys et al. 2001).

Salmonella vector combined with NR and CD has shown successful results *in vivo* and both are currently undergoing phase I clinical trials in cancer patients. *Salmonella* has been combined with carboxypeptidase G2 (CPG2), an enzyme that converts a range of mustard prodrugs to DNA cross-linking agents and has shown high levels of activity *in vivo*. TAPET (Tumour Amplified Protein Expression Therapy) uses VNP20009, an attenuated strain of *S. typhimurium* as a bacterial vector and expresses an *E. coli* CD for preferentially delivering anticancer drugs to solid tumours (Luo et al. 2001). The expression of the prodrug converting enzyme HS-thymidine kinase (TK) in a purine auxotroph has demonstrated enhanced antitumour activity upon the addition of ganciclovir, the corresponding prodrug (Pawelek et al. 1997). Likewise the expression of HSV-TK in VNP20009 has demonstrated its selective accumulation in subcutaneously implanted murine colon tumours (Tjuvajev et al. 2001). *Bifidobacterium longum* transfected with pBLES100-S-eCD produces cytosine deaminase in the hypoxic tumour, and studies have confirmed this as an effective prodrug-enzyme therapy (Fujimori et al. 2003).

Recently a strain of *E. coli,* DH5α-lux/βG has been developed which expresses β-glucuronidase (βG) as well as the luxCDABE gene cluster enzymes for selective prodrug activation and non-invasive imaging in tumours respectively. These bacteria emit light for imaging and hydrolyze the glucuronide prodrug 9ACG to the topoisomerase I inhibitor 9-aminocamptothecin (9AC). Optical imaging, colony-forming units (CFUs) and staining for βG activity have indicated that this bacterial strain localizes and replicates in human tumour xenografts thereby producing antitumour activity along with systemic 9ACG prodrug therapy (Cheng et al. 2008). These results have generated interest in prodrug activating bacteria as useful selective cancer chemotherapy.

8.3.6 Bacteria as Immunotherapeutic Agents

Since tumours are immunogenic, the immunotherapeutic strategy employs stimulation of the immune system to destroy cancerous cells. Immunotherapy for cancer offers great promise as an emerging and effective approach. But the major drawback is the ability of tumours to escape the immune system due to development of tolerance as they are weakly immunogenic and sometimes body takes them as self antigens. Thus one of the novel immunotherapeutic strategies employs bacteria to enhance the antigenicity of tumour cells (Xu et al. 2009). It has been reported that attenuated but still invasive, *S. typhimurium* infects malignant cells both *in vitro* and *in vivo* which in turn triggers the immune response. Attenuated *S. typhimurium* has demonstrated successful invasion of melanoma cells that can present antigenic determinants of bacterial origin and become targets for anti-Salmonella-specific T cells. However, better outcomes were achieved after vaccinating tumour bearing mice with *S. typhimurium* before intratumoral *Salmonella* injection (Avogadri et al. 2005). Genetically engineered attenuated strains of *S. typhimurium* expressing murine cytokines have exhibited the capacity to modulate immunity to infection and have retarded the growth of experimental melanomas. Results have suggested that IL-2 encoding *Salmonella* organisms are superior in suppressing tumour growth as compared to the parental non-cytokine expressing strain (Al-Ramadi et al. 2008). Tumour antigen DNA sequences have been introduced into bacteria too such as *Salmonella* and *Listeria*, resulting in protective immunity in animal models. A xenogenic DNA vaccine encoding human tumour endothelial marker 8 (TEM8) carried by attenuated *S. typhimurium* has been reported to generate TEM8-specific CD8 cytotoxic T-cell response after oral administration. Suppression of angiogenesis in the tumours along with protection of mice from lethal challenges against tumour cells and reduced tumour growth support the potential of anti-angiogenesis immunotherapy (Ruan et al. 2009). Recently, a recombinant strain of attenuated *S. typhimurium* expressing a gene encoding LIGHT, a cytokine known to promote tumour rejection has been reported to inhibit growth of primary tumours, as well as the dissemination of pulmonary metastases, in various mouse tumour models employing murine carcinoma cell lines in immunocompetent mice. Antitumor activity was achieved without significant toxicity (Loeffler et al. 2007).

Because of its ability to stimulate strong innate and cell-mediated immunity, recombinant forms of the facultative intracellular bacterium, *Listeria monocytogenes*, have been used as vector for cancer vaccine. A recombinant *L. monocytogenes* vaccine strain (Lm-NP) expressing nucleoprotein (NP) from influenza strain A/PR8/34 has shown great therapeutic potential pre-clinically by regressing growth of macroscopic tumours of all types. Treatment with another recombinant Listeria strain Lm-LLO-E7 has demonstrated effective cure of the majority of tumour bearing mice. And clinical trials are currently underway for the use of Lm-LLO-E7 as a cancer immunotherapeutic for cervical cancer (Wood et al. 2008). An attenuated *L. monocytogenes* (LM)-based vaccine expressing truncated listeriolysin O (LLO) has demonstrated the eradication of all metastases and almost the entire primary tumour in the syngeneic, aggressive mouse breast tumour model 4T1 (Kim et al. 2009).

C. novyi has been reported to induce massive leukocytosis and inflammation. Furthermore, the antitumor effects of inflammation are well known too. Systemic administration of *C. novyi-NT* spores destroys adjacent cancer cells and triggers an inflammatory reaction by producing cytokines such as IL-6, MIP-2, G-CSF, TIMP-1, and KC that attract inflammatory cells i.e. neutrophils followed by monocyte and lymphocytes. The inflammatory reaction restricts the bacterial infection and directly contributes to the destruction of tumour cells through the production of reactive oxygen species, proteases, and other degradative enzymes. Finally, it stimulates a potent cellular immune response leading to destruction of residual tumour cells. A phase I clinical trial combining spores of a *C. novyi-NT* with an anti-microtubuli agent has been initiated (Xu et al. 2009).

The cell wall skeleton of *Mycobacterium bovis Bacillus Calmette-Guérin* (BCG-CWS) has been used as an effective adjuvant for immunotherapy of a variety of cancer patients (Hayashi et al. 2009). Recently it has been demonstrated that BCG/CWS has a radiosensitizing effect on colon cancer cells through the induction of autophagic cell death. *In vitro* as well as *in vivo* studies have revealed that BCG/CWS in combination with ionizing radiation (IR) is a promising therapeutic strategy for enhancing radiation therapy in colon cancer cells (Yuk et al. 2010). All these findings indicate the promising potential of non-virulent bacteria as cancer immunotherapeutic agents. A summary of relevant clinical trials using bacteria is shown in Table 8.1.

8.4 Limitations of Bacterial Therapy

The major limitation of bacterial therapy is toxicity. As cancer cells are generally insensitive, reducing the dose results in diminished efficacy and increasing the dose to achieve the therapeutic efficacy may produce toxicity to non-cancerous cells. Another challenge is the lack of specificity, which is being tried to overcome by bacterially directed enzyme prodrug therapy. Moreover, systemic administration of bacteria is rather inconvenient and carries higher risk of obvious toxicity. Furthermore, even removal of the toxin genes like in COBALT therapy led to ~15–45% mortality in mice (Dang et al. 2001). A more difficult problem is the treatment of small non-necrotic metastases of large primary tumours as metastasis is the major cause of mortality from cancer. Due to small hypoxic regions of these metastases, targeting by bacteria is difficult to achieve. Other problem is incomplete tumour lysis as bacteria do not consume all parts of the malignant tissue which necessitates the combination of therapy with chemotherapeutic treatments. In case of bacteria based vector therapy, the major hurdle is the inaccessibility because most of the times an intratumoral injection is required (Hatefi and Canine 2009). Another major challenge faced by bacterial therapy is the potential for DNA mutations i.e. any loss of functionality due to mutations may lead to wide variety of problems like failure of therapy or exaggerated infection. Although

Table 8.1 Important clinical studies involving bacterial intervention in cancer. Bacteria are well recognized as pathogenic or disease causing agents since their discovery. Scientists were working on the pathogenic properties rather than focusing on their beneficial effects. Decades back, it is surprised to abide by their beneficial effects in the treatment of cancer. Thereafter, number of preclinical studies have been done and found beneficial effects of anaerobic bacteria. Thereafter, several clinical trials have been done to justify their role in humans, they are as follows:

Author (year)	Intervention compounds[a]	Clinical study (type)	Cancer type	Outcome/purpose of study
Mussai et al. (2010)	HA22 (CAT-8015)	Cell culture	Acute lymphoblastic leukaemia	Provide a strong diagnostic rationale in children with drug-resistant ALL
King et al. (2009)	VNP20009 and TAPET-CD	Preclinical and Phase I	Solid tumour	Induced direct tumoricidal activity or to deliver tumoricidal agents directly to tumours
Sampson et al. (2008)	TP-38	Open trial	Malignant brain tumours/ Glioblastoma multiforme	Dose escalation study Potential efficacy of drugs
Vogelbaum et al. (2007)	Interleukin-13-PE38QQR	Open trial	Malignant gliomas	Assessed the safety of CB administered by convection Well tolerated in adults
Kunwar et al. (2007)	IL13-PE38QQR	Phase I clinical trial	Glioblastoma multiforme	Favorable risk-benefit profile
Powell et al. (2007)	LMB-2	Open trial	metastatic melanoma	CD25-directed immunotoxin to selectively mediate antitumor responses
Kunwar et al. (2006)	IL13-PE38QQR	Open trial	Malignant glioma	Symptomatic windows identified, particularly those in which chimeric cytotoxins are used
Shimamura et al. (2006)	IL4-PE and IL13-PE	In vitro (biopsy and cell culture)	Glioma	Demonstrated that the phenotype of these cytokine receptors on tumour cells is different from that found on normal immune cells
	IL13-PE	Phase I/II clinical trial Phase III	Glioblastoma multiforme	Effectiveness of intracranial administration by using convection-enhanced delivery (CED) Found good tolerance and Benefit/risk profile for treatment of patients with glioma Patients being monitored for safety, duration of overall survival, and quality of life

Kreitman et al. (2005)	RFB4(dsFv)-PE38 (BL22)	Phase I trial Phase II (underway)	B-cell malignancies (CD22+ non-Hodgkin's lymphoma, chronic lymphocytic leukemia, and hairy cell leukemia)	Well tolerated and highly effective in hairy cell leukemia
Garland et al. (2005)	NBI-3001	Phase I	Glioblastoma multiforme	NBI-3001 at 0.016 mg/m^2 was well tolerated
Parney et al. (2005)	Interleukin 13-PE38QQR	Phase I	Malignant glioma	Determined efficacy with surrogate end points and MR imaging
Sampson et al. (2003)	TP-38	Phase I	Malignant brain tumours	Patients survival increased
Weber et al. (2003)	NBI-3001	Open trial	Malignant glioma	Observed acceptable safety and toxicity
Posey et al. (2002)	BR96 sFv-PE40	Phase I	Advanced solid tumours	Determined maximal tolerated dose (MTD) with limited toxicity
Kreitman et al. (2001)	BL22	Dose-escalation	Chemotherapy-resistant hairy-cell leukemia	Efficacious in complete patients, resistant to chemotherapy
Cunningham and Nemunaitis (2001)	TAPET-CD, VNP20029	Phase I Clinical trial	Advanced or metastatic cancer	Applied direct intratumoral injection Potential antitumor activity with 5-FU
Rand et al. (2000)	IL-4(38–37)-PE38KDEL	Open trial	High-grade glioma	Safe without systemic toxicity
Kreitman et al. (2000)	LMB-2	Phase I	Hematologic malignancies	First recombinant immunotoxin efficacious in cancer
Puri (1999)	IL4-PE	Cell culture	Glioblastoma/malignant astrocytoma	Indicated that localized administration can produce nontoxic levels of IL4-PE that may have significant activity against astrocytoma
Kreitman et al. (1999)	LMB-2	Phase I	Refractory hairy cell leukemia	Proof of principal study which targeted therapy with recombinant Fv-containing proteins; found efficacious
Pai-Scherf et al. (1999)	erb-38	Phase I	Breast cancer and esophageal cancer patient	Unexpected organ toxicities (Hepatotoxicity)

(continued)

Table 8.1 (continued)

Author (year)	Intervention compounds[a]	Clinical study (type)	Cancer type	Outcome/purpose of study
Pai et al. (1996)	LMB-1	Phase I	Metastatic breast and colon cancer	First reported antitumor activity in epithelial tumours with vascular leak syndrome manifested by hypoalbuminemia, fluid retention, hypotension and, in one case, pulmonary edema
Goldberg et al. (1995)	TP40	Phase I	bladder cancer	Well-tolerated and efficacious in bladder cancer
Pai et al. (1991)	OVB3gPE	Dose escalating study	Ovarian cancer	No clinical antitumor responses were observed and have Neurologic toxicity and encephalopathy

[a]Refer list of abbreviations for details about compounds

recombinant DNA technology has solved some of these safety concerns yet more developments are needed.

8.5 Conclusion

Various strategies employing bacteria have been investigated so far as anti-cancer modalities. Of all these, live/attenuated bacteria as antitumor agents and vectors for gene-directed enzyme prodrug therapy have emerged as promising strategies. IL-4 fused with *Pseudomonas* exotoxin is in Phase I clinical trials in patients with glioblastoma. VNP20009 and TAPET-CD have been investigated successfully in Phase I clinical trials in cancer patients. Chimeric toxins are also being investigated as future toxin-based anticancer therapies. However more investigations and studies are needed to establish this therapy.

8.6 Future Directions

Recently it has been hypothesized that bacteria growing with cancer cells of certain types (like the metastatic cells) in the presence of anticancer drugs such as DNA replication inhibitors undergo the *SOS response* and lead to generation of novel beneficial phenotypes which acquire the capacity to invade the cancer cells in a bid to escape drug pressure. Thus, further research is required to implicate these phenotypes as novel bacterial anti-metastasis regimens. Although bacteria have shown promising and significant potency in eradicating established tumours found in pre-clinical mouse tumour models yet the successful translation of these pre-clinical strategies into clinical practice will depend on the outcome of clinical trials. Of all the strategies discussed above, anaerobic bacteria vector-mediated cancer therapy and immunotherapy are very promising. But as we know cancer is a multifactorial disease, no single therapy is completely suitable for it. The combination of recombinant DNA technology along with immunotherapy applied to the anaerobic bacteria is being considered as the foundation for the multimodality therapeutic strategies for cancer.

References

Abd El-Aal H, Habib E, Mishrif M (2005) Wilms' tumor: the experience of the pediatric unit of Kasr El-Aini Center of Radiation Oncology and Nuclear Medicine (NEMROCK). J Egypt Natl Canc Inst 17(4):308–311

Al-Ramadi BK, Fernandez-Cabezudo MJ, El-Hasasna H et al (2008) Attenuated bacteria as effectors in cancer immunotherapy. Ann N Y Acad Sci 1138(1):351–357

Ansiaux R, Gallez B (2007) Use of botulinum toxins in cancer therapy. Expert Opin Investig Drugs 16(2):209–218

Avogadri F, Martinoli C, Petrovska L et al (2005) Cancer immunotherapy based on killing of salmonella-infected tumor cells. Cancer Res 65(9):3920–3927

Ballard TE, Melander C (2008) Kinamycin-mediated DNA cleavage under biomimetic conditions. Tetrahedron Lett 49:3157

Bermudes D, Zheng L, King IC (2002) Live bacteria as anticancer agents and tumor-selective protein delivery vectors. Curr Opin Drug Discov Devel 5(2):194–199

Bettegowda C, Dang LH, Abrams R et al (2003) Overcoming the hypoxic barrier to radiation therapy with anaerobic bacteria. Proc Natl Acad Sci USA 100(25):15083–15088

Carey R, Holland J, Whang H et al (1967) Clostridial oncolysis in man. Eur J Cancer 3:37–46

Carswell EA, Old LJ, Kassel RL et al (1975) An endotoxin induced serum factor that causes necrosis of tumors. Proc Natl Acad Sci USA 72:3666–3670

Cheng CM, Lu YL, Chuang KH et al (2008) Tumor-targeting prodrug-activating bacteria for cancer therapy. Cancer Gen Ther 15:393–401

Cheong I, Huang X, Bettegowda C et al (2006) A bacterial protein enhances the release and efficacy of liposomal cancer drugs. Science 314(5803):1308–1311

Cunningham C, Nemunaitis J (2001) A phase I trial of genetically modified *Salmonella typhimurium* expressing cytosine deaminase (TAPET-CD, VNP20029) administered by intratumoral injection in combination with 5-fluorocytosine for patients with advanced or metastatic cancer. Protocol no: CL-017. Version: April 9, 2001. Hum Gene Ther 12(12):1594–1596

Dang LH, Bettegowda C, Huso DL et al (2001) Combination bacteriolytic therapy for the treatment of experimental tumors. Proc Natl Acad Sci USA 98(26):15155–15160

Dang LH, Bettegowda C, Agrawal N et al (2004) Targeting vascular and avascular compartments of tumors with *C. novyi*-NT and antimicrotubule agents. Cancer Biol Ther 3(3):326–337

Diaz LA Jr, Cheong I, Foss CA et al (2005) Pharmacologic and toxicologic evaluation of *C.novyi*-NT spores. Toxicol Sci 88(2):562–575

Falnes PO, Ariansen S, Sandwig K et al (2000) Requirement for prolonged action in the cytosol for optimal protein synthesis inhibition by diphtheria toxin. J Biol Chem 275:4363–4368

Fan D, Yano S, Shinohara H et al (2002) Targeted therapy against human lung cancer in nude mice by high affinity recombinant antimesothelin single chain Fv immunotoxin. Mol Cancer Ther 1:595–600

Fiorentini C, Matarrese P, Straface E et al (1998) Toxin induced activation of Rho GTP-binding protein increases Bcl-2 expression and influences mitochondrial homeostasis. Exp Cell Res 242:341–350

Frankel AE, Rossi P, Kuzel TM et al (2002) Diphtheria fusion protein therapy of chemoresistant malignancies. Curr Cancer Drug Targets 2:19–36

Fujimori M, Amano J, Taniguchi S (2003) The genus Bifidobacterium for cancer gene therapy. Curr Opin Drug Discov Devel 5:200–203

Garland L, Gitlitz B, Ebbinghaus S et al (2005) Phase I trial of intravenous IL-4 pseudomonas exotoxin protein (NBI-3001) in patients with advanced solid tumors that express the IL-4 receptor. J Immunother 28(4):376–381

Goldberg MR, Heimbrook DC, Russo P et al (1995) Phase I clinical study of the recombinant oncotoxin TP40 in superficial bladder cancer. Clin Cancer Res 1(1):57–61

Greenfield L, Johnson VG, Youle RJ (1987) Mutations in diphtheria toxin separate binding from entry and amplify immunotoxin selectivity. Science 238:536–539

Hagihara N, Walbridge S, Olson AW et al (2000) Vascular protection by chloroquine during brain tumor therapy with Tf-CRM 107. Cancer Res 60:230–234

Hatefi A, Canine BF (2009) Perspectives in vector development for systemic cancer gene therapy. Gene Ther Mol Biol 13(A):15–19

Hayashi A, Nishida Y, Yoshii S et al (2009) Immunotherapy of ovarian cancer with cell wall skeleton of *Mycobacterium bovis* Bacillus Calmette-Guérin: effect of lymphadenectomy. Cancer Sci 100(10):1991–1995

Hoption Cann SA, van Netten JP, van Netten C (2003) Dr. William Coley and tumour regression: a place in history or in the future. Postgrad Med J 79:672–680

Hough CD, Sherman Baust CA, Pizer ES (2000) Large scale serial analysis of gene expression reveals genes differentially expressed in ovarian cancer. Cancer Res 60:6281–6287

Jain RK (2001) New approaches for the treatment of cancer. Adv Drug Deliv Rev 46:149–168

Julien B, Shah S (2002) Heterologous expression of epothilone biosynthetic genes in *Myxococcus xanthus*. Antimicrob Agents Chemother 46(9):2772

Karsten V, Pike J, Troy K et al (2001) A strain of *Salmonella typhimurium* VNP20009 expressing an anti-angiogenic peptide from platelet factor-4 has enhanced anti-tumor activity. Proc Annu Meet Am Assoc Cancer Res 42:3700

Keller R, Keist R, Joller P et al (1995) Coordinate up and down modulation of inducible nitric oxide synthase, nitric oxide production and tumoricidal activity in bone marrow derived mononuclear phagocytes by lipopolysaccharides and gram negative bacteria. Biochem Biophys Res Commun 211:183–189

Khatua S, Nair C, Ghosh K (2004) Immune-mediated thrombocytopenia following dactinomycin therapy in a child with alveolar rhabdomyosarcoma: the unresolved issues. J Pediatr Hematol Oncol 26(11):777–779

Kim SH, Castro F, Paterson Y et al (2009) High efficacy of a Listeria based vaccine against metastatic breast cancer reveals a dual mode of action. Cancer Res 69(14):5860–5866

King I, Itterson M, Bermudes D (2009) Tumor-targeted *Salmonella typhimurium* overexpressing cytosine deaminase: a novel, tumor-selective therapy. Methods Mol Biol 542:649–659

Kokai Kun JF, Mcclane BA (1997) Determination of functional regions of *Clostridium perfringens* enterotoxin through deletion analysis. Clin Infect Dis 25:S165–S167

Kokai Kun JF, Benton K, Wieckowski EU et al (1999) Identification of a *Clostridium perfringens* enterotoxin region required for large complex formation and cytotoxicity by random mutagenesis. Infect Immun 67:5634–5641

Kominsky SL, Vali M, Korz D (2004) *Clostridium perfringens* enterotoxin elicits rapid and specific cytolysis of breast carcinoma cells mediated through tight junction proteins claudin 3 and 4. Am J Pathol 164:1627–1633

Kreitman RJ, Wilson WH, Robbins D et al (1999) Responses in refractory hairy cell leukemia to a recombinant immunotoxin. Blood 94(10):3340–3348

Kreitman RJ, Wilson WH, White JD et al (2000) Phase I trial of recombinant immunotoxin anti-Tac(Fv)-PE38 (LMB-2) in patients with hematologic malignancies. J Clin Oncol 18(8):1622–1636

Kreitman RJ, Wilson WH, Bergeron K et al (2001) Efficacy of the anti-CD22 recombinant immunotoxin BL22 in chemotherapy-resistant hairy-cell leukemia. N Engl J Med 345(4):241–247

Kreitman RJ, Squires DR, Stetler-Stevenson M et al (2005) Phase I trial of recombinant immunotoxin RFB4(dsFv)-PE38 (BL22) in patients with B-cell malignancies. J Clin Oncol 23(27):6719–6729

Kunwar S, Chang SM, Prados MD et al (2006) Safety of intraparenchymal convection-enhanced delivery of cintredekin besudotox in early-phase studies. Neurosurg Focus 20(4):E15

Kunwar S, Prados MD, Chang SM, Cintredekin Besudotox Intraparenchymal Study Group et al (2007) Direct intracerebral delivery of cintredekin besudotox (IL13-PE38QQR) in recurrent malignant glioma: a report by the Cintredekin Besudotox Intraparenchymal Study Group. J Clin Oncol 25(7):837–844

Lanzerin M, Sand O, Olsnes S (1996) GPI-anchored diphtheria toxin receptor allows membrane translocation of the toxin without detectable ion channel activity. EMBO J 15:725–734

Li X, Fu GF, Fan YR et al (2003) *Bifidobacterium adolescentis* as a delivery system of endostatin for cancer gene therapy: selective inhibitor of angiogenesis and hypoxic tumor growth. Cancer Gene Ther 10:105–111

Lin SL, Spinka TL, Le TX et al (1999) Tumor directed delivery and amplification of tumor-necrosis factor-α (TNF) by attenuated *Salmonella typhimurium*. Clin Cancer Res 5:3822

Liong MT (2008) Roles of probiotics and prebiotics in colon cancer prevention postulated mechanisms and in-vivo evidence. Int J Mol Sci 9(5):854–863

Liu SC, Minton NP, Giaccia AJ et al (2002) Anticancer efficacy of systemically delivered anaerobic bacteria as gene therapy vectors targeting tumor hypoxia/necrosis. Gene Ther 9:291–296
Loeffler M, Le'Negrate G, Krajewska M et al (2007) Attenuated Salmonella engineered to produce human cytokine LIGHT inhibit tumor growth. Proc Natl Acad Sci USA 104(31):12879–12883
Louie GV, Yang W, Bowman ME et al (1997) Crystal structure of the complex of diphtheria toxin with an extracellular fragment of its receptor. Mol Cell 1:67–68
Low KB, Ittensohn M, Lin S et al (1999) VNP20009, a genetically modified *Salmonella typhimurium* for treatment of solid tumors. Proc Am Assoc Cancer Res 40:851
Luo X, Ittensohn M, Low B et al (1999) Genetically modified *Salmonella typhimurium* inhibited growth of primary tumors and metastase. Proc Annu Meet Am Assoc Cancer Res 40
Luo X, Li Z, Shen SY et al (2001) Genetically armed *Salmonella typhimurium* delivered therapeutic gene and inhibited tumor growth in preclinical models. Proc Annu Meet Am Assoc Cancer Res 42
Malmgren RA, Flanigan CC (1955) Localization of the vegetative form of *Clostridium tetani* in mouse tumors following intravenous spore administration. Cancer Res 15:473–478
Mengesha A, Dubois L (2009) Clostridia in anti-tumor therapy. In: Bruggemann H, Gottschalk G (eds) Clostridia: molecular biology in the post-genomic era, 3rd edn. Caister Academic Press, Norfolk
Michl P, Buchholz M, Rolke M (2001) Claudin-4: a new target for pancreatic cancer treatment using *Clostridium perfringens* enterotoxin. Gasrtoenterology 121:678–684
Minton NP (2003) Clostridia in cancer therapy. Nat Rev Microbiol 1:237–242
Mussai F, Campana D, Bhojwani D et al (2010) Cytotoxicity of the anti-CD22 immunotoxin HA22 (CAT-8015) against paediatric acute lymphoblastic leukaemia. Br J Haematol 150(3):352–358
Nauts HC (1980) The beneficial effects of bacterial infections on host resistance to cancer: end result in 449 cases, 2nd edn, Monograph no. 8. Cancer research institute, New York
Nauts HC, McLaren JR (1990) Coley's toxins the first century. Adv Exp Med Biol 267:483–500
Nauts HC, Fowler G, Bogatko F (1953) A review of the influence of bacterial infection and of bacterial products (Coley's toxins) on malignant tumors in man. Acta Med Scand 276:1–103
Nougayrede JP, Taieb F, De Rycke J et al (2005) Cyclomodulins: bacterial effectors that modulate the eukaryotic cell cycle. Trends Microbiol 13:103–110
Oswald E, Sugai M, Labigne A et al (1994) Cytotoxic necrotizing factor type 2 produced by virulent *Escherichia coli* modifies the small GTP-binding proteins Rho involved in assembly of actin stress fibers. Proc Natl Acad Sci USA 91:3814–3818
Pai LH, Bookman MA, Ozols RF et al (1991) Clinical evaluation of intraperitoneal Pseudomonas exotoxin immunoconjugate OVB3-PE in patients with ovarian cancer. J Clin Oncol 9(12):2095–2103
Pai LH, Wittes R, Setser A et al (1996) Treatment of advanced solid tumors with immunotoxin LMB-1: an antibody linked to Pseudomonas exotoxin. Nat Med 2(3):350–353
Pai-Scherf LH, Villa J, Pearson D et al (1999) Hepatotoxicity in cancer patients receiving erb-38, a recombinant immunotoxin that targets the erbB2 receptor. Clin Cancer Res 5(9):2311–2315
Parney IF, Kunwar S, McDermott M et al (2005) Neuroradiographic changes following convection-enhanced delivery of the recombinant cytotoxin interleukin 13-PE38QQR for recurrent malignant glioma. J Neurosurg 102(2):267–275
Pastan I (1997) Targeted therapy of cancer with recombinant immunotoxins. Biochim Biophys Acta 1333:C1–C6
Pawelek JM, Low KB, Bermudes D (1997) Tumor-targeted Salmonella as a novel anticancer vector. Cancer Res 57:4537–4544
Posey JA, Khazaeli MB, Bookman MA et al (2002) A phase I trial of the single-chain immunotoxin SGN-10 (BR96 sFv-PE40) in patients with advanced solid tumors. Clin Cancer Res 8(10):3092–3099
Powell DJ Jr, Felipe-Silva A, Merino MJ et al (2007) Administration of a CD25-directed immunotoxin, LMB-2, to patients with metastatic melanoma induces a selective partial reduction in regulatory T cells *in vivo*. J Immunol 179(7):4919–4928
Prakash V, Winston WT (2010) Update on options for treatment of metastatic castration-resistant prostate cancer. Oncol Targets Ther 3:39–51

Punj V, Bhattacharya S, Saint-Dic D et al (2004) Bacterial cupredoxin azurin as an inducer of apoptosis and regression in human breast cancer. Oncogene 13:2362–2374

Puri RK (1999) Development of a recombinant interleukin-4-Pseudomonas exotoxin for therapy of glioblastoma. Toxicol Pathol 27(1):53–57

Rafter J (2004) The effects of probiotics on colon cancer development. Nutr Res Rev 17:277–284

Rand RW, Kreitman RJ, Patronas N et al (2000) Intratumoral administration of recombinant circularly permuted interleukin-4-Pseudomonas exotoxin in patients with high-grade glioma. Clin Cancer Res 6(6):2157–2165

Reisser O, Pance A, Jeanin JF (2002) Mechanism of anti-tumor effect of lipid A. Bioassays 24:284–289

Renault J, Baron M, Mailliet P et al (1981) Heterocyclic quinones.2.Quinoxaline-5,6-(and 5–8)-diones – Potential antitumoral agents. Eur J Med Chem 16(6):545–550

Ribi EE, Granger DL, Milner KC et al (1975) Tumor regression caused by endotoxins and mycobacterial fractions. J Natl Cancer Inst 55:1253–1257

Richardson MA, Ramirez T, Russell NC et al (1999) Coley toxins immunotherapy: a retrospective review. Altern Ther Health Med 5:42–47

Roller M, Clune Y, Collins K et al (2007) Consumption of prebiotic inulin enriched with oligofructose in combination with the probiotics *Lactobacillus rhamnosus* and *Bifidobacterium lactis* has minor effects on selected immune parameters in polypectomised and colon cancer patients. Br J Nutr 97(4):676–684

Ruan Z, Yang Z, Wang Y et al (2009) DNA vaccine against tumor endothelial marker 8 inhibits tumor angiogenesis and growth. J Immunother 32(5):486–491

Saltzman DA, Heise CP, Hasz DE et al (1996) Attenuated *Salmonella typhimurium* containing interleukin – 2 decreases MC-38 hepatic metastases: a novel anti-tumor agent. Cancer Biother Radiopharm 11:145–153

Saltzman DA, Katsanis E, Heise CP et al (1997) Patterns of hepatic and splenic colonization by an attenuated strain of *Salmonella typhimurium* containing the gene for human interleukin-2: a novel anti-tumor agent. Cancer Biother Radiopharm 12:37–45

Sampson JH, Akabani G, Archer GE et al (2003) Progress report of a Phase I study of the intracerebral microinfusion of a recombinant chimeric protein composed of transforming growth factor (TGF)-alpha and a mutated form of the Pseudomonas exotoxin termed PE-38 (TP-38) for the treatment of malignant brain tumors. J Neurooncol 65(1):27–35

Sampson JH, Akabani G, Archer GE et al (2008) Intracerebral infusion of an EGFR-targeted toxin in recurrent malignant brain tumors. Neuro Oncol 10(3):320–329

Seow SW, Cai S, Rahmat JN et al (2010) *Lactobacillus rhamnosus* GG induces tumor regression in mice bearing orthotopic bladder tumors. Cancer Sci 101(3):751–758

Shimamura T, Husain SR, Puri RK (2006) The IL-4 and IL-13 Pseudomonas exotoxins: new hope for brain tumor therapy. Neurosurg Focus 20(4):E11

Takimoto CH, Calvo E (2008) Principles of oncologic pharmacotherapy. In: Pazdur R, Wagman LD, Camphausen KA et al (eds) Cancer management: a multidisciplinary approach, 11th edn. Cmp United Business Media, New York

Theys J, Landuyt W, Nuyts S et al (2001) Specific targeting of cytosine deaminase to solid tumors by engineered *Clostridium acetobutylicum*. Cancer Gene Ther 8:294–297

Thiele E, Arison R, Boxer G (1963) Oncolysis by Clostridia IV effect of nonpathogenic Clostridial spores in normal and pathological tissues. Cancer Res 24:234–238

Tjuvajev J, Blasberg R, Luo X et al (2001) Salmonella based tumor-targeted cancer therapy: tumor amplified protein expression therapy (TAPET) for diagnostic imaging. J Control Release 74:313–315

Turan T, Karacay O, Tulunay G et al (2006) Results with EMA/CO (etoposide, methotrexate, actinomycin D, cyclophosphamide, vincristine) chemotherapy in gestational trophoblastic neoplasia. Int J Gynecol Cancer 16(3):1432–1438

Vogelbaum MA, Sampson JH, Kunwar S et al (2007) Convection-enhanced delivery of cintredekin besudotox (interleukin-13-PE38QQR) followed by radiation therapy with and without temozolomide in newly diagnosed malignant gliomas: phase 1 study of final safety results. Neurosurgery 61(5):1031–1037

Weber FW, Floeth F, Asher A et al (2003) Local convection enhanced delivery of IL4-Pseudomonas exotoxin (NBI-3001) for treatment of patients with recurrent malignant glioma. Acta Neurochir Suppl 88:93–103

Wei MQ, Ellem KAO, Dunn P et al (2007) Facultative or obligate anaerobic bacteria have the potential for multimodality therapy of solid tumours. Eur J Cancer 43:490–496

Wood LM, Guirnalda PD, Seavey MM et al (2008) Cancer immunotherapy using *Listeria monocytogenes* and listerial virulence factors. Immunol Res 42:233–245

Xu J, Liu XS, Zhou SF et al (2009) Combination of immunotherapy with anaerobic bacteria for immunogene therapy of solid tumours. Genet Ther Mol Biol 13:36–52

Yuhua L, Kunyuan G, Hui C et al (2001) Oral cytokine gene therapy against murine tumor using attenuated *Salmonella typhimurium*. Int J Cancer 94:438–443

Yuk JM, Shin DM, Song KS et al (2010) Bacillus calmette-guerin cell wall cytoskeleton enhances colon cancer radiosensitivity through autophagy. Autophagy 6(1):46–60

Zacharski LR, Sukhatme VP (2005) Coley's toxin revisited: immunotherapy or plasminogen activator therapy of cancer? J Thromb Haemost 3:424

Chapter 9
Targeting Cancer with Amino-Acid Auxotroph *Salmonella typhimurium* A1-R

Robert M. Hoffman

Abstract *Salmonella, Clostridium* and *Bifidobacterium* have been shown to control tumour growth and promote survival in animal models. However, *Clostridium* and *Bifidobacterium* are obligate anaerobes which limits their growth to the necrotic region of tumour and thus limits their effectiveness. In contrast, *Salmonella* is a facultative anaerobe which can grow in the viable as well as necrotic regions of tumours giving it greater potential as an anti-tumour agent. However, previous experiments with *Salmonella*, including clinical trials, have used over-attenuated mutants, limiting the anti-tumour efficacy. We have developed an effective bacterial cancer therapy strategy using *Salmonella typhimurium* auxotrophs which grow in viable as well as necrotic areas of tumours. The auxotrophy severely restricts growth of these bacteria in normal tissue. The *S. typhimurium* A1-R mutant, which is auxotrophic for leu-arg, has high anti-tumour virulence. *In vitro*, A1-R infects tumour cells and causes nuclear destruction. A1-R was initially used to treat metastatic human prostate and breast tumours that had been orthotopically implanted in nude mice. Forty percent of treated mice were cured completely and survived as long as non-tumour-bearing mice. A1-R administered i.v. to nude mice with primary osteosarcoma and lung metastasis was highly effective, especially against metastasis. A1-R was also targeted to both axillary lymph and popliteal lymph node metastasis of human pancreatic cancer and fibrosarcoma, respectively, as well as lung metastasis of the fibrosarcoma in nude mice. The bacteria were delivered via a lymphatic channel to target the lymph node metastases and systemically via the tail vein to target the lung metastasis. The metastases were cured without the need of chemotherapy or any other treatment. A1-R was administered intratumorally to nude mice with an orthotopically

R.M. Hoffman (✉)
AntiCancer, Inc., 7917 Ostrow Street, San Diego, CA 92111, USA

Department of Surgery, University of California San Diego,
200 West Arbor Drive, San Diego, CA 92111-8220, USA
e-mail: all@anticancer.com

A.A. Khan (ed.), *Bacteria and Cancer*, DOI 10.1007/978-94-007-2585-0_9,

transplanted human pancreatic tumour. The primary pancreatic cancer regressed without additional chemotherapy or any other treatment. A1-R was also effective against pancreatic cancer liver metastasis when administered intrasplenically to nude mice. A1-R was also highly effective against spinal cord glioma in orthotopic nude mouse models, a highly treatment-resistant disease. Tumour vascularity positively correlates with susceptibility to A1-R therapy. Substrains with tumour-specific promoters and mutants which enhance selective tumour targeting have been identified. The approach described here, where bacterial monotherapy effectively treats primary and metastatic tumours, is a significant improvement over previous bacterial tumour therapy strategies that require combination with toxic chemotherapy.

Keywords *S. typhimurium* • Leucine-arginine auxotrophs • Cancer therapy • Green fluorescent protein • Red fluorescent protein • Fluorescence • Imaging • Mice • Nude mice • Tumour targeting

Abbreviations

5-FU	5-fluorouracil
FACS	Fluorescence-activated cell sorting
FNR	Fumarate and nitrate reduction global regulator
GFP	Green fluorescent protein
HIP-1	Hypoxia-inducible promoter
LLC-RFP	Lewis lung carcinoma expressing red fluorescence protein
ND-GFP	Nestin-driven GFP
NTG	Nitrosoguanidine
RFP	Red fluorescent protein
VEGF	Vascular endothelial growth factor

9.1 Introduction

Recently Forbes (Forbes 2010) has outlined advantages bacteria have over other types of cancer treatment. Bacteria have flagella that enable tumour penetration (Dang et al. 2001) and chemotactic receptors that direct them to tumours by sensing molecules in the tumour microenvironment (Kasinskas and Forbes 2006, 2007). For example, the TAR receptor detects aspartate secreted by viable cancer cells, and the TRG receptor promotes migration towards ribose in necrotic tissue (Kasinskas and Forbes 2007; Forbes 2010). Since bacteria can migrate far from the vasculature, bacteria can penetrate tumours to a greater degree than small molecules that diffuse only passively (Forbes 2010). Forbes described five potential mechanisms that influence the accumulation of facultative anaerobes in tumours: entrapment of bacteria in tumours vasculature (Forbes et al. 2003); bacterial invasion of tumours following inflammation (Leschner et al. 2009); chemotaxis towards compounds produced by tumours (Kasinskas and Forbes 2006, 2007); preferential growth in tumour-specific microenvironments (Kasinskas and Forbes 2006; Zhao et al. 2005); and protection from clearance by the immune system (Sznol et al. 2000; Forbes 2010). *Salmonella* identify, penetrate tumours and cancer cells themselves, and by sensing and migrating towards small molecule gradients of serine, aspartate and ribose (Kasinskas and Forbes 2006, 2007). The growth rate of *Salmonella* is greater in tumours when dying cells are present (Kasinskas and Forbes 2006; Forbes et al. 2003; Leschner et al. 2009; Low et al. 1999). We have shown increased tumour specificity of auxotrophic *Salmonella* that require leucine and arganine, which are nutrients derived from dying tumour tissue (Zhao et al. 2005, 2006; Forbes 2010), which is the theme of the present review. Depleting host neutrophils increases and enables bacterial invasion of viable tumour tissue (Westphal et al. 2008). The ease of genetically manipulating bacteria gives it, its greatest potential for cancer therapy (Forbes 2010).

Bacterial therapy of cancer has a long anecdotal history going back to the early nineteenth century. Coley (1906) was among the first to put bacteria treatment of cancer on a scientific basis. He observed, more than a century ago, that some cancer patients were cured of their tumours following post-operative bacterial infection. Coley treated cancer patients with bacteria, and later treated cancer patients with extracts of bacteria called Coley's Toxins, with significant success. However, after Coley's death, Coley's Toxins fell out of favor. In the middle of the last century, Malmgren and Flanigan (1955) showed that anaerobic bacteria had the ability to survive and replicate in necrotic tumour tissue with low oxygen content. Several approaches aimed at utilizing bacteria for cancer therapy have subsequently been described (Gericke and Engelbart 1964; Moese and Moese 1964; Thiele et al. 1964; Kohwi et al. 1978; Kimura et al. 1980; Fox et al. 1996; Lemmon et al. 1997; Brown and Giaccia 1998; Low et al. 1999; Clairmont et al. 2000; Sznol et al. 2000; Yazawa et al. 2000, 2001).

Bifidobacterium longum has been shown to selectively grow in hypoxic regions of tumours following intravenous administration and has been shown to be an effective antitumour agent in combination with drug treatment by delivering the cysteine deaminase gene whose product converts the low-toxicity agent fluorocytosine to 5-fluorouracil (5-FU) an often effective cancer drug (Taniguchi et al. 2010).

Vogelstein et al. created a strain of *Clostridium novyi*, an obligate anaerobe, which was depleted of its lethal toxin (Dang et al. 2001). This strain of *C. novyi* was termed *C. novyi* NT. Following intravenous administration, the *C. novyi* NT spores germinated in the avascular regions of tumours in mice, causing damage to the surrounding viable tumour (Dang et al. 2001). Combined with conventional chemotherapy or radiotherapy, intravenous *C. novyi* NT spores caused extensive tumour damage within 24 h (Dang et al. 2001).

However, *Clostridium* and *Bifidobacterium* are obligate anaerobes which limits their growth to the necrotic region of tumour and thus limits their effectiveness. In contrast, *Salmonella* is a facultative anaerobe which can grow in the viable as well as necrotic regions of tumours giving it greater potential as an anti-tumour agent. However, previous experiments with *Salmonella*, including clinical trials, have used over-attenuated mutants, limiting the anti-tumour efficacy.

Following attenuation by purine and other auxotrophic mutations, the facultative anaerobe *Salmonella typhimurium* was used for cancer therapy (Low et al. 1999; Hoiseth and Stocker 1981; Pawelek et al. 1997). These genetically modified bacteria replicated in tumours to levels more than 1,000-fold greater than in normal tissue (Low et al. 1999). *S. typhimurium* was further modified genetically by disrupting the msbB gene to reduce the incidence of septic shock (Low et al. 1999).

The *msbB* mutant of *S. typhimurium* has been tested in a Phase I clinical trial to determine its efficacy on metastatic melanoma (Toso et al. 2002). To raise the therapeutic index, *S. typhimurium* was further attenuated by deletion of the *purI* as well as *msbB* gene. The new strain of *S. typhimurium*, termed VNP20009, could then be safely administered to patients (Toso et al. 2002*)*. More studies are needed to completely characterize the safety and efficacy of these bacteria and to improve its therapeutic index.

Mengesha et al. (2006) utilized *S. typhimurium* as a vector for gene delivery by developing a hypoxia-inducible promoter (HIP-1) to limit gene expression to hypoxic tumours. HIP-1 was able to drive gene expression in bacteria infecting human tumour xenografts implanted in mice. Genes linked to the HIP-1 promoter showed selective expression in tumours.

Yu et al. (2003, 2004) used green fluorescent protein (GFP)-labeled bacteria to visualize tumour targeting abilities of three pathogens: *Vibrio cholerae*, *S. typhimurium* and *Listeria monocytogenes.* Nguyen et al. (2010) targeted tumours with *S. typhimurium* engineered to express luciferase and cytotoxic genes that are induced by arabinose. However, more effective bacteria are needed for cancer therapy.

9.2 Development of Tumour-Targeting Amino Acid Auxotrophic Strain A1 of *S. typhimurium*

We initially developed a strain of *S. typhimurium*, termed A1, which selectively grew in tumour xenografts (Zhao et al. 2005). In contrast, normal tissue rapidly cleared infecting bacteria, even in immunodeficient athymic mice. *S. typhimurium* A1 is auxotrophic (leu/arg dependent), but receives sufficient support from tumour tissue. The steps used to develop *S. typhimurium* A1 are described below:

9.2.1 GFP Transfection and Stable Expression in S. typhimurium

Initially *S. typhimurium* 14028 was transfected with the pGFP gene by electroporation. The transformed *S. typhimurium* expressed GFP over 100 passages with GFP expression monitored at each passage (Zhao et al. 2005).

9.2.2 Cancer Cell Killing by S. typhimurium In Vitro

To observe the intracellular replication and virulence of *S. typhimurium*-GFP in a human prostate cancer cell line *in vitro*, PC-3 human prostate cancer cells were labeled with retroviral red fluorescent protein (RFP) in the cytoplasm, and with GFP in the nucleus by means of a fusion of GFP with histone H2B. This has allowed the interaction between bacteria and cancer cells to be visualized by dual color spatial-temporal imaging. The quantitative ability of *S. typhimurium* to kill prostate cancer cells was determined with the MTT method and observed to be dose dependent (Zhao et al. 2005).

9.2.3 Mutation, Isolation and Identification of Auxotrophs of S. typhimurium-GFP

S. typhimurium-GFP auxotrophic strains were obtained after nitrosoguanidine (NTG) mutagenesis. Twelve of 300 isolates tested were identified as auxotrophic mutants (4%) using minimal medium supplemented with various amino acids. Nude mice were inoculated i.v. with 10^7 cfu of each mutant. After inoculation with wild-type *S. typhimurium*, the mice died within 2 days. The mice which lived longest were those inoculated with auxotroph A1 and survived as long as control uninfected mice. A1 required leu and arg and was chosen for efficacy studies (Zhao et al. 2005).

9.3 Efficacy of *S. typhimurium* Amino Acid Auxotrophs on Cancer

9.3.1 In Vivo Efficacy Testing of S. typhimurium A1

To observe the interaction of prostate cancer cells with bacteria, we used a PC-3 human prostate cancer cell line expressing RFP, so their response to the bacteria could be visualized *in vivo* (Zhao et al. 2005).

To evaluate the efficacy of *S. typhimurium* A1, ten NCR nude mice, 6–8 weeks old, were implanted subcutaneously (s.c.) on the mid-right side with 2×10^6 RFP-labeled PC-3 human prostate cancer cells. Bacteria were grown and harvested at late-log phase and then diluted in PBS and injected directly into the tail vein (5×10^7 cfu/100 μl PBS). Tumour size was determined from fluorescence imaging at each time point after infection. *S. typhimurium* A1 selectively colonized the PC-3 tumour and suppressed its growth.

9.3.2 Isolation of High-Tumour-Virulence Variant *S. typhimurium A1*

To enhance tumour virulence, *S. typhimurium* GFP A1 was passaged by injection in nude mice transplanted with the HT-29 human colon tumour. Bacteria, expressing GFP, isolated from the infected tumour were then cultured. The re-isolated A1 was termed A1-R. The ability of A1-R to adhere to tumour cells was evaluated in comparison with the parental A1 strain *in vitro*. The number of A1-R bacteria attached to HT-29 human colon cancer cells was approximately six times higher than parental A1 (Zhao et al. 2006).

9.3.3 Enhanced Tumour Virulence of *S. typhimurium* A1-R in PC-3 Human Prostate Cancer Cells In Vitro

Virulence of GFP-labeled *S. typhimurium* A1 and A1-R bacteria was compared *in vitro* under fluorescence microscopy. Both strains infected dual color PC-3 cancer cells expressing RFP in the cytoplasm and GFP in the nucleus. Whereas almost all cells were infected and died after 2 h with A1-R, it took 24 h to obtain the same result with A1. Thus, the tumour virulence of A1-R was greatly increased (Zhao et al. 2007).

9.3.4 Enhanced Tumour Targeting of *S. typhimurium* A1-R in Nude Mice with the PC-3 Tumour

GFP-labeled *S. typhimurium* A1 and A1-R (5×10^7 cfu/100 μl) were administered (i.v.) to nude mice with the human PC-3 prostate tumour. The biodistribution of the bacteria in tumour tissue was determined at day 4. A1-R had 100× greater cfu in PC-3 tumour tissue than A1. This result suggested that A1-R has greater tumour targeting efficacy than A1 (Zhao et al. 2007).

9.3.5 Efficacy of A1-R in Orthotopic Metastatic Human Prostate Tumour Models

A1-R was used to treat metastatic PC-3 human prostate tumours that had been orthotopically implanted in nude mice. A1-R could eradicate tumours in the orthotopic nude mouse models of PC-3. Of ten mice with the PC-3 tumours that were injected weekly with A1-R, seven were alive at the time the last untreated mouse died. Four of the tumour-bearing mice were apparently cured by weekly bacterial treatment.

9.3.6 Dose Response of A1-R

Weekly dosing of A1-R was much more effective than two doses only (Zhao et al. 2007). In contrast to 12 weekly doses, with only two doses, only one of ten mice was cured (Zhao et al. 2007).

9.3.7 Comparison of Biodistribution Between S. typhimurium Strains A1 and A1-R in Normal Tissue In Vivo

GFP-labeled *S. typhimurium* A1 and A1-R bacteria (5×10^7 cfu/100 μl) were administered i.v. to nude mice with the PC-3 tumour. The biodistribution of the bacteria in liver and spleen tissue was determined at day 2. Fewer A1-R cfu (6×10^6) were recovered from the liver than A1 (4×10^7) indicating greater tumour selectivity of A1-R (Zhao et al. 2007).

9.3.8 Tumour Targeting by A1-R

To compare bacterial infection in the tumour with infection in normal tissue, A1-R bacteria (5×10^7 cfu/100 μl) were administered i.v. in PC-3-bearing nude mice. On day 4 after injection, the tumour, liver and spleen were removed. The tissues were homogenized and plated on LB agar plates. After overnight growth at 37°C, the cfu were counted. The ratio of tumour to normal tissue was approximately 10^6, indicating a very high degree of tumour targeting by A1-R (Zhao et al. 2007).

9.3.9 Efficacy of A1-R on Breast Tumour Growth

Treatment with A1-R resulted in significant tumour shrinkage in nude mice with s.c. MARY-X human breast cancer. Bacteria (5×10^7 cfu/100 μl) were inoculated i.v. in MARY-X-bearing nude mice. Tumour growth was monitored by caliper measurement in two dimensions. The infected tumours regressed by day 5 after infection and complete regression occurred by day 25. In orthotopic models of MARY-X, A1-R treatment also led to tumour regression following a single i.v. injection of A1-R. The regression of the tumour in treated mice was visualized by whole-body imaging. The difference in tumour volume between the treated group, which showed quantitative regression, and the control was statistically significant ($P < 0.05$) (Zhao et al. 2006).

9.3.10 Survival Efficacy with A1-R Treatment in Orthotopic Breast Cancer Models

The survival of the A1-R-treated mice with orthotopic MARY-X tumours was prolonged with a 50% survival time of 13 weeks compared with 5 weeks of control animals. Forty percent of the treated mice survived as long as control non-tumour-bearing mice. In the cured animals, tumours were completely eradicated with no regrowth. The parental *S. typhimurium* A1 was less effective than A1-R. Tumour growth was only slowed after A1 i.v. injection and not eradicated (Zhao et al. 2006).

9.3.11 Efficacy of A1-R on Primary Pancreatic Cancer

9.3.11.1 Intracellular Growth of *S. typhimurium* A1-R

A1-R GFP could invade and replicate intracellularly in the XPA1 human pancreatic cancer cell line expressing GFP in the nucleus and RFP in the cytoplasm. Intracellular bacterial infection led to cell fragmentation and cell death (Nagakura et al. 2009).

9.3.11.2 A1-R Treatment Schedule *In Vivo*

On day 0, a tumour piece of the dual color XPA1 tumour was transplanted on the pancreas of nude mice. On day 7, the tumour was exposed and observed with the Olympus OV100 Small Animal Imaging System. The size of the tumour (fluorescent area, mm^2) was measured. Three mice were treated with a low concentration of A1-R (10^7 cfu/ml); three were treated with a high concentration (10^8 cfu/ml); and three were used as untreated controls. The bacteria were injected into the tumour. Tumour volume (mm^3) was calculated with the formula $V = 1/2 \times (\text{length} \times \text{width}^2)$. On day 14, the tumour was exposed again and the size was measured as described above to determine the efficacy of treatment (Nagakura et al. 2009).

9.3.11.3 Efficacy of A1-R on Pancreatic Cancer

Before treatment, the average tumour size (fluorescent area) on day 7 was 3.2 ± 1.9 mm^2 in the untreated group, 3.1 ± 1.4 mm^2 in the high-bacteria-dose group and 3.5 ± 0.75 mm^2 in the low-bacteria-dose group. On day 14, after 7 days treatment, the tumour fluorescence area was 19.9 ± 4.3 mm^2 in the untreated group; 2.2 ± 0.89 mm^2 in the high-bacteria-concentration treatment group; and 12.7 ± 6.5 mm^2 in the low-bacteria-concentration treatment group (Nagakura et al. 2009).

9.4 Targeting Metastasis with A1-R

9.4.1 Targeting A1-R to Pancreatic Cancer Liver Metastasis

We have demonstrated the efficacy of locally as well as systemically administered A1-R on liver metastasis of pancreatic cancer expressing RFP. Mice treated with A1-R given locally via intrasplenic injections or systemically via tail vein injections had a much lower hepatic and splenic tumour burden as compared to untreated control mice. Systemic treatment with intravenous A1-R also increased survival time. All results were statistically significant (Yam et al. 2010).

9.4.2 Experimental Lymph Node Metastasis Cured by Specific Targeting of a A1-R

A new experimental model of lymph node metastasis was developed for this study. To obtain experimental metastasis in the axillary lymph node, XPA1-RFP human pancreatic cancer cells were injected into the inguinal lymph node in the nude mice. Just after injection, cancer cells were imaged trafficking in the efferent lymph duct to the axillary lymph node. Metastasis in the axillary lymph node was subsequently formed. A1-R bacteria were then injected into the inguinal lymph node to target the axillary lymph node metastasis. Just after bacterial injection, a large amount of bacteria were visualized around the axillary lymph node metastasis. By day 7, all lymph node metastases had been eradicated in contrast to growing metastases in the control group. There were very few bacteria in the lymph node by day 7 and no bacteria were detected after day 10. This route of administration was, therefore, able to deliver sufficient A1-R bacteria to eradicate the lymph node metastasis after which the bacteria became undetectable. The average tumour size (fluorescent area) in the axillary lymph nodes on day 0 was 0.4 ± 0.19 mm^2 in the treatment group and 0.46 ± 0.08 mm^2 in the untreated group. On day 7, it was 0 mm^2 in the treatment group and 0.98 ± 0.17 mm^2 in the untreated group (Hayashi et al. 2009a).

9.4.3 A1-R Targeted Therapy of Spontaneous Lymph Node Metastasis

We then tested bacterial therapy strategy for spontaneous lymph node metastasis from a fibrosarcoma tumour growing in the footpad. At first, only A1-R bacteria were injected in the footpad in nude mice in order to determine any adverse effects. No infection, skin necrosis, or body weight loss or fatality was detected. Then double-labeled HT 1080-GFP-RFP human fibrosarcoma cells were injected into the footpad of additional nude mice. The presence of popliteal lymph node metastasis was determined by weekly imaging. Once the metastasis was detected, A1-R bacteria were injected subcutaneously in the footpad (Hayashi et al. 2009a).

Bacteria are small particles and when injected subcutaneously, the lymph system immediately collects them from the site of injection. The lymph system is well known as a drainage route for bacterial infection.

We observed the injected bacteria trafficking in the lymphatic channel. The popliteal region was exposed just after bacteria injection and a large amount of GFP A1-R bacteria targeting the popliteal lymph node metastasis was observed by fluorescence imaging. Dual color labeling of the cancer cells with RFP in the cytoplasm and GFP in the nucleus distinguished them from the GFP bacteria. After treatment, the popliteal lymph node was observed every week by fluorescence imaging. One mouse was used to image the bacteria by exposing the popliteal lymph node on day 7. GFP bacteria invading the lymph node metastasis were observed. All lymph node metastases shrank and five out of six were eradicated within 7–21 days after treatment in contrast to growing metastases in the untreated control group (Hayashi et al. 2009a).

9.4.4 A1-R Therapy for Experimental Lung Metastasis

To obtain lung metastasis, dual-color RFP-GFP-HT1080 cells were injected into the tail vein of nude mice (day 0). On days 4 and 11, A1-R bacteria were injected into the tail vein. On day 16, all animals were killed and the lungs were imaged to determine the efficacy of bacteria therapy on lung metastasis. To observe the lung metastasis at lower magnification, an RFP filter was used (excitation 545 nm, emission 570–625 nm). In the A1-R treatment group, only a few cancer cells were observed in contrast to multiple metastases in the control (untreated) group. The number of metastases on the surface of the lung was significantly lower in the treatment group than in the control group ($P<0.005$). There were no significance differences between the treated and untreated groups in body weight (Hayashi et al. 2009a).

9.4.5 Targeting of Primary Bone Tumour and Lung Metastasis of High Grade Osteosarcoma in Nude Mice with A1-R

Mice were transplanted with 143B-RFP osteosarcoma cells in the tibia and developed primary bone tumour and lung metastasis. Seven days after tumour injection, the RFP tumour was confirmed inside the tibia. After three times weekly injections of A1-R, the bone tumour size and lung metastasis were examined on day 28. The bone tumour size (RFP area) was 231.7 ± 69.7 mm^2 in the untreated group and 94.6 ± 22.7 mm^2 in the treated group ($P<0.05$). The lung was excised and the metastases on the surface were counted. The number of metastasis was 52 ± 29.6 in the untreated group and 2.3 ± 2.1 in the treated group ($P<0.05$). A1-R therapy was, therefore, effective for primary and metastatic osteosarcoma (Hayashi et al. 2009b).

9.5 Targeting of Spinal Cord with A1-R

S. typhimurium A1-R were administered systemically or intrathecally, to spinal cord cancer in orthotopic nude mouse models. Tumour fragments of human U87-RFP glioma were implanted by surgical orthotopic implantation into the dorsal site of the spinal cord. Five and 10 days after transplantation, eight mice in each group were treated with A1-R (2×10^7 CFU/200 μl i.v. injection or 2×10^6 CFU/10 μl intrathecal injection) (Kimura et al. 2010).

9.6 Tumour-Positive Vascularity Correlates with Susceptibility to A1-R

S. typhimurium A1-R destroys tumour blood vessels and this is enhanced in tumours with high vascularity. RFP-expressing Lewis lung cancer cells (LLC-RFP) were transplanted subcutaneously in the ear, back skin and footpad of nestin-driven GFP (ND-GFP) transgenic nude mice, which selectively express GFP in nascent blood vessels. Color-coded *in vivo* imaging demonstrated that the LLC-RFP ear tumour had the highest cell density and the footpad tumour had the least. The ear tumour had more abundant blood vessels than tumour on the back or footpad. The tumour-bearing mice were treated with A1-R via tail-vein injection. Tumours in the ear were the earliest responders to bacterial therapy and hemorrhaged severely the day after A1-R administration. Tumours growing in the back were the second fastest responders to bacterial treatment and appeared necrotic 3 days after A1-R administration. Tumours growing in the footpad had the least vascularity and were the last responders to A1-R. Therefore, tumour vascularity correlated positively with efficacy of A1-R. The present study suggests that A1-R efficacy on tumours involves vessel destruction which depends on the extent of vascularity of the tumour (Liu et al. 2010).

Leschner et al. (2009) observed a rapid increase of TNF-á in blood, in addition to other pro-inflammatory cytokines, after *S. typhimurium* treatment of tumours. Bacterial treatment induced a great influx of blood into the tumours by vascular disruption and bacteria were flushed into the tumour along with the blood. Our results, discussed above, suggest the degree of vascularity is most important when bacteria target tumours and destroy tumour blood vessels (Liu et al. 2010).

9.7 Screening for *Salmonella* Promoters Differentially Activated in the PC-3 Prostate Tumour

We have used a high-throughput method to screen for *S. typhimurium* promoters that are selectively activated in tumours in the mouse. A random library of *S. typhimurium* with DNA cloned upstream of a promoter less GFP, was injected intravenously in nude mice with s.c. human PC-3 prostate tumours as well as in control nude mice. GFP-positive *S. typhimurium* clones from tumour, spleen, liver and from *in vitro* growth in LB medium were isolated by fluorescence-activated cell sorting (FACS). Active promoters in all environments were amplified by PCR and identified by DNA-microarray hybridization. Among promoters identified as preferentially induced in tumours, and not induced in any of the other environments (spleen, liver, or *in vitro*), were those of at least five genes known to be controlled by the fumarate and nitrate reduction global regulator (FNR). At least five other genes with unknown regulation were also enriched in tumours. The natural tendency of *S. typhimurium* to target tumours preferentially over other tissues, combined with the use of promoters preferentially induced in the tumour environment versus other environments, may allow the exquisitely tumour-specific expression of fusion proteins on the surface or secreted by *S. typhimurium* for highly selective tumour therapy (Arrach et al. 2008).

9.8 Screening for *S. typhimurium* Mutants That Enhance Selective Tumour Targeting

To select *Salmonella* strains that are avirulent in normal tissues and yet efficient in tumour targeting, the relative fitness of 41,000 *Salmonella* transposon insertion mutants growing in mouse models of human prostate and breast cancer were evaluated. Two classes of potentially safe mutants were identified. Class 1 mutants showed reduced fitness in normal tissues and unchanged fitness in tumours (e.g., mutants in htrA, SPI-2, and STM3120). Class 2 mutants showed reduced fitness in tumours and normal tissues (e.g., mutants in aroA and aroD). In a competitive fitness assay in human PC-3 tumours growing in mice, class 1 mutant STM3120 had a fitness advantage over class 2 mutants aroA and aroD, validating the findings of the initial screening of a large pool of transposon mutants and indicating a potential advantage of class 1 mutants for delivery of cancer therapeutics. In addition, an STM3120 mutant successfully targeted tumours after intragastric delivery, (please also see below) (Arrach et al. 2010).

9.9 Oral Administration of Tumour-Targeting Bacteria

Following oral administration in mice, *Salmonella* preferentially accumulated in tumours and maintained its anticancer effects (Jia et al. 2007) with very low toxicity (Chen et al. 2009). Oral delivery might be different in humans, in which bacterial escape from the gut to the circulation occurs less often than in mice (Bermudes et al. 2000; Forbes 2010).

9.10 Conclusion

Our goal is to develop tumour-targeting *S. typhimurium* strains that can kill primary and metastatic cancer without toxic effects to the host and without the need for combination with toxic chemotherapy. Toward this goal, a new substrain of *S. typhimurium*, A1-R was developed, that has greatly increased anti-tumour efficacy but maintains its original auxotrophy for leu-arg that prevents it from mounting a continuous infection in normal tissues. A1-R was able to effect cures in monotherapy on mouse models of metastatic human cancer. Candidate *S. typhimurium* tumour-specific promoters have been identified that may enhance the anti-tumour efficacy of A1-R by driving expression of toxins that could be selectively expressed in the tumours. Mutants with enhanced selectivity of tumour targeting have also been identified. *Salmonella* could also be used for DNA delivery to tumours (Darji et al. 1997; Weiss and Chakraborty 2001; Forbes 2010). For example, transfer of the endostatin gene from *Salmonella* reduced microvessel density, decreased vascular endothelial growth factor (VEGF) expression and slowed tumour growth in mice (Lee et al. 2004). Bacterial transfer of the genes encoding TRAIL and SMAC into tumour cells from *Salmonella* has also shown to be effective (Fu et al. 2008; Forbes 2010). Gene silencing in tumours has also been achieved by transferring plasmids encoding small hairpin RNAs (shRNA) from *Salmonella* into cancer cells (Zhang et al. 2007; Yang et al. 2008; Forbes 2010). Future studies will be aimed to bring bacterial treatment of cancer back to the clinic.

References

Arrach N, Zhao M, Porwollik S et al (2008) *Salmonella* promoters preferentially activated inside tumors. Cancer Res 68:4827–4832

Arrach N, Cheng P, Zhao M et al (2010) High-throughput screening for *Salmonella* avirulent mutants that retain targeting of solid tumors. Cancer Res 70:2165–2170

Bermudes D, Low B, Pawelek J (2000) Tumor-targeted *Salmonella*. Highly selective delivery vectors. Adv Exp Med Biol 465:57–63

Brown JM, Giaccia AJ (1998) The unique physiology of solid tumors: opportunities (and problems) for cancer therapy. Cancer Res 58:1408–1416

Chen G, Wei DP, Jia LJ et al (2009) Oral delivery of tumor-targeting *Salmonella* exhibits promising therapeutic efficacy and low toxicity. Cancer Sci 100:2437–2443

Clairmont C, Lee KC, Pike J et al (2000) Biodistribution and genetic stability of the novel antitumor agent VNP20009, a genetically modified strain of *Salmonella typhimurium*. J Infect Dis 181:1996–2002

Coley WB (1906) Late results of the treatment of inoperable sarcoma by the mixed toxins of erysipelas and *Bacillus prodigiosus*. Am J Med Sci 131:375–430

Dang LH, Bettegowda C, Huso DL et al (2001) Combination bacteriolytic therapy for the treatment of experimental tumors. Proc Natl Acad Sci USA 98:15155–15160

Darji A, Guzmán CA, Gerstel B et al (1997) Oral somatic transgene vaccination using attenuated *S. typhimurium*. Cell 91:765–775

Forbes NS (2010) Engineering the perfect (bacterial) cancer therapy. Nat Rev Cancer 10:785–794

Forbes NS, Munn LL, Fukumura D et al (2003) Sparse initial entrapment of systemically injected *Salmonella typhimurium* leads to heterogeneous accumulation within tumors. Cancer Res 63:5188–5193

Fox ME, Lemmon MJ, Mauchline ML et al (1996) Anaerobic bacteria as a delivery system for cancer gene therapy: *in vitro* activation of 5-fluorocytosine by genetically engineered *Clostridia*. Gene Ther 3:173–178

Fu W, Chu L, Han XW et al (2008) Synergistic antitumoral effects of human telomerase reverse transcriptase-mediated dual-apoptosis-related gene vector delivered by orally attenuated *Salmonella enterica* serovar *typhimurium* in murine tumor models. J Gene Med 10:690–701

Gericke D, Engelbart K (1964) Oncolysis by *Clostridia*. II. Experiments on a tumor spectrum with a variety of clostridia in combination with heavy metal. Cancer Res 24:217–221

Hayashi K, Zhao M, Yamauchi K et al (2009a) Cancer metastasis directly eradicated by targeted therapy with a modified *Salmonella typhimurium*. J Cell Biochem 106:992–998

Hayashi K, Zhao M, Yamauchi K et al (2009b) Systemic targeting of primary bone tumor and lung metastasis of highgrade osteosarcoma in nude mice with a tumor-selective strain of *Salmonella typhimurium*. Cell Cycle 8:870–875

Hoiseth SK, Stocker BA (1981) Aromatic-dependent *Salmonella typhimurium* are non-virulent and effective as live vaccines. Nature 291:238–239

Jia LJ, Wei DP, Sun QM et al (2007) Oral delivery of tumor-targeting *Salmonella* for cancer therapy in murine tumor models. Cancer Sci 98:1107–1112

Kasinskas RW, Forbes NS (2006) *Salmonella typhimurium* specifically chemotax and proliferate in heterogeneous tumor tissue *in vitro*. Biotechnol Bioeng 94:710–721

Kasinskas RW, Forbes NS (2007) *Salmonella typhimurium* lacking ribose chemoreceptors localize in tumor quiescence and induce apoptosis. Cancer Res 67:3201–3209

Kimura NT, Taniguchi S, Aoki K et al (1980) Selective localization and growth of *Bifidobacterium bifidum* in mouse tumors following intravenous administration. Cancer Res 40:2061–2068

Kimura H, Zhang L, Zhao M, et al (2010) Targeted therapy of spinal cord glioma with a genetically-modified *Salmonella typhimurium*. Cell Prolif 43:41–48

Kohwi Y, Imai K, Tamura Z et al (1978) Antitumor effect of *Bifidobacterium infantis* in mice. Gann 69:613–618

Lee CH, Wu CL, Shiau AL (2004) Endostatin gene therapy delivered by *Salmonella choleraesuis* in murine tumor models. J Gene Med 6:1382–1393

Lemmon MJ, Van Zijl P, Fox ME et al (1997) Anaerobic bacteria as a gene delivery system that is controlled by the tumor microenvironment. Gene Ther 4:791–796

Leschner S, Westphal K, Dietrich N et al (2009) Tumor invasion of *Salmonella enterica* serovar *typhimurium* is accompanied by strong hemorrhage promoted by TNF-alpha. PLoS One 4:e6692

Liu F, Zhang L, Hoffman RM et al (2010) Vessel destruction by tumor-targeting *Salmonella typhimurium* A1-R is enhanced by high tumor vascularity. Cell Cycle 9:4518–4524

Low KB, Ittensohn M, Le T et al (1999) Lipid A mutant *Salmonella* with suppressed virulence and TNFα induction retain tumor-targeting *in vivo*. Nat Biotechnol 17:37–41

Malmgren RA, Flanigan CC (1955) Localization of the vegetative form of *Clostridium tetani* in mouse tumors following intravenous spore administration. Cancer Res 15:473–478

Mengesha A, Dubois L, Lambin P et al (2006) Development of a flexible and potent hypoxia-inducible promoter for tumor-targeted gene expression in attenuated *Salmonella*. Cancer Biol Ther 5:1120–1128

Moese JR, Moese G (1964) Oncolysis by Clostridia. I. Activity of *Clostridium butyricum* (M-55) and other nonpathogenic clostridia against the Ehrlich carcinoma. Cancer Res 24:212–216

Nagakura C, Hayashi K, Zhao M et al (2009) Efficacy of a genetically-modified *Salmonella typhimurium* in an orthotopic human pancreatic cancer in nude mice. Anticancer Res 29:1873–1878

Nguyen VH, Kim HS, Ha JM et al (2010) Genetically engineered *Salmonella typhimurium* as an imageable therapeutic probe for cancer. Cancer Res 70:18–23

Pawelek JM, Low KB, Bermudes D (1997) Tumor-targeted *Salmonella* as a novel anticancer vector. Cancer Res 57:4537–4544

Sznol M, Lin SL, Bermudes D et al (2000) Use of preferentially replicating bacteria for the treatment of cancer. J Clin Invest 105:1027–1030

Taniguchi S, Fujimori M, Sasaki T et al (2010) Targeting solid tumors with non-pathogenic obligate anaerobic bacteria. Cancer Sci 101(9):1925–1932

Thiele EH, Arison RN, Boxer GE (1964) Oncolysis by *Clostridia*. III. Effects of clostridia and chemotherapeutic agents on rodent tumors. Cancer Res 24:222–233

Toso JF, Gill VJ, Hwu P et al (2002) Phase I study of the intravenous administration of attenuated *Salmonella typhimurium* to patients with metastatic melanoma. J Clin Oncol 20:142–152

Weiss S, Chakraborty T (2001) Transfer of eukaryotic expression plasmids to mammalian host cells by bacterial carriers. Curr Opin Biotechnol 12:467–472

Westphal K, Leschner S, Jablonska J et al (2008) Containment of tumor-colonizing bacteria by host neutrophils. Cancer Res 68:2952–2960

Yam C, Zhao M, Hayashi K et al (2010) Monotherapy with a tumor-targeting mutant of *S. typhimurium* inhibits liver metastasis in a mouse model of pancreatic cancer. J Surg Res 164: 248–255

Yang N, Zhu X, Chen L et al (2008) Oral administration of attenuated *S. typhimurium* carrying shRNA-expressing vectors as a cancer therapeutic. Cancer Biol Ther 7:145–151

Yazawa K, Fujimori M, Amano J et al (2000) *Bifidobacterium longum* as a delivery system for cancer gene therapy: selective localization and growth in hypoxic tumors. Cancer Gene Ther 7:269–274

Yazawa K, Fujimori M, Nakamura T et al (2001) *Bifidobacterium longum* as a delivery system for gene therapy of chemically induced rat mammary tumors. Breast Cancer Res Treat 66:165–170

Yu YA, Timiryasova T, Zhang Q et al (2003) Optical imaging: bacteria, viruses, and mammalian cells encoding light emitting proteins reveal the locations of primary tumors and metastases in animals. Anal Bioanal Chem 377:964–972

Yu YA, Shabahang S, Timiryasova TM et al (2004) Visualization of tumors and metastases in live animals with bacteria and vaccinia virus encoding light-emitting proteins. Nat Biotechnol 22: 313–320

Zhang L, Gao L, Zhao L et al (2007) Intratumoral delivery and suppression of prostate tumor growth by attenuated *Salmonella enterica* serovar *typhimurium* carrying plasmid-based small interfering RNAs. Cancer Res 67:5859–5864

Zhao M, Yang M, Li XM et al (2005) Tumor-targeting bacterial therapy with amino acid auxotrophs of GFP-expressing *Salmonella typhimurium*. Proc Natl Acad Sci USA 102:755–760

Zhao M, Yang M, Ma H et al (2006) Targeted therapy with a *Salmonella typhimurium* leucine-arginine auxotroph cures orthotopic human breast tumors in nude mice. Cancer Res 66:7647–7652

Zhao M, Geller J, Ma H et al (2007) Monotherapy with a tumor-targeting mutant of *Salmonella typhimurium* cures orthotopic metastatic mouse models of human prostate cancer. Proc Natl Acad Sci USA 104:10170–10174

Chapter 10
Bacterial Asparaginase: A Potential Antineoplastic Agent for Treatment of Acute Lymphoblastic Leukemia

Abhinav Shrivastava, Abdul Arif Khan, S.K. Jain, and P.K. Singhal

Abstract Among the pediatric cancer in developed countries, acute leukemia constitutes major part with affecting 30–45 per 1,000,000 children each year. Although one thirds of acute lymphoblastic leukemia cases are curable, but the effect of treatment varies with differences in patients clinical, immunologic and genetic characteristics. L-asparaginase is among the main drugs used for the treatment of acute lymphoblastic leukemia and certain non Hodgkin lymphoma. L-asparaginase has been isolated from various sources including bacteria, algae, fungi, plant and mammals. Among these, bacterial asparaginase is most commonly used for treatment of ALL. But the efficacy of every asparaginase preparation varies in their activity, efficacy and side effects. This difference is reflected between the different marketed products even in bacterial asparaginase obtained from two different bacteria. Currently, native asparaginase obtained from *E. coli* and *Erwinia chrysanthemi* are most frequently used, but other products like PEG-asparaginase are also gaining popularity for management of ALL. Similarly, development of asparaginase in erythrocytes, asparaginase obtained from sources other than bacteria, and recombinant asparaginases are also a subject of research. Present chapter reviews the current status of bacterial asparaginase in ALL management, including its potential

A. Shrivastava (✉)
College of Life Sciences, Cancer Hospital and Research Institute, Cancer Hills, Gwalior, Madhya Pradesh 474009, India
e-mail: abhi.shri@gmail.com

A.A. Khan
College of Life Sciences, Cancer Hospital and Research Institute, Cancer Hills, Gwalior, Madhya Pradesh 474009, India

Microbiology Unit, Department of Pharmaceutics, College of Pharmacy, King Saud University, Riyadh 11451, Saudi Arabia

S.K. Jain
Department of Microbiology, Vikram University, Ujjain, Madhya Pradesh 496010, India

P.K. Singhal
Deparment of Biosciences, Rani Durgavati University, Jabalpur, Madhya Pradesh, India

A.A. Khan (ed.), *Bacteria and Cancer*, DOI 10.1007/978-94-007-2585-0_10,

benefits, side effects and future research going in the direction of development of L-asparaginase as potential antineoplastic agent.

Keywords ALL • Leukemia • Bacterial enzymes • Asparaginase • Chemotherapy • *Fusarium* • Antineoplastic • Pegaspargase

Abbreviations

ALL	Acute lymphoblastic leukemia
AML	acute myeloid leukemia
BLLOQ	below the lower limit of quantification
CLL	chronic lymphocytic leukemia
CML	chronic myelogenous (or myeloid) leukemia
PEG	poly (ethylene glycol)
SPR	surface plasmon resonance

10.1 Introduction

Cancer remains a major public health problem through out the word. In the year 2008, the American Cancer Society estimated that 12.4 million cases of cancer are diagnosed and 7.6 million deaths occurred due to cancer around the world. With these figures cancer accounts for one in every eight deaths throughout the world (ACS 2010).

Nearly all cancers are caused by abnormalities in the genetic material of the transformed cells (Vogelstein and Kinzler 2002). These abnormalities may be due to the effects of carcinogens, such as tobacco smoke (Clayson 2001), radiation, chemicals or infectious agents (Trichopoulos et al. 1996). Other cancer-promoting genetic abnormalities may be randomly acquired through errors in DNA replication, or are inherited, and thus present in all cells from birth. The heritability of cancer is usually affected by complex interactions between carcinogens and the host's genome (Emery et al. 2001). New aspects of the genetics of cancer pathogenesis, such as DNA methylation and microRNAs are increasingly recognized as important. Genetic abnormalities associated with cancer typically affect two types of genes. First: *cancer-promoting oncogenes*, which are typically activated in cancer cells, giving those cells new properties, such as hyperactive growth and division, protection against programmed cell death, loss of respect for normal tissue boundaries, and the ability to become established in diverse tissue environments (Hanahan and Weinberg 2000). Second: *tumour suppressor genes*, which are inactivated in cancer cells, result in the loss of normal functions in those cells, such as accurate DNA replication, control over the cell cycle, orientation and adhesion within tissues, and interaction with protective cells of the immune system (Sherr 2004).

Among these cancers, leukemia is a type of cancer which involves blood and bone marrow. It is estimated by National Cancer Institute that, 43,050 cases of leukemia will be diagnosed and 21,840 patients will die due to this disease in 2010 (Howlader et al. 2010). Leukemia is found to be a most common cause of cancer death in males under age 40 and females under age 20 (Jemal et al. 2008). There are four most common types of leukemia including acute lymphoblastic (or lymphoid) leukemia (ALL), acute myeloid leukemia (AML), chronic lymphocytic leukemia (CLL), chronic myelogenous (or myeloid) leukemia (CML). Other forms of leukemia include hairy cell leukemia, chronic myelomonocytic leukemia (CMML) and juvenile myelomonocytic leukemia (JMML).

Cancer is ranked as second most common cause of death in children aged between 1 and 14 years (Jemal et al. 2008). Among the cases of pediatric cancer, acute lymphoblastic leukemia (ALL) is most prevalent cancer accounting for 25% of all malignancies diagnosed among children below age 15 years. ALL is a major scourge in pediatric oncology, and cause considerable catastrophe (Pui 2010). As the name suggest, ALL is a malignant disorder of lymphoid progenitor cells. The clinical outcome of ALL has been improved remarkably since last 50 years with the aid of knowledge in chemotherapy of ALL.

10.2 Chemotherapy

The word chemotherapy was coined by Paul Ehrlich in twentieth century while searching for chemical agent effective in treatment of Syphlis. His observation pertaining to selective incorporation of certain chemicals in bacterial cells paved the way for discovery of many therapeutic modalities for management of bacterial infection. Selective incorporation of certain chemical agents was thought to cause toxicity in bacterial cells. Ehrlich also warned about the possible human toxicity of his Silver bullet (Salvarsan), which was arsenic by chemical nature and used for treatment of syphilis (Thorburn 1983). During world war first, nitrogen mustard was used as a potential chemical warfare, which was further studied during world war second and used for cytotoxicity against cancer cells on the basis of results obtained during accidental exposure of mustard gas on normal human beings and resulted low WBC count (Hirsch 2006). Consequently, after of discovery of potential cytotoxicity of mustard gas against human being and its use in cancer chemotherapy, this idea ushered in to a new era of anticancer discovery, where many chemical substances otherwise discarded due to their potential toxicities on human being, were tested for cancer chemotherapy.

Survival in ALL has improved significantly due to specific diagnosis overcoming the problem of biologic heterogeneity of ALL, risk assessment, and optimization of chemotherapy combinations, proper CNS prophylaxis, Treatment of ALL should be very urgent and require to be given within few days and sometimes the same day as the diagnosis is made. The initial treatment of chemotherapy called induction chemotherapy. During this treatment, it is necessary that patients should be admitted in the hospital for 1 month.

10.2.1 Remission or Induction Chemotherapy

The general medicines used for induction treatment of ALL are asparaginase, daunorubicin, vincristine, prednisone, and sometimes cyclophosphamide (Cytoxan). Intensive supportive care is very necessary during the chemotherapy, including transfusion of red blood cells and platelets. Uses of antibiotics are required for both prevention and treatment of bacterial and fungal infections. A repeat bone marrow biopsy is necessary when blood counts have returned to normal range. A complete remission or induction is achieved when the blood cells and bone marrow does not show any symptom of persistent leukemia and counting of blood has returned to normal scale.

10.2.2 Consolidation or Intensification Chemotherapy

Consolidation chemotherapy generally includes many cycles of intensive chemotherapy given over a 6–9 month period. Frequent hospitalizations are necessary and intensive supportive care is still required, including red blood cell and platelet transfusions.

10.2.3 Maintenance Chemotherapy

When patient have completed intensive chemotherapy, they required to take oral drugs for about 18–24 months. These oral drugs are usually well-tolerated with only minimal side effects. It is necessary to regular check up of patient blood once a month while taking chemotherapy drugs. During this stage of maintenance therapy most of the patients with ALL can return to their routine work.

10.2.4 CNS Prophylaxis

ALL frequently can transfer in the spinal fluid. At this step, it is necessary that a needle is inserted between the vertebrae of the lower back and infusing chemotherapy directly into the clear spinal fluid, it is called intrathecal chemotherapy. Regularly 6–12 injections of intrathecal chemotherapy are given to prevent recurrence of ALL in patients. Most of the patients complete intrathecal therapy within 2–4 months of starting their treatment.

During chemotherapy there are various drugs, which are used for treatment in cancer. L-asparaginase is known chemotherapeutic agents against cancer, such as acute lymphoblastic leukemia and lymphosarcoma. It is used mainly in the treatment of cancer in children.

L-Asparaginase (E.C.3.5.1.1) is an enzyme which catalyzes breakdown of extra cellular L-asparagines into L-aspartate and ammonia in following way (Asselin et al. 1989, 1991)

$$HOOCCH \cdot NH_2 \cdot CH_2 \cdot CONH_2 + H_2O \rightarrow HOOCCH \cdot NH_2 \cdot CH_2COOH + NH_3 -$$

This enzyme is generally used for the treatment of cancer like ALL (Acute Lymphoblastic Leukemia) and NHL (Non-Hodgkin's Lymphoma) (Capizzi et al. 1970).

10.3 Biomedical Application of L-Asparaginase

L-asparaginase is a known antineoplastic agent, which is used mainly in the treatment of children. Several example of recent research are available concerning the use of L-asparaginase in cancer therapy (Muller and Boos 1998; Avramis and Panosyan 2005; Pieters et al. 2011). Leukaemic cells have deficiency or absence of mammalian asparagine synthetase enzyme (Keating et al. 1993). Due to the absence of asparagine synthetase, these cells can not synthesize their own L-asparagine, thus completely dependent on external L-asparagine for protein synthesis and survival. So, required L-asparagine should be absorbed from out side for survival. For these reasons this enzyme is injected intravenously in order to decrease the blood concentration of L-asparagine, affecting selectively the new neoplastic cells (Mitchell et al. 1994). Due to decrease in serum asparagine concentration, leukemic blast cells are restricted in protein synthesis. In consequence, the cells not expressing sufficient amount of asparagine synthetase are killed (Capizzi and Holcenberg 1993; Chabner 1990).

L-asparaginase has two isozymes (Schwartz et al. 1966): asparaginase I, which is found in the cytoplasm and has a low affinity for L-asparagine (EC1; Km = 3.5 mM) and L-asparaginase II, that is a periplasmic enzyme used in the treatment of acute lymphoblastic leukemia and in contrast, it is a high-affinity enzyme (EC2; Km = 10 pM). L-asparaginase II is widely distributed in both prokaryotic and eukaryotic cells and has been intensively studied over the past few years.

The removal of amino group of serum asparagine selectively kills leukaemic cells, leaving normal cells having capability to synthesize intracellular asparagines (Broome 1968). Moreover, studies of the action of asparaginase upon neoplastic cells, led to the introduction of new drugs as well as the combination of asparaginase with drugs of similar modes of action in the clinical treatment of lymphoblastic leukemia (Ronghe et al. 2001).

L-asparaginases are present in many plants, mammalian and bacterial species, but only the enzymes from *E. coli* and *Erwinia chrysanthemi* have been produced on industrial scale. Although, the enzyme from these sources have identical mechanisms of action and toxicities, their pharmacokinetic properties are different, and patients allergic to one drug have frequently resistant to the other. Though important in cancer therapy, the clinical applications of L asparaginase are often limited by three factors (Wang et al. 2003; Woo et al. 2000; Chakrabarti and Schuster 1997). *First*, many side effects are associated with L-asparaginase treatment, including immuno suppression and pancreatitis (Wang et al. 2003). There are two types of toxic effects, first one which is related to immunologic sensitization to a foreign protein and another related to the inhibition of protein synthesis (Wang et al. 2003). *Second*, about 10% of successfully treated patients suffer a relapse with the appearance of tumours that have resistance against further L-asparaginase therapy. *Lastly*, long duration treatment with L-asparaginase may improve the growth and survival of resistant tumours and increase their metastatic property (Woo et al. 2000). There are two possible ways for L-asparaginase resistance have been proposed (Woo et al. 2000). The first way is related to an increase concentration in asparagine synthetase, which has been found in the blasts cells of patients with acute lymphoblastic leukemia showing resistance against the drug (Asselin 1999). Another way appears to be the induction of a host response leading to the production of anti-asparaginase antibodies, which neutralize L-asparaginases impeding their enzymatic activity (Chakrabarti and Schuster 1997). A PEG conjugated or PEG *E. coli* asparaginase is a better option in patients with prior clinical hypersensitivity to native *E. coli* asparaginase (Inada et al. 1995). PEG-asparaginase has the capability to decrease the immunogenicity of the protein, increases its stability in plasma and it is possible to use this drug in heavily pretreated patients (Inada et al. 1995). L-asparaginase drug therapy may be used alone or in combination with other treatment like–surgery, radiation therapy and chemotherapy.

10.3.1 L-Asparaginase as Anti-tumour Drug

Enzymatic conversion of L-asparagine to L-aspartate and ammonia was first detected by Lang (1904) who observed that L-asparaginase activity is present in several beef tissues. Lang's results were confirmed by Fürth and Friedmann in 1910, who studied the asparagine hydrolysis in horse and pig organs and confirmed that asparaginase activity exist in all animal tissues. Contradicting results were observed by Clementi in 1922, who showed that asparaginase activity is present only in the liver of omnivorous animals (like pig), while organs of carnivorous mammals, amphibians and reptiles do not contain L-asparaginase at all, where as enzymatic activity can be found practically in all tissues only in herbivores. In 1953, Kidd found that guinea pig serum inhibited a number of transplantable lymphomas in mice and rats as well as certain spontaneous and radiation-induced leukemia in mice. Clementi's discovery of L-asparaginase activity in the blood of guinea pig was brought to light again after 40 years. Broome (1961) also presented some evidence that the anti cancerous element in guinea pig serum was L-asparaginase. Sensitive tumour cells are unable to synthesize L-asparagine (Broome 1963; Boyse et al. 1967; Sobin and Kidd 1965) and so there is inhibition of protein synthesis in sensitive cells in the absence of L-asparagine due to L-asparaginase (Haskell and Canellose 1969). Normal cells are able to synthesize and do not dependent upon exogenous supply of

L-asparagine (Broome 1963). Sensitive cells have very low activity of asparagine synthetase, whereas resistant tumours have high enzyme levels (Haskell and Canellose 1969; Keefer et al. 1985). In man, there is no much difference in L-asparagine synthetase content between sensitive and resistant leukaemic cells. However, when L-asparaginase is injected, the amount of asparagine synthetase is increased in resistant cells but in sensitive cells there is no change (Haskell and Canellose 1969). So, there is generation of differential cytotoxicity after administration of L-asparaginase (Worton et al. 1991). In sensitive cells, very rapid inhibition of protein synthesis take place during L-asparaginase treatment (Bartalena et al. 1986; Villa et al. 1986) as well as delayed inhibition of DNA and RNA synthesis (Sobin and Kidd 1965; Becker and Broome 1967). Inhibition of incorporation of L-asparagine (Asn), which caused alterations in protein and nucleic acid metabolism of the 6C3HED lymphoma cells by the guinea pig asparaginase, was found to be responsible for the tumour growth inhibition (Sobin and Kidd 1966; Kidd and Sobin 1966). The final proof that asparaginase was the tumour-inhibitory agent of guinea pig serum was furnished by other investigators who isolated the enzyme to homogeneity as judged by immunoelectrophoresis and demonstrated that it was strongly inhibitory to lymphoma tumours (Yellin and Wriston 1966). Furthermore, other researchers found that *E. coli* yielded preparations inhibited tumours, but other bacterial asparaginases were either less active or completely inactive (Mashburn and Wriston 1964; Broome 1965). Subsequently, the native *E. coli* asparaginase was then developed as a drug for use in patients.

L-asparaginase has capability to block the cell cycle in human leukaemic cell line of Jurkat T cell at the phase between G1 and S but did not inhibit the cell cycle of resistant cell line (Takase et al. 1985). Some workers suggested a link between L-asparaginase and L-asparaginyl t-RNA concentration (Waye and Stenner 1981). Some sensitive cells have an asparaginyl t-RNA that is not found in resistant cells, but these data remained inconclusive (Waye and Stenner 1981).

10.3.2 General Mechanism of Reaction Catalyzed by L-Asparaginase

The study of catalytic reaction of L-asparaginases has been compared with classic serine proteases, in which activity depends on a set of amino acid residues, typically Ser-His-Asp, known as the "catalytic triad" (Carter and Wells 1988). This set contains a nucleophilic amino acid (Ser), a general base (His), and an additional acidic residue (Asp), all linked by a chain of hydrogen bonds. It is two step reactions. In the first step, the enzyme's nucleophile become activated through a strong O-H...B hydrogen bond to an adjacent basic residue, attacks on the Carbon atom of the amide group of substrate, leading through a tetrahedral transition state which form an acyl-enzyme intermediate product. A negative charge develops on the "Oxygen" atom of the amide group during the transition state which is stabilized by interactions with adjacent hydrogen bond donors. These conditions lead to formation of

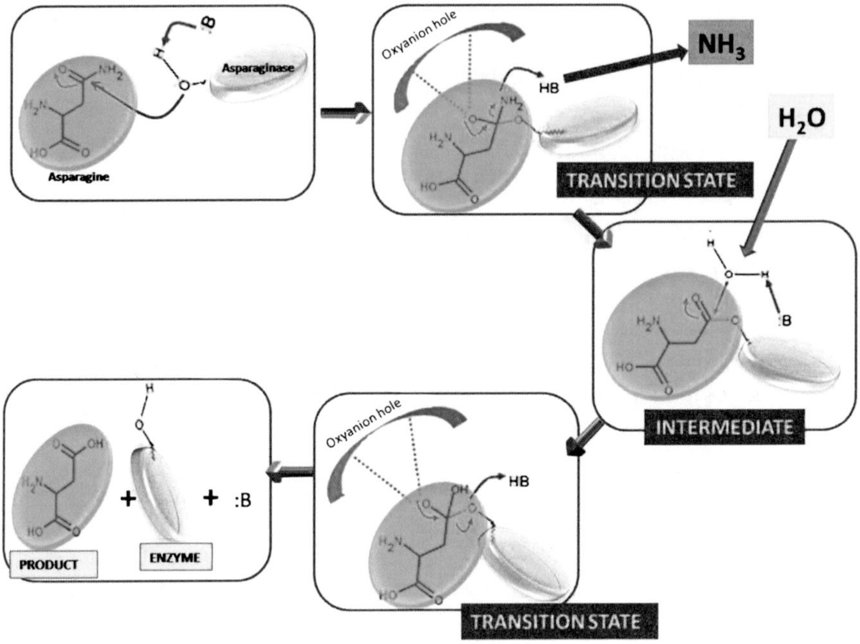

Fig. 10.1 Schematic representation of L-asparaginase activity

celestial sphere or constellation of those donors (which typically are main-chain N–H groups) and it is known as the "oxyanion hole". During the second step of the reaction C atom of the ester is attacked by a nucleophile activated water molecule. All these condition has been shown in an imaginary figure (Fig. 10.1). A suitable general base is also required for the activation of the nucleophilic residue.

10.3.2.1 Biochemistry and Mechanism of Action Catalyzed by L-Asparaginase

Enzymes are the natural, ideal catalysts, much more efficient and specific in their reaction for a given substrate in comparison to any artificial man made catalyst. However, enzymes of pharmaceuticals importance have unique requirements, such as purity of bacterial enzymes. These bacterial proteins must be purified extensively to eliminate toxicities and to minimize immunogenic reactions, and these proteins should have limited bio distribution and rapid elimination from circulation (Capizzi and Holcenberg 1993). The L-asparaginase from *E. coli* and *Erwinia* were purified and used as an antileukaemic agent experimentally in patients (Clavell et al. 1986; Story et al. 1993). This enzyme has high potential against children's acute lymphoblastic leukemia (Hill et al. 1967; Oettgen et al. 1967). Several researchers have studied asparaginase production and purification in attempt to minimize impurities that produce allergenic reactions and other side effects (Campbell et al. 1967; Boss 1997; Gallagher et al. 1989). *E. coli* contains two enzymes, EC1 expressed constitutively (Km 5 mM) and another EC2 induced by anaerobiosis (Km 12.5 μM); only the second show antineoplastic activity (Schwartz et al. 1966). Due to the low Km of EC 2 the enzyme can persist sufficiently long in the circulation of the recipient animal for effective anti-tumour action. Now there is a clear understanding that asparaginase was attacking cancerous cells on the basis of affecting nutritional requirement caused by the lack of asparagine. Thus, use of L-asparaginase combining with the other drugs like cytosine arabinoside (ara-C) and 6-mercaptopurine (6-MP) or thioguanine (6-TG) and daunomycin with vinca alkaloids was helpful to achieve the 50 days cures in mice (Broome 1981; Burchenal and Karnofsky 1970). Asparagine depletion or nutritional deprivation after asparaginase treatment may cause the significant changes in the absolute pool sizes, especially of ATP, UTP, and CTP. Some experiments suggested that consumption of the growth medium after one hour of asparaginase action inhibited the conversion of external uridine to CTP by the cells. In 6C3HED lymphoma cells, the uridine nucleotide pool, which forms the immediate precursors of RNA, provided a system which is in rapid equilibrium with externally supplied nucleosides (Goody and Ellem 1975). Intact bacterial native enzyme has molecular weight of 140 kDa ± 3.3 kDa with no signs of association or dissociation, or polymerization through analytical sedimentation equilibrium (Holcenberg and Teller 1976; Chabner and Loo 1996). Similar molecular weight (134 kDa) was also calculated for asparaginase with glutaminase activity from *E. coli* and other microbes. Glutaminase-asparaginase of *Pseudomonas* 7A contain four subunits with a molecular weight of 36 kDa ± 0.5 kDa using sedimentation equilibrium and 34 kDa through amino acid analysis.

10.3.2.2 More on the Mechanism of Action of L-Asparaginase

As mentioned earlier, asparaginase shows its antileukemic effect on neoplastic cells through rapidly completing conversion of asparagine to aspartic acid and ammonia in blood. Although asparagine is not an essential amino acid, but thymus, and T-cell types of leukemia depend on extracellular sources of asparagine for their protein synthesis due to lack of this amino acid. Approximately 50 μM asparagine present in the serum at steady state level which is produced from ingested food nutrients and via *de novo* pathway in the liver through the reaction of aspartic acid and ammonia (from glutamine, Gln) by the enzyme asparagine synthetase (AS) in mammalian cells (Avramis et al. 2002; Panosyan et al. 2004). Some ALL patients show the drug resistance against L-asparaginase due to increase in concentration of AS in the liver as well as in the leukemia cells, so that ultimate result is sufficient levels of asparagines for protein synthesis. The high concentrations of asparagines in serum are responsible for higher rates of relapsed ALL in children (Gaynon 2005; Jarrar et al. 2006).

10.4 Micro Organisms Associated with L-Asparaginase Production

L-asparaginase has been found in variety of bacteria, fungi, yeast, plants and mammals. L-asparaginase II produced by a large number of microorganisms such as *E. coli* (Khushoo et al. 2005; Derst et al. 1994), *Erwinia carotovora* (Aghaiypour et al. 2001; Borisova et al. 2003), *Erwinia chrysanthemi* (Kotzia and Labrou 2007), *Enterobacter aerogenes* (Mukharjee et al. 2000) *Pseudomonas aeruginosa* (El-Bessoumy et al. 2004), and a large number of gram positive and gram negative bacteria (Law and Wriston 1971). L-asparaginase with antineoplastic activity have been reported from *E. coli* (Bagert and Rohm 1989), *Erwinia aroideae* (Peterson and Ciegler 1972), *Erwinia carotovora* (Lee et al. 1989), *Proteus vulgaris* (Tosa et al. 1971), *Citrobacter freundii* (Davidson et al. 1977), *Vibrio succinogenes* (Distasio et al. 1976), *V. proteus* (Sinha et al. 1991), *Azotobacter vinelandii* (Gaffar and Shethna 1975) and *Pseudomonas aurantiaca* (Lebedeva et al. 1988).

A large number of substrates were screened for the better source of enzyme like plant leaves (Sieciechowicz et al. 1988a, b; Sieciechowicz and Ireland 1989), Yeast (Imada et al. 1973) and fungi (De-Angeli et al. 1970).

Although, L-asparaginase is a potential antineoplastic agent (Mashburn and Wriston 1964), only few purified asparaginase showed antineoplastic activity (Wriston and Yellin 1973). It has been observed that anticancerous activity of asparaginase is present in few gram negative bacteria (Law and Wriston 1971) and fungi (Raha et al. 1990). Recently L-asparaginase from *Fusarium solani* was also purified which shows anticancerous activity (Shrivastava et al. 2010). Although these experiments are in initial stage, further advancement is necessary to use this agent for therapeutic purpose.

10.4.1 Bacterial Asparaginase

Bacterial asparaginase has been studied in detail, because other microbes did not show appreciable enzyme activity for clinical trials (Selvakumar 1979). Among these, asparaginase from *E. coli* is most extensively studied and commonly used for therapeutic purposes.

In many microorganisms there is existence of more than one form of asparaginase, which differs in their physicochemical properties like heat stability, Km value, pH maxima and inactivation by inhibitor (Whelan and Wriston 1969). For example *E. coli* have two type of asparaginase, known as EC1 and EC2 (Campbell et al. 1967). Only EC2, which is present in periplasmic space, show relatively heat stability and high affinity for substrate as well as anti-neoplastic activity (Campbell et al. 1967). Similarly, two form of asparaginase are present in *Mycobacterium tuberculosis* and three forms exist in *Citrobacter freundii,* but only one form of both microorganisms show anti-cancerous activity.

E. coli asparaginase has molecular weight about 1,20,000–1,50,000 Da (Hellman et al. 1983) and it consist of four subunit with 30,000–35,000 molecular weight of each (Holcenberg et al. 1978).

In *E.coli* asparaginase, tyrosine and histidine residues are reported to be present in active site (Bagert and Rohm 1989). Lebedeva et al. (1988) showed that presence of carboxyl group in active site is essential for binding of substrate in *Pseudomonas aurantiaca* asparaginase.

For therapeutic purpose, the most important requirement of L-asparaginase is that, its isolation should be relatively easy, the enzyme should be stable at physiological pH and temperature, it should have low Km value and enzyme should not show feed back inhibition. The affinity of enzyme toward its substrate is considered as most important factor, which is responsible for anti-neoplastic activity of L-asparaginase (Lebedeva et al. 1988). L-asparaginase from *E. coli* and other microbes have high anti-leukemic activity with very low K_m value in the range of $1–5\times10^{-5}$ M, whereas L-asparaginase with non antineoplstic activity present in *Bacillus coagulans* has a very high K_m value with 4.4×10^{-3} M (Law and Wriston 1971). Relationship between antitumour activity and half life of L-asparaginase has been proven in various experiments (Alpar and Lewis 1985; Ho et al. 1988). It has been demonstrated in mice, that L-asparaginases from *E. coli* and guinea pig serum have half life 2.5 and 19 h respectively. Though, both contain high antineoplastic activity but L-asparaginase from other sources with low antitumour activity are rapidly disappeared from plasma (Adamson and Fabro 1968). Immobilized enzyme increases its serum half lives (Alpar and Lewis 1985; Ho et al. 1988).

The use of inhibitor of glutamine or asparagine synthesis along with asparaginase treatment increases its therapeutic efficiency (Prager et al. 1982; Capizzi and Cheng 1981). Nowadays, L-asparaginase is administered intravenously at a dose of 200–1,000 IU/Kg/Day for 1 month for the treatment of ALL and other related cancer (Capizzi et al. 1970).

10.4.1.1 Side Effect of Bacterial Asparaginase

There are various side effects associated with asparaginase treatment like hypersensitivity reactions, hepatotoxicity, pancreatitis, coagulation disorders, and hyperglycemia (Cairo 1982). The toxic effects are related to immune reactions against the bacterial protein and to the effects of depletion of L-asparagine, and later on inhibition of protein synthesis in major glands like the liver and pancreas. These allergic reactions are the major toxicities, which can be produced primarily due to formation of anti-asparaginase antibodies in circulation. During the study of high risk ALL patients, it was observed that clinical allergy and high titer antibodies were very common with the augmented regimen on CCG-1961 (Avramis et al. 2002; Panosyan et al. 2004).

Hypersensitivity

Immunologic reactions caused by exposure to bacterial proteins can be quite common and may range in severity from localized, transient erythema and rash at the site of injection to urticaria, respiratory distress, and acute life-threatening anaphylaxis. Clinical symptoms of asparaginase hypersensitivity reactions includes, allergic reactions, anaphylaxis (rare), serum sickness, itching and swelling of extremities, edema, urticaria and rash, broncospasm, localized or generalized erythema and other clinically related reactions. Other toxicities related to asparaginases treatment include, organ toxicities, liver dysfunction, pancreatitis and related hyperglycemia, ketoacidosis, glucosuria, cerebral dysfunction, decreased protein synthesis, hypofibrinonemia, hypercoagulable state-coagulopathies, hypoalbuminemia (Avramis and Tiwari 2006).

Development of hypersensitivity reaction and inactivation of L-asparaginase by antibodies is evident in up to 60% cases with *E. coli* asparaginase. In fact, the pegylated form of *E. coli* L-asparaginase is not free from such adverse reaction. Consequently, the patients developing hypersensitivity reaction for one asparaginase formulation is shifted to another (Pieters et al. 2011). But the options for such shift are very limited, and necessitate the development of better asparaginase formulation with little adverse effect. This requirement urged for searching asparaginase potential in fungi, plant and mammals (Shrivastava et al. 2010).

Coagulation Disorder

Asparaginase therapy leads to severe acquired deficiency of serpin (inhibitor of serine protease) class of proteins including, antithrombin and $\alpha\alpha 1$ antitrypsin. It has been estimated that, 2.1–15% children and adults develop these complication after receiving native asparaginase preparation, while this risk is limited to 1.1–4% with pegylated form of asparaginase (Nowak-Göttl et al. 2003; Holle 1997). Antithrombin is a major protein involved in physiological inhibition of thrombin and other coagulation factor like IXa, Xa, XIa as well as factor VII. Antithrombin forms irreversible link with enzyme inhibiting its proteolytic activity (Jakubas et al. 2008). It has been reported that L-asparaginase induces conformational changes in antithrombin molecule leading to loss of its activity and formation of protein aggregate accumulated in ER cisterns (Hernandez-Espinoza et al. 2006). The adverse side events related to coagulation disorders are also produced due to effect of drug on protein synthesis, including reduction in antithrombin, fibrinogen, plasminogen, and factors IX and X with prolongation of activated partial thromboplastin time (Mitchell et al. 1994; Miniero et al. 1986). These complications are believed to be due to several mechanisms, including vessel wall damage by the catheter itself, as well as the chemotherapeutic agent.

Deficiencies of protein C and S have also been observed as a result of L-asparaginase treatment (Athale and Chan 2003: Nowak-Gottl et al. 2001). These abnormalities cause the excess thrombin levels and may increase the risk for bleeding or thrombosis. Hypofibrinogenemia has been reported to occur anywhere from 15% to 65%. Asparaginase therapy may affect the hemostatic system indirectly through the effect of supportive care (Athale and Chan 2003), for example, central venous line catheters used during treatment may cause thrombotic complications (Revel-Vilk 2006).

Pancreatitis, Hyperglycemia, Hepatotoxicity

The causes of these side effects are not well defined but may possibly arise due to any abnormality in protein synthesis. Hepatotoxicity may involve glutamine deficiency, oxidative stress, decreased hepatic protein synthesis, and later on impairment of beta-oxidation in mitochondria (Bodmer et al. 2006; Fromenty and Pessayre 1995, 1997). Histological results are rarely reported in patients who develop liver abnormalities while being treated with asparaginase. However, instances of macro- and microvesicular liver steatosis have been described (Bodmer et al. 2006; Sahoo and Hart 2003). Microvesicular steatosis can lead to liver failure and coma and may be fatal in certain cases.

L-asparaginase therapy cause significant increase in pancreatic amylase and lipase. L-asparagine is involved in synthesis inhibition of above mentioned enzymes. In the absence of L-asparagine, severe complications emerge in pancreas (Jakubas et al. 2008). L-asparaginase can adversely affect both the endocrine (insulin-secreting) and exocrine (digestive enzyme-secreting) cells of the pancreas. Some patients develop signs and symptoms of diabetes due to decreased synthesis of insulin.

Beside above mentioned complications, L-asparaginase therapy may also causes CNS related symptoms. These include drowsiness, hallucination, and amentia in some individuals. The reduced level of L-asparagine and/or L-glutamine in cerebral tissues is proposed to be a reason behind these complications (Chabner and Loo 1996).

10.4.1.2 Adverse Effects by Asparaginase Formulations

Three types of asparaginase preparations are available, first one native L-asparaginase from *Escherichia coli* (native L-asparaginase, Elspar, Merck), second *E. coli* L-asparaginase conjugated with polyethylene glycol (PEG) (pegaspargase, Oncaspar, Enzon) and third is L-asparaginase produced from the plant bacteria *Erwinia chrysanthemi* (*Erwinia* L-asparaginase, Erwinase).These L-asparaginase are approved and available in Europe and United States manufactured by EUSA Pharma (Earl 2009).

The Kyowa-Hakko native asparaginase preparation is also available in market under different brand names, like Crasnitin™ and Medac™ in Europe and Asia. All these asparaginases have bacterial origin, which act by deamenating the amino acids asparagines and Glutamine (Holcenberg and Teller 1976; Capizzi and Holcenberg 1993; Avramis et al. 2002; Panosyan et al. 2004).

Comparison of Hypersensitivity Reaction with Different Aspararaginase Preparation

L-asparaginase is covalently bind with PEG, blocks potentially immunogenic epitopes without obstructing the substrate-interaction site, and so that reduce the tendency for dose-limiting hypersensitivity without functional cooperation

(Abuchowski et al. 1984). This property makes it as a better substituent of native L-asparaginase. Similar results were obtained while comparing L-asparaginase isolated from *Erwinia* and *E. coli.* Native *E. coli* L-asparaginase was associated with more allergic reactions than *Erwinia* L-asparaginase (14% vs 6% respectively, $P=0.03$) according to the protocol of Dana-Farber Cancer Institute (Moghrabi et al. 2007).

Comparison of Hepatotoxicity Reaction with Different Asparaginase Preparation

In the study of Children's Cancer Group CCG-1962 trial, no patients receiving pegaspargase exhibited abnormal liver function, compared with 7% of patients receiving native L-asparaginase (e.g., aspartate aminotransferase, alanine aminotransferase, or alkaline phosphatase >1.5 times the normal value, or total bilirubin >1.5 times the normal value) (Avramis et al. 2002). While *Erwinia* L-asparaginase also showed low frequency of liver dysfunction (24%) in comparison to native (50%) and PEG (58%)- asparaginase (Dhall et al. 2008).

10.4.2 Recent Improvement in Bacterial L-Asparaginase

L-asparaginase is an antileukemic agent. These proteins awake the immune response due to bacterial origin of this enzyme, so observation of anti-asparaginase antibodies and its antigen-neutralizing activity may be helpful in prognosis but it is not very sensitive method (Panosyan et al. 2004). Recently enzyme-linked immunosorbent assay (ELISA) was developed and used for therapeutic monitoring to support clinical trials and for the evaluation of anti-asparaginase antibodies in patients (Panosyan et al. 2004; Avramis et al. 2009). But one drawback of this method is the uncertainty of below the lower limit of quantification (BLLOQ) by ELISA alone in patients with evident clinical allergy symptoms and this limitation need to be clarified. At the same time, one SPR (surface plasmon resonance) chip based bioassay method was developed. In this CM5 sensor chip human *E. coli* asparaginase, pegaspargase and Erwinase enzymes were covalently attached to the carboxy-methylated dextran matrix. This method shows high specificity, sensitivity and reproducibility of the RU (resonance units). The bioassay method was ten times more sensitive than the ELISA antibodies assay. The results of some experiments demonstrated that the SPR assay could detect approximately 10-fold lower concentration of weak affinity antibodies, at which the antibodies could be washed away in the ELISA assay. SPR method is very fast occurring in 300 s whereas ELISA which is an end-point analysis needs 24 h. Furthermore, the SPR method also inform about the subtype of antibodies (IgG).

In the last, due to the very high sensitivity of the SPR bioassay chip, the probability of false-negative results was decreased. Overall these studies inform that SPR assay method detects very rapidly, with high specificity and more accuracy against asparaginase than the ELISA method and it is now used to determine immunogenicity in cancerous samples (Avramis et al. 2009).

10.5 Conclusion

L-asparaginases are an essential element of treatment protocols for acute lymphoblastic leukemia for nearly 40 years. It is used in remission induction and intensification steps of ALL management. The effect of L-asparaginase is to decrease concentration of amino acid L-asparagine, which is necessary for cancer cells due to their inability to synthesize such amino acid. Currently three L-asparaginase preparations are available for clinical use. *E. coli* native asparaginase, pegylated form of *E. coli* asparaginase (Pegasparaginase) and the last one in the L-asparaginase isolated from *Erwinia chrysanthemi*. The efficacy and adverse reaction associated with different formulations of L-asparaginase varies with different factors. Although, this enzyme has been a cornerstone in the management of ALL, but the side effects associated with these formulations should be controlled. Present research in the direction of finding better substitute of asparaginase led to the detection of this activity in fungi. Fungi are evolutionary close in comparison to bacteria, perhaps the development of fungal asparaginase may involve less adverse reactions in comparison to bacterial asparaginase. But as far as present situation is concerned, the bacterial asparaginases are an integral element for management of ALL. It is the influx of asparaginase and other suitable chemotherapeutic modalities in cancer management, due to which the life expectancy of ALL patients has significantly increased.

References

Abuchowski A, Kazo GM, Verhoest CR et al (1984) Cancer therapy with chemically modified enzymes. I. Antitumor properties of polyethylene glycol asparaginase conjugates. Cancer Biochem Biophys 7(2):175–186

Adamson RH, Fabro S (1968) Antitumor activity and other biological properties of L-asparaginase (NSC – 109229)- a review. Cancer Chemother Rep 52(6):617–626

Aghaiypour K, Wlodawer A, Lubkowski J (2001) Structural basis for the activity and substrate specificity of *Erwinia chrysanthemi* L-asparaginase. Biochemistry 40(19):5655–5664. doi:dx.doi.org

Alpar HO, Lewis DA (1985) Therapeutic efficacy of asparaginase encapsulated in intact erythrocytes. Biochem Pharmacol 34(2):257–261

American Cancer Society (2010) Cancer facts and figure 2010. http://www.cancer.org/acs/groups/content/@nho/documents/document/acspc-024113.pdf

Asselin BL (1999) The three asparaginases. Comparative pharmacology and optimal use in childhood leukemia. Adv Exp Med Biol 457:621–629

Asselin BL, Ryan D, Frantz CN et al (1989) *In vitro* and *in vivo* killing of acute lymphoblastic leukemia cells by L-asparaginase. Cancer Res 49(15):4363–4368

Asselin BL, Lorenson MY, Whitin JC et al (1991) Measurement of serum L-asparagine in the presence of L-asparaginase requires the presence of an L-asparaginase inhibitor. Cancer Res 51(24):6568–6573

Athale UH, Chan AKC (2003) Thrombosis in children with acute lymphoblastic leukemia: Part II. Pathogenesis of thrombosis in children with acute lymphoblastic leukemia: effects of the disease and therapy. Thromb Res 111(4–5):199–212

Avramis VI, Panosyan EH (2005) Pharmacokinetic/pharmacodynamic relationships of asparaginase formulations: the past, the present and recommendations for the future. Clin Pharmacokinet 44(4):367–393

Avramis VI, Tiwari PN (2006) Asparaginase (native ASNase or pegylated ASNase) in the treatment of acute lymphoblastic leukemia. Int J Nanomedicine 1(3):241–254

Avramis VI, Sencer S, Periclou AP et al (2002) A randomized comparison of native *Escherichia coli* asparaginase and polyethylene glycol conjugated asparaginase for treatment of children with newly diagnosed standard-risk acute lymphoblastic leukemia: a Children's Cancer Group study. Blood 99(6):1986–1994

Avramis VI, Avramis EV, Hunter W et al (2009) Immunogenicity of native or pegylated *E. coli* and *Erwinia* asparaginases assessed by ELISA and surface plasmon resonance (SPR-Biacore) assays of IgG antibodies (Ab) in sera from patients with acute lymphoblastic leukemia (ALL). Anticancer Res 29(1):299–302

Bagert U, Rohm KH (1989) On the role of histidine and tyrosine residues in *E. coli* asparaginase. Chemical modification and ^{1}H-nuclear magnetic resonance studies. Biochim Biophys Acta 999(1):36–41

Bartalena L, Martino E, Antonelli A et al (1986) Effect of the antileukemic agent L-asparaginase on thyroxine-binding globulin and albumin synthesis in cultured human hepatoma (HEP G2) cells. Endocrinology 119(3):1185–1188

Becker FF, Broome JD (1967) L-asparaginase: inhibition of early mitosis in regenerating rat liver. Science 156(3782):1602–1603

Bodmer M, Sulz M, Stadlmann S et al (2006) Fatal liver failure in an adult patient with acute lymphoblastic leukemia following treatment with L-asparaginase. Digestion 74(1):28–32

Borisova AA, El'darov MA, Zhgun AA et al (2003) Purification and properties of recombinant *Erwinia carotovora* L-asparaginase expressed in *E. coli* cells. Biomed Khim 49(5):502–507

Boss J (1997) Pharmacokinetics and drug monitoring of L-asparaginase treatment. Int J Clin Pharmacol Ther 35(3):96–98

Boyse EA, Old LJ, Campbell HA et al (1967) Suppression of murine leukemias by L-asparaginase. Incidence of sensitivity among leukemias of various types: comparative inhibitory activities of guinea pig serum L-asparaginase and Escherichia coli L-asparaginase. J Exp Med 125(1):17–31

Broome JD (1961) Evidence that the L-asparaginase activity of guinea pig serum is responsible for its antilymphoma effects. Nature 191:1114–1115

Broome JD (1963) Evidence that the L-asparaginase activity of guinea pig serum is responsible for its anti-lymphoma effects. I. Properties of the L-asparaginase of guinea pig serum in relation to those of the antilymphoma substance. J Exp Med 118:99–120

Broome JD (1965) Antilymphoma activity of L-asparaginase in vivo: clearance rates of enzyme preparations from guinea pig serum and yeast in relation to their effects on tumor growth. J Natl Cancer Inst 35(6):967–974

Broome JD (1968) L-asparaginase: the evolution of a new tumor inhibitory agent. Trans NY Acad Sci 30(5):690–704

Broome JD (1981) L-asparaginase: discovery and development as a tumor inhibitory agent. Cancer Treat Rep 65(Suppl 4):111–114

Burchenal JH, Karnofsky DA (1970) Clinical evaluation of L-asparaginase. Introduction. Cancer 25(2):241–243

Cairo MS (1982) Adverse reactions of L-asparaginase. Am J Pediatr Hematol Oncol 4(3):335–339

Campbell HA, Mashburn LT, Boyse EA et al (1967) Two L-asparaginases from *Escherichia coli* B. Their separation, purification and antitumor activity. Biochemistry 6(3):721–730

Capizzi RL, Cheng YC (1981) Therapy of neoplasia with asparaginase. In: Holcenberg J, Roberts J (eds) Enzymes as drugs. Wiley, New York

Capizzi RL, Holcenberg JS (1993) Asparaginases. In: Holland JF, Frei E, Bast RC et al (eds) Cancer medicine, 3rd edn. Lea & Febiger, Philadelphia

Capizzi RL, Bertino JR, Handschumacher RE (1970) L-asparaginase. Ann Rev Med 21:433–444

Carter P, Wells JA (1988) Dissecting the catalytic triad of a serine protease. Nature 332(6164): 564–568

Chabner B (1990) Enzyme therapy, L-asparaginase. In: Chabner B, Collins J (eds) Cancer chemotherapy: principles and practice. Lippincott, Philadelphia

Chabner BA, Loo TL (1996) Enzyme therapy: L-asparaginase. In: Chabner BA, Longo DL (eds) Cancer chemotherapy and biotherapy, principles and practice, 2nd edn. Lippincott-Raven Publishers, Philadelphia

Chakrabarti R, Schuster SM (1997) L-asparaginase: perspectives on the mechanisms of action and resistance. Int J Pediatr Hematol Oncol 4:597–611

Clavell LA, Gelber RD, Cohen HJ et al (1986) Four agent induction and intensive asparaginase therapy for treatment of childhood acute lyphobalastic leukemia. New Engl J Med 315(11):657–663

Clayson DB (2001) Toxicological carcinogenesis. Lewis Publishers, Boca Raton

Clementi A (1922) La desamidation enzymatique de l'asparagine chez les differentes especes animals et la signification physiologique de sa presence dans l'organisme. Arch Intern Physiol 19:369–398

Davidson L, Burkom M, Ahn S et al (1977) l-Asparaginases from Citrobacter freundii. Biochim Biophys Acta 480(1):282–294

De-Angeli CL, Pocchiari F, Russi S et al (1970) Effect of L-asparaginase from *Aspergillus terreus* on ascites sarcoma in the rat. Nature 225(5232):549–550

Derst C, Wehner A, Specht V et al (1994) State and function of tyrosine residues in *Escherichia coli* asparaginase II. Eur J Biochem 224(2):533–540

Dhall G, Robison NJ, Rubin JI et al (2008) Incidence of adverse reactions to post-induction asparaginase (ASP) therapy in children and adolescents with high risk acute lymphoblastic leukemia (ALL): a report from the children's oncology group study CCG-1961. J Clin Oncol 26:abstr 10021

Distasio JA, Nredrerman RA, Kafkewitz D et al (1976) Purification and characterization of L-asparaginase with anti lymphoma activity from *Vibrio succinogenes*. J Biol Chem 251(22): 6929–6933

Earl M (2009) Incidence and management of asparaginase-associated adverse events in patients with acute lymphoblastic leukemia. Clin Adv Hematol Oncol 7(9):600–606

El-Bessoumy AA, Sarhan M, Mansour J (2004) Production, isolation, and purification of L-asparaginase from *Pseudomonas aeruginosa* 50071 using solid-state fermentation. J Biochem Mol Biol 37(4):387–393

Emery J, Lucassen A, Murphy M (2001) Common hereditary cancers and implications of primary care. Lancet 358(9275):56–63

Fromenty B, Pessayre D (1995) Inhibition of mitochondrial beta-oxidation as a mechanism of hepatotoxicity. Pharmacol Ther 67(1):101–154

Fromenty B, Pessayre D (1997) Impaired mitochondrial function in microvesicular steatosis. Effects of drugs, ethanol, hormones and cytokines. J Hepatol 26(suppl 2):43–53

Furth O, Friedmann M (1910) Uber die verbreitung asparaginspaltender organfermente. Biochem Z 26:435–440

Gaffar SA, Shethna YI (1975) Partial purification and antitumour activity of L-asparaginase from *Azotobacter vinelandi*. Curr Sci 44:727–729

Gallagher MP, Marshall RD, Wilson R (1989) Asparaginase as a drug for treatment of acute lymphoblastic leukaemia. Essays Biochem 24:1–40

Gaynon PS (2005) Childhood acute lymphoblastic leukaemia and relapse. Br J Haematol 131(5): 579–587

Goody HE, Ellem KA (1975) Nutritional effects on precursor uptake and compartmentalization of intracellular pools in relation to RNA synthesis. Biochim Biophys Acta 383(1):30–39

Hanahan D, Weinberg RA (2000) The hallmarks of cancer. Cell 100(1):57–70
Haskell CM, Canellos GP (1969) L-asparaginase resistance in human leukemia-asparagine synthetase. Biochem Pharmacol 18(10):2578–2580
Hellman K, Miller DS, Cammack KA (1983) The effect of freeze-drying on the quarternary structure of L-asparaginase from *Erwinia carotovora*. Biochim Biophys Acta 749(2):133–142
Hernandez-Espinoza D, Minano A, Martinez C et al (2006) L-Asparaginase-induced antithrombin type I deficiency: implication for conformational diseases. Am J Pathol 169(1):142–153
Hill JM, Roberts J, Loeb E et al (1967) L-asparaginase therapy for leukemia and other malignant neoplasms. Remission of human leukemia. J Am Med Assoc 202(9):882–888
Hirsch J (2006) An anniversary for cancer chemotherapy. JAMA 296(12):1518–1520
Ho DH, Wang CY, Lin JR et al (1988) Polyethylene glycol-L-asparaginase and L-asparaginase studies in rabbits. Drug Meta Dispos 16(1):27–29
Holcenberg JS, Teller DC (1976) Physical properties of antitumour glutaminase-asparaginase from *Pseudomonas* 7A. J Biol Chem 251(17):5375–5380
Holcenberg JS, Ericsson L, Roberts J (1978) Amino acid sequence of the diazooxonorleucine binding site of *Acinetobacter* and *Pseudomonas* 7A glutaminase-asparaginase enzymes. Biochemistry 17(3):411–417
Holle LM (1997) Pegaspargase: an alternative? Ann Pharmacother 31(5):616–624
Howlader N, Noone AM, Krapcho M et al (eds) (2010) SEER cancer statistics review, 1975–2008. National Cancer Institute, Bethesda
Imada A, Igarasi S, Nakahama K et al (1973) Asparaginase and glutaminase activities of micro organisms. J Gen Microbiol 76(1):85–99
Inada Y, FurukawaM SH et al (1995) Biomedical and biotechnological applications of PEG and PM modified proteins. Trends Biotechnol 13(3):86–91
Jakubas BP, Kuli MK, Giebel S et al (2008) Use of L-asparaginase in acute lymphoblastic leukemia: recommendations of the Polish Adult Leukemia Group. Pol Arch Med Wewn 118(11):664–669
Jarrar M, Gaynon PS, Periclou AP et al (2006) Asparagine depletion after pegylated *E. Coli* asparaginase treatment and induction outcome in children with acute lymphoblastic leukemia in first bone marrow relapse: a children's oncology group study (CCG-1941). Pediatr Blood Cancer 47(2):141–146
Jemal A, Siegel R, Ward E et al (2008) Cancer statistics, 2008. CA Cancer J Clin 58(2):71–96
Keating MJ, Holmes R, Lerner SH et al (1993) L-asparaginase and PEG asparaginase–past, present and future. Leuk Lymphoma 10:153–157
Keefer JF, Moraga DA, Schuster SM (1985) Amino acid content of L5178Y and L5178Y/L-ASE cells after L-asparaginase treatment. Biochem Med 34(2):135–142
Khushoo A, Pal Y, Mukherjee KJ (2005) Optimization of extracellular production of recombinant asparaginase in *Escherichia coli* in shake flask and bioreactor. Appl Microbiol Biotechnol 68(2):189–197
Kidd JG (1953) Regression of transplanted lymphomas induced *in vivo* by means of normal guinea pig serum. I. Course of transplanted cancers of various kinds in mice and rats given guinea pig serum, horse serum, or rabbit serum. J Exp Med 98(6):565–582
Kidd JG, Sobin LH (1966) The incorporation of L-asparagine-14C by lymphoma 6C3HED cells: its inhibition by guinea pig serum. Cancer Res 26(2):208–211
Kotzia GA, Labrou NE (2007) L-asparaginase from *Erwinia chrysanthemi* 3937: cloning, expression and characterization. J Biotechnol 127(4):657–669
Lang S (1904) Uber desamidierung im Tierkorper. Beitr chem Physiol Pathol 5:321–345
Law AS, Wriston JC (1971) Purification and properties of *Bacillus coagulans* L-asparaginase. Arch Biochem Biophys 147(2):4744–4752
Lebedeva ZI, Kabanova EA, Berezov TT (1988) The nature of functional group of active centers of anti-tumour glutamine-(asparagine-)ase. Bull Exp Biol Med 105(4):426–429
Lee SM, Wroble MH, Ross JT (1989) L-asparaginase from *Erwinia carotovora*. An improved recovery and purification process using affinity chromatography. Appl Biochem Biotechnol 22(1):1–11

Mashburn LT, Wriston JC (1964) Tumor inhibitory effect of L-asparaginase from *Escherichia coli*. Arch Biochem Biophys 105:450–452

Miniero R, Pastore G, Saracco P et al (1986) Hemostatic changes in children with acute lymphoblastic leukemia treated according to two different L-asparaginase schedules. Am J Pediatr Hematol Oncol 8(2):116–120

Mitchell LG, Halton JM, Vegh PA et al (1994) Effect of disease and chemotherapy on hemostasis in children with acute lymphoid leukemia. Am J Pediatr Hematol Oncol 16(2):120–126

Moghrabi A, Levy DE, Asselin B et al (2007) Results of the Dana-Farber Cancer Institute ALL consortium protocol 95–01 for children with acute lymphoblastic leukemia. Blood 109(3): 896–904

Mukharjee J, Majumdar S, Scheper T (2000) Studies on nutritional and oxygen requirements for production of L-asparaginase by *Enterobacter aerogenes*. Appl Microbiol Biotechnol 53(2):180–184

Müller HJ, Boos J (1998) Use of L-asparaginase in childhood ALL. Crit Rev Oncol Hematol 28(2):97–113

Nowak-Gottl U, Heinecke A, von Kries R et al (2001) Thrombotic events revisited in children with acute lymphoblastic leukemia: impact of concomitant *Escherichia coli* asparaginase/prednisone administration. Thromb Res 103(3):165–172

Nowak-Göttl U, Ahlke E, Fleischhack G et al (2003) Thromboembolic events in children with acute lymphoblastic leukemia (BFM protocols): prednisone versus dexamethasone administration. Blood 101(7):2529–2533

Oettgen HF, Old LJ, Boyse EA et al (1967) Inhibition of leukemias in man by L-asparaginase. Cancer Res 27(12):2619–2631

Panosyan EH, Grigoryan RS, Avramis IA et al (2004) Deamination of glutamine is a prerequisite for optimal asparagine deamination by asparaginases in vivo (CCG-1961). Anticancer Res 24(2C):1121–1125

Peterson RE, Ciegler A (1972) Factors influencing L-asparaginase production by *Erwinia aroideae*. Appl Microbiol 23(3):671–673

Pieters R, Hunger SP, Boos J et al (2011) L-asparaginase treatment in acute lymphoblastic leukemia: a focus on *Erwinia* asparaginase. Cancer 117(2):238–249

Prager Morton D, Baechtel Samuel F, Heifetza A (1982) In: Jeljaszewicz J, Pulverer G, Roszkowski W (eds) Bacteria and cancer. Academic, London

Pui CH (2010) Recent research advances in childhood acute lymphoblastic leukemia. J Formos Med Assoc 109(11):777–787

Raha SK, Roy SK, Dey S et al (1990) Purification and properties of an L-asparaginase from *Cylindrocarpon obtusisporum* MB-10. Biochem Int 21(6):987–1000

Revel-Vilk S (2006) Central venous line-related thrombosis in children. Acta Haematol 115(3–4):201–206

Ronghe M, Burke GA, Lowis SP et al (2001) Remission induction therapy for childhood acute lymphoblastic leukaemia: clinical and cellular pharmacology of vincristine, corticosteroids, L asparaginase and anthracyclines. Cancer Treat Rev 27(6):327–337

Sahoo S, Hart J (2003) Histopathological features of L-asparaginase-induced liver disease. Semin Liver Dis 23(3):295–299

Schwartz JH, Reeves JY, Broome JD (1966) Two L-asparaginases from *E. coli* and their action against tumors. Proc Natl Acad Sci USA 56(5):1516–1519

Selvakumar N (1979) Studies on L-asparaginase (antileukaemic agent) from coastal area of Porto novo. PhD thesis, Annamalai University, Chidambaram, India

Sherr CJ (2004) Principles of tumor suppression. Cell 116(2):235–246

Shrivastava A, Khan AA, Jain SK et al (2010) Biotechnological advancement in isolation of antineoplastic compounds from natural origin: a novel source of L-asparaginase. Acta Biomed 81(2): 104–108

Sieciechowicz KA, Ireland RJ (1989) Isolation and properties of an asparaginase from leaves of *Pisum sativum*. Phytochemistry 28(9):2275–2279

Sieciechowicz KA, Joy KW, Ireland RJ (1988a) Diurnal changes in asparaginase activity in pea leaves. I. The requirement for light for increased activity. J Exp Bot 39(6):695–706

Sieciechowicz KA, Ireland RJ, Joy KW (1988b) Diurnal changes in asparaginase activity in pea leaves. II. Regulation of activity. J Exp Bot 39:707–721

Sinha A, Manna S, Roy SK et al (1991) Induction of L-asparaginase synthesis in *Vibrio proteus*. Indian J Med Res 93:289–292

Sobin LH, Kidd JG (1965) A metabolic difference between two lines of lymphoma 6C3HED cells in relation to asparagine. Proc Soc Exp Biol Med 119:325–327

Sobin LH, Kidd JG (1966) Alterations in protein and nucleic acid metabolism of lymphoma 6C3HED-OG cells in mice given guinea pig serum. J Exp Med 123(1):55–74

Story MD, Voehring DW, Stephens LC et al (1993) L-asparaginase kills lymphoma cells by apoptosis. Cancer Chemother Pharmacol 32(2):129–133

Takase K, Takagi S, Okawa H et al (1985) Analysis of effect of L-asparaginase on the cell cycle of leukemic T cells by flow cytometry. Nippon Ketsueki Gakkai Zasshi 48(1):159–163

Thorburn AL (1983) Paul Ehrlich: pioneer of chemotherapy and cure by arsenic (1854–1915). Br J Vener Dis 59(6):404–405

Tosa T, Sano R, Yamamota K et al (1971) L-Asparaginase from *Proteus vulgaris*. Appl Microbiol 22(3):387–392

Trichopoulos D, Li FP, Hunter DJ (1996) What causes cancer? Sci Am 275(3):80

Villa P, Corada M, Bartošek I (1986) L-asparaginase effects on inhibition of protein synthesis and lowering of the glutamine content in cultured hepatocytes. Toxicol Lett 32(3):235–241

Vogelstein B, Kinzler KW (2002) The genetic basis of human cancer, 2nd edn. McGraw Hill, Toronto

Wang B, Relling MV, Storm MC et al (2003) Evaluation of immunologic crossreaction of antiasparaginase antibodies in acute lymphoblastic leukemia (ALL) and lymphoma patients. Leukemia 17(8):1583–1588

Waye MM, Stanners CP (1981) Role of asparagine synthetase and asparagyl-transfer RNA synthetase in the cell-killing activity of asparaginase in Chinese hamster ovary cell mutants. Cancer Res 41(8):3104–3106

Whelan HA, Wriston JC (1969) Purification and properties of asparaginase from *Escherichia coli* B. Biochemistry 8(6):2386–2393

Woo MH, Hak LJ, Storm MC et al (2000) Hypersensitivity or development of antibodies to asparaginase does not impact treatment outcome of childhood acute lymphoblastic leukemia. J Clin Oncol 18(7):1525–1532

Worton KS, Kerbel RS, Andrulis IL (1991) Hypomethylation and reactivation of the asparagine synthetase gene induced by L asparaginase and ethyl methanesulfonate. Cancer Res 51(3):985–989

Wriston JC, Yellin TO (1973) L-asparaginase: a review. Adv Enzymol Relat Areas Mol Biol 39:185–248

Yellin TO, Wriston JC (1966) Antagonism of purified asparaginase from guinea pig serum towards lymphoma. Science 151(713):998–999

Chapter 11
Can Bacteria Evolve Anticancer Phenotypes?

Navya Devineni, Reshma Maredia, and Tao Weitao

Abstract Can bacteria evolve anticancer phenotypes when growing together with cancer cells? Three hypotheses have been developed to address this question. First, based on an analogy of bacterial biofilm development from planktonic cells to carcinogenesis from metastatic cells, there seems to be an evolutionary relationship between bacteria and cancer. Such an analogy in lifestyle suggests that bacteria and cancer may compete with each other while encountering nutrient needs or anti-proliferation drug treatment. These thoughts lead to the second hypothesis that bacteria can form biofilms on cancer cells under treatment of DNA replication inhibitory anti-cancer drugs, thereby impairing metastasis. Under such treatment, bacteria may be able to adhere to and invade into cancer cells—thus resulting in advantageous bacterial phenotypes—so that bacteria survive the drug attack. This hypothesis proposes that bacteria, growing with cancer cells and replication inhibition drugs, undergo the SOS response through which the evolution of the anticancer phenotypes is facilitated.

Keywords Bacteria • Biofilms • Infections • SOS • Bacterial attachment • Bacterial invasion • Cancer • Anticancer drug • Anticancer therapy • Metastasis • DNA replication inhibitors • DNA damage • Proteins • Proteomics • Molecular evolution

Abbreviations

HRLY	High growth rate but low yield
LRHY	Low growth rate but high yield
dNTP	deoxyribonucleoside triphosphates
c-di-GMP	bis-(3′-5′)-cyclic dimeric guanosine monophosphates

N. Devineni • R. Maredia • T. Weitao (✉)
Biology Department, The University of Texas at San Antonio,
One UTSA Circle, San Antonio, TX 78249-0662 USA
e-mail: tao.wei@utsa.edu

A.A. Khan (ed.), *Bacteria and Cancer*, DOI 10.1007/978-94-007-2585-0_11,

11.1 Introduction

Would bacteria always be a human's deadly enemy? Bacteria are very small organisms, usually consisting of single cell, lacking a nucleus, and organelles such as mitochondria and chloroplasts, yet having an intriguing life cycle (Weitao and Nordstrom 2010). Many bacteria are so tiny that a million of them, laid end-to-end, would measure no more than about 5 cm (2 in.). Bacteria are found everywhere, in air, soil, water, and inside our body and on our skin, though we cannot visualize them with naked eye. It is amazing to know that microbes cover most of our body; in fact, the ratio of bacterial cells to our body cells is 10:1 approximately. When we hear the word bacteria, often negative side of falling sick comes to our mind, because certain bacteria act as pathogens, causing infectious diseases, such as tetanus, typhoid fever, pneumonia, syphilis, cholera, food-borne illness and tuberculosis. Also, harmful bacterial toxins in food cause botulism, which can lead to paralysis or even death. However, not all bacteria are harmful. Certain types of bacteria live in the intestines and on the skin of human beings as normal flora, not only providing essential vitamins, but also protecting humans from pathogenic bacteria. Importantly, bacteria have been exploited by industries, to produce a variety of beneficial things including antibiotics, cheese, ethanol etc. Could bacteria be further exploited and turned into anticancer therapeutics? This question will make a new and interesting turning point in the journey to fight cancer.

Cancer as we all know burdens the healthcare system significantly. For example, the population of breast cancer survivors currently is estimated to be 2.5 millions in the United States; of these patients, 65% have survived for more than 5 years since their initial diagnosis (Ries et al. 2009). Such intimidating lethality was well documented to be attributed to cancer invasion and metastasis. Metastasis, which is the cause of 90% of deaths from solid tumours, is a process in which cancer cells migrate to distant sites and adapt to the tissue microenvironment from the primary cancer. The cancer prognosis becomes unpredictable if invasion and metastases have occurred before surgical removal of the primary cancer. However, if invasion and metastases are inhibited or blocked, cancer can be no longer life threatening.

Understandably, a variety of strategies using live agents to curb metastasis and cancer progression have been the focus of intense anticancer research. Among the strategies, viruses have been administered for cancer treatment over decades (Liu et al. 2007), and oncolytic viruses are being developed as anticancer drugs (Russell and Peng 2007). Yet, the key hurdles are low specificity of the microbial agents for cancer cells, immunity barrier to the clinical efficacy, and high-risk infections for patients. These hurdles have hindered the clinical advancement towards cancer treatment (Power and Bell 2007). Attempts were also made to administer live bacteria to control cancer progression (Coley 1991; Nauts et al. 1946). Later on, attenuated bacteria have been used as vehicles to deliver a variety of protective agents into cancer hosts (Chakrabarty 2003; Roland et al. 2005). Additionally, certain bacteria have tumour finding nature defined as the ability to replicate inside tumour cells preferentially where the intracellular bacteria may evade host immunity (Yu et al. 2004). These approaches with bacteria so far have relied mostly on host immunity to indirectly impair cancer cells. However, such strategies that depend on host immunity and tumour finding nature are unlikely to effectively control cancer metastasis and progression, because the immune defense of cancer patients is often impaired (Lehrnbecher et al. 2008), and infections are usually fatal, and the innate immune response though elicited by the bacteria is unspecific and short lived. Due to the fatal infections together with the deficient immunity, research to advance bacterial anti-cancer approach has been neglected.

Present chapter aims to address the neglected issue and to set a new direction towards bacterial evolution against cancer. We begin this chapter by discussing how bacteria form biofilms in response to the replication stress which leads to a hypothesis of evolutionary relationship between bacteria and cancer (Weitao 2009a). Although we focus mainly on the response of bacteria to an anti-proliferative drug hydroxyurea, the concept discussed applies to other replication inhibitors as well. We then describe how bacterial macromolecules released during anticancer treatment can result in the formation of biofilms against cancer metastasis (Weitao 2009b). Finally, we shall confront the riveting question whether bacteria can evolve anticancer phenotypes by using our understanding of the bacterial SOS response against DNA replication inhibitors (Dallo and Weitao 2010). In the later part of the chapter we present the hypothesis: bacteria when grown along with cancer cells, generate novel anticancer phenotypes by undergoing SOS response induced by anticancer drugs (Dallo and Weitao 2010).

11.2 Multicellularity of a Unicellular Organism in Response to DNA Replication Stress

11.2.1 Bacterial Biofilm vs. Cancer

Bacteria were traditionally thought of as unicellular organisms that grow strictly as free living (planktonic) cells. However, under certain conditions bacteria behave like multi-cellular organisms, forming multi-cellular communities—biofilms, which involve active communication and cooperation among cells. Such behavior may underscore their evolutionary relationship to multi-cellular organisms; this thought has led to a futuristic hypothesis of whether biofilm development from planktonic cells is analogous to carcinogenesis from metastatic cells.

11.2.2 A New Mechanism for Biofilm Development in Response to DNA Replication Inhibitors

The aforementioned analogy could be derived from computer simulations of biofilms. Two growth strategies have been suggested *in silico* for bacteria (Kreft 2004)—high growth rate but low yield (HRLY) and low growth rate but high yield (LRHY)—where the growth rate is defined as biomass formation per unit of time, and the yield as biomass generation per amount of resources consumed. Planktonic cells tend to grow at maximum growth rates during exponential phase whenever nutrients are available, characteristic of HRLY cells that consume resource egoistically to gain the short-term advantage. In contrast, LRHY subpopulations would utilize resources ecologically for the long-term advantage. Based on such reasoning, it can be predicted that cells in biofilms may grow more slowly than cells in a planktonic subpopulation; in fact, *Pseudomonas aeruginosa* grows planktonically at a higher rate than stagnantly in biofilms. Because of the long-term advantage, LRHY may render biofilms predominant in nature (Kreft 2004). While comprehensive empirical evidence for this model is lacking, we hypothesized that selectively impairing HRLY favors LRHY biofilm formation.

To test this hypothesis, we used DNA replication inhibitor hydroxyurea. Hydroxyurea inhibits ribonucleotide reductases, which are essential for *de novo* synthesis of deoxyribonucleoside triphosphates (dNTPs), a rate-limiting step in DNA biosynthesis. In *Escherichia coli*, hydroxyurea treatment results in intracellular dNTP depletion, which arrests DNA replication. This arrest may be differentially harmful to highly proliferating HRLY cells for the following reasons. HRLY bacteria can grow so quickly that their doubling time is shorter than the time needed to complete chromosomal replication. Consequently, ongoing rounds of replication cannot be completed at cell division and must be carried over to the next generation, leading to multiple replication forks in a single cell. If a cell over-initiates replication and generate multiple forks, arrest would occur at the multiple forks under dNTP depletion conditions causing DNA damage beyond the repair capacity and eventually growth arrest. This notion is evidenced by a study with *dnaA* (cos) mutants that over-initiate replication and therefore become susceptible to hydroxyurea (Nordman et al. 2007). Hence, HRLY cells would be susceptible to hydroxyurea.

We therefore tested the hypothesis with hydroxyurea, asking whether hydroxyurea suppresses HRLY planktonic cells of *P. aeruginosa* but induces LRHY biofilm formation. This hypothesis receives support from the work of Gotoh et al. (2008), showing inhibition of planktonic growth of *P. aeruginosa* but induction of biofilm formation in the presence of hydroxyurea and involvement of a second messenger bis-(3′–5′)-cyclic dimeric guanosine monophosphate (c-di-GMP) signaling. Treatment of bacteria with the replication inhibitor also induces a transcriptional response of bacteria to DNA damage known as the SOS response. While evidence for the link of c-di-GMP signaling to the SOS response is lacking, it is worth future investigation as to whether biofilms are induced through a network of c-di-GMP signaling and the SOS response.

11.2.3 Analogous Behaviors of Biofilms and Cancer Encountering DNA Replication Inhibitors

Moreover, one of the far-reaching implications from the study above seems to be the analogous effects of hydroxyurea treatment of bacteria and of cancer. Analogies can be drawn between planktonic bacteria and metastatic cells, and between biofilm and solid tumour development. Like planktonic bacteria, some types of cancer cells that vigorously proliferate can be viewed as HRLY in a broader sense. Cancer cells that have escaped from the proliferation controls are characteristic of the HRLY bacteria. Cancer cells compete with the surrounding normal cells for nutrient uptake and waste removal, and their uncontrolled growth behavior enables them to expand and invade, much like invasion of the LRHY communities by the HRLY (Kreft 2004). Understandably, "hyper-tumours" would gain a short-term advantage by growing quickly and egoistically through exploiting the developed vasculature system until they die due to the lack of further angiogenesis—an analogy to the short-term advantage of the HRLY bacteria.

Hydroxyurea has been historically used to treat myeloproliferative diseases. In fact, the drug has been generally recommended as the first line of treatment for essential thrombocythemia and polycythemia vera, rather than for advanced slowly growing solid tumours. While the mechanisms behind the recommendation are poorly understood, the slowly growing tumours can be viewed as LRHY and accordingly are refractory to hydroxyurea, but a fraction of cells in certain hematopoietic diseases and metastatic populations can have planktonic HRLY behavior in bloodstream and consequently are susceptible. This assumption is evidenced by a study showing that treatment of such a slowly growing tumour as meningioma with hydroxyurea has only a marginal benefit, but the treatment of migrating and proliferating murine fibrosarcoma shows greater inhibitory effects (Haug et al. 1993).

11.2.4 Evolutionary Relationship Between Bacterial Biofilms and Cancer

In conclusion, hydroxyurea, an antiproliferative drug for tumour treatment, induces multicellularity of *P. aeruginosa* via impairing planktonic proliferation. Apparently, the planktonic cells are susceptible while the biofilm cells are refractory to this drug. Analogously, metastatic cells like planktonic bacteria are susceptible (Haug et al. 1993), but slowly-growing tumours like biofilms seem refractory (Loven et al. 2004). The differential responses suggest that both bacterial and certain tumour cells might be proliferating egoists under certain conditions that make them susceptible to the replication inhibitor, implying their evolutionary relationship as generally proposed for the uni- and multi-cellular organisms. The cells from the two distinct life domains would be expected to compete with each other in the ecosystem of a eukaryotic host. Such competition may be due to the mechanisms either underlying nosocomial chronic biofilm related infections for individuals under chemotherapies or underlying anti-cancer therapeutic regimens.

We now know that *P. aeruginosa* forms multicellular communities when treated with anticancer drugs, such as hydroxyurea. Also, the analogy between planktonic cells and metastatic cells leads us to the thought that bacterial biofilms can be generated against cancer cells. In the next section of this chapter we will discuss whether and how the formed biofilms can disrupt the process of metastasis.

11.3 Bacteria Form Biofilms Against Cancer Metastasis

11.3.1 A Hypothesis About Bacterial Anti-metastasis Therapy

Sarcomas in patients was regressed with acute *Streptococcal* infections (Coley 1991; Nauts et al. 1946), a nineteenth century observation that has stimulated research communities in multi-facets; one of which is metastasis. While mechanisms behind the sarcoma regression are unclear, we hypothesize that metastasis can be impaired by antagonist bacterial infections or the bacterial macromolecules during treatment with the type of anticancer drugs—DNA replication inhibitors.

11.3.2 Can Bacteria Form Biofilms on Cancer Cells?

The hypothesis above predicts that since formation of bacterial biofilms is induced by hydroxyurea (Gotoh et al. 2008), biofilms can form on cancer cells, disrupting metastasis. This prediction is evidenced by the finding that *P. aeruginosa* not only acquires biofilm like properties during growth within airway epithelial cells (Garcia-Medina et al. 2005) but also attaches to and penetrates the human lung epithelial cells derived from a human bronchus alveolar carcinoma (Carterson et al. 2005). Specific attachment of bacteria to cancer cells may allow the bacteria to grow on the surface of cancer cells so that the bacteria invade to acquire nutrients and interfere with cancer metastasis and other cellular functions.

11.3.3 Can Bacteria Produce Anticancer Macromolecules?

This hypothesis also predicts that the bacterial macromolecules can coat cancer cells to block metastasis. Such macromolecules as proteins and DNA are released from the bacteria treated with the anticancer drug and mediate the biofilm formation (our unpublished data). Released proteins and DNA are indeed thought to be potential countermeasures against cancer (Mahfouz et al. 2007). Additionally, polysaccharides from *Streptococcus agalactiae* inhibit adhesion of cancer cells to endothelial cells—an essential step in cancer metastasis (Miyake et al. 1996). Conceivably, encaged by bacteria and their macromolecules, cancer should not be able to metastasize and colonize, and the coated cancer cells are immuno-reactive so that the protective immunity can be enhanced.

This hypothesis is testable. Bacterial adhesion to cancer cells in the presence of anticancer drugs can be analyzed with *in vitro* and *in vivo* imaging. The bacterial proteins required for specific bacterial attachment and for formation of biofilms on the cancer surface can be identified by proteomic analysis. The impact is profound in control and treatment of cancer, as metastasis leads to major deaths from solid tumours (Gupta and Massague 2006).

In conclusion, disruption of the metastasis by formation of biofilms over cancer cells is a very good start which would lead to effective treatment or prevention of cancer in future. As to be discussed next, the formed biofilm and the bacterial proteins are the products of the SOS response triggered by anti-proliferative drugs, such as hydroxyurea (Gotoh et al. 2010). Clearly, this work may pave a new multi disciplinary avenue by bringing together bacteriology and cancer research to development of the evolution-based therapeutics against cancer.

11.4 Bacteria Under SOS Evolve Anticancer Phenotypes

11.4.1 Background

11.4.1.1 Live Bacteria as Anti-cancer Therapeutics

Attempts with live bacteria to control cancer progression were tried over a century ago (Coley 1991; Nauts et al. 1946). Although undesired infections raised a concern, hypotheses have been proposed and tested to address the problem with promising progress. As reviewed by Chakrabarty (2003), antitumour treatment with *Clostridium novyi* was proposed, based on propensity of the anaerobe to grow in anaerobic core of the tumours and to deprive tumours of oxygen and essential nutrients. *Salmonella*, a facultative anaerobe, was also found to have tumour propensity that appears to be encoded by the pathogenicity island. Furthermore, bacteria could be engineered for selective destruction of tumours and for bacterial gene-directed prodrug therapy; in fact, such bacteria appeared to kill tumours selectively but not the normal tissue (Jain 2001). While these data support the notion of bacterial tropism and cancer killing, it remains unclear how they are developed and what evolutionary relationship of bacteria is with cancer.

11.4.1.2 Analogy of Bacterial Lifestyle to Cancer Cell Behaviors

We previously proposed the analogy of bacterial lifestyle to cancer cell behaviors, projecting the evolutionary relationship (Weitao 2009a, b). The shared features are reflected by observations that bacteria and cancer cells respond similarly to such anticancer drugs as DNA replication inhibitors (Weitao 2009a). These common lifestyles imply that they may compete with each other under certain conditions (Weitao 2009a, b). Bacteria growing under competition and drug influence are highly likely to evolve new phenotypes against cancer.

11.4.1.3 Bacterial SOS Response as Evolution Driving Force

Replication inhibitors induce the SOS response (Walker 1984) during which generation of new phenotype may be facilitated. SOS is a transcriptional response, in which

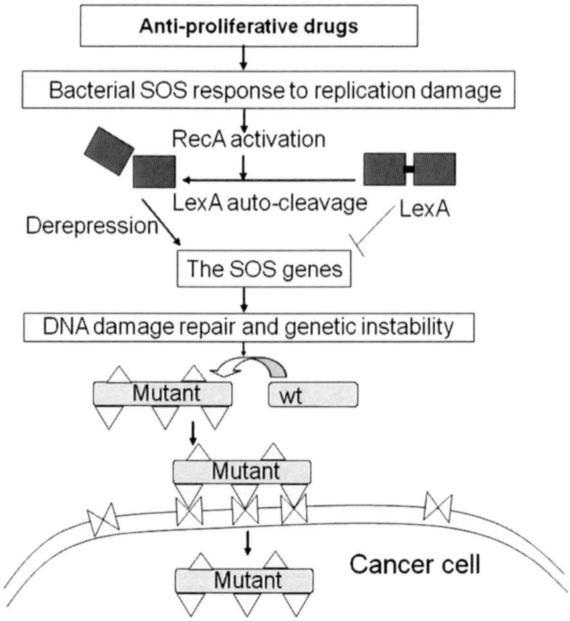

Fig. 11.1 SOS-facilitated evolution of bacterial cancer capture-invasion phenotypes. The bacterial SOS response is triggered by DNA damage caused by treatment with replication inhibition anticancer drugs. This response is controlled by the interplay of LexA and RecA. These players regulate the SOS genes that encode functions required for DNA damage repair. LexA represses these genes. DNA damage activates RecA to stimulate autocatalytic cleavage of LexA so that the SOS genes are derepressed and expressed for repair. Cell division is inhibited and delayed resulting filamentation to allow repair before cell division. If the DNA damage is so extensive that the cells cannot directly repair, the lesions of damage can be bypassed, leading to mutagenesis and genetic instability. A variety of bacterial mutants can be generated consequently and selected for adhesion to cancer cells and invade them to evade the drug attack. The mutants produce proteins (*triangles*) that recognize the cancer cells surface (*double triangles*) and mediate bacterial adhesion to the cancer cells

at least 40 SOS genes in *E. coli* (Courcelle et al. 2001; Fernandez de Henestrosa et al. 2000; Khil and Camerini-Otero 2002) and 15 in *P. aeruginosa* (Cirz et al. 2006) are induced through interplay of the SOS regulators LexA and RecA (Fig. 11.1). In the presence of single stranded DNAs that are generated during replication inhibition, RecA coprotease senses the signals and binds to the single stranded DNA to assume an active conformation (Sassanfar and Roberts 1990) and to stimulate auto-cleavage of LexA (Little 1991). Consequently, LexA repression of the SOS genes is prevented by this cleavage leading to a global induction of the SOS response.

These SOS gene products are involved in cytogenesis, DNA recombination, DNA replication, DNA damage repair, and segregation of chromosomes during cell division (Cox 1998; Sherratt 2003). For instance, the SOS gene, *sulA,* is induced to inhibit and delay cell division transiently leading to cell filamentation (Fig. 11.1) until DNA damage is ameliorated. The SOS controlled *umu* operon is involved in the error-prone translesion DNA synthesis (Kitagawa et al. 1985). If damage is so extensive that it cannot be directly repaired, the lesions of damage can be bypassed by the translesion synthesis with aid of the *umu* encoded proteins (Bridges and Woodgate 1985), leading to mutagenesis and genetic instability. A variety of bacterial mutants can be generated consequently. If bacteria grow with cancer and anticancer drugs, pools of these bacterial mutants are, in fact, selected for new phenotypes (Fig. 11.1).

This section aims to present a creative hypothesis as below. This hypothesis predicts that bacteria can evolve the cancer adhesion-invasion phenotypes, to challenge the limitation of anticancer treatment arising from bacterial natural propensity to cancer. The outcomes should help develop novel bacterial anticancer regimens to deal with the safety and specificity issues poised over a century ago.

11.4.2 Presentation of the Hypothesis

11.4.2.1 Can Bacteria Be Tamed to Produce Anticancer Phenotypes?

Our hypothesis states that bacteria, growing with cancer cells and replication inhibition drugs, evolve advantageous phenotypes. This hypothesis suggests that if treated with drugs, bacteria can be induced to adhere and invade cancer cells so that bacteria can survive drug attack. These features are defined as the cancer adhesion-invasion phenotypes. Obviously, our hypothesis is not based on the bacterial natural antitumor propensity but on the SOS-induced molecular evolution of new phenotypes. This hypothesis will be tested with *P. aeruginosa* as a starter. While its ecological niche may not be necessarily tumour, this bacterium could attach to and penetrate human lung epithelial cells derived from a human bronchus alveolar carcinoma (Carterson et al. 2005). These antitumor activities may be mediated by bacterial proteins; in fact, *P. aeruginosa* does have such an antitumor potential since it has genes encoding antitumor proteins. Azurin is a periplasmic antitumor protein in *P. aeruginosa* (reviewed in Mahfouz et al. 2007). Release of Azurin depends on contact with cancer cells, and Azurin targets preferentially cancer cells but marginally normal cells (Yamada et al. 2005). Additionally, Laz and Pa-CARD display cytotoxic activity against leukemia cells (Kwan et al. 2009). Our hypothesis suggests that such proteins and new candidates would emerge when bacteria undergoing the anticancer drugs induced SOS mutagenesis interact with cancer cells.

11.4.2.2 Generation of Anticancer Phenotypes During SOS

Furthermore, bacterial adhesion to cancer may be induced as we proposed previously (Weitao 2009b). *P. aeruginosa* can be induced to attach on abiotic surface and can form biofilms in response to hydroxyurea (Gotoh et al. 2008; Weitao

2009a). Indeed, such stress inducible biofilm formation is SOS-dependent (Gotoh et al. 2010). While historically it is an antiproliferative drug for tumour treatment (Gotoh et al. 2008; Weitao 2009a), hydroxyurea is also a replication inhibitor targeting at ribonucleotide reductases that are a good anticancer target (Shao et al. 2006). This drug inhibits growth of proliferating planktonic bacterial cells but stimulates bacterial adhesion (Gotoh et al. 2008; Weitao 2009a), likely to cancer if the bacterial cells grow with the cancer cells. Such replication inhibitors induce the bacterial SOS response (Fig. 11.1) (Walker 1984) during which generation of the advantageous phenotypes may be facilitated (Fig. 11.1). For instance, error prone DNA replication generates mutagenesis and genetic instability during SOS, yielding a variety of bacterial mutants. Since bacterial entry into cancer cells can evade the drug attack, these mutants can be selected for cancer invasion for bacterial survival.

11.4.3 Testing the Hypothesis

11.4.3.1 Experimental Design

To test this hypothesis, we will first examine attachment of *P. aeruginosa* to cancer cells and cancer invasion (Fig. 11.1). We will use the *recA* mutant as a bacterial SOS control because RecA initiates SOS that may facilitate development of the adhesion-invasion phenotypes. We will incubate these bacterial cells with cancer cells in the presence of the anticancer drugs that inhibit DNA replication enzymes including DNA polymerases (Berdis 2008), DNA helicases (Sharma et al. 2005), ribonucleotide reductases (Shao et al. 2006), and topoisomerases (Teicher 2008). We will harvest the invaded bacteria from the cancer cells. Second, we will use proteomic analysis to reveal proteins that exhibit distinct changes in expression in the bacterial cells that adhere to and invade into cancer cells. Lastly, according to the proteomic results, we will identify the mutated genes encoding cancer adhesion and invasion. We will construct the deletion mutants by deletion-insertion of the genes encoding the induced proteins. Then, we will test the mutants for cancer adhesion and invasion. We will also over produce these proteins and examine adhesion-invasion phenotypes against the cancer cells and the non cancer control.

11.4.3.2 Bacterial Proteins Differentially Induced During Bacteria-Cancer Interaction

Bacteria under the anticancer drug induced SOS are expected to produce proteins that mediate the cancer adhesion-invasion phenotypes (Fig. 11.1). The protein induction is generally reflected by alterations in the intensities of the

2-D gel protein spots from the invaded and non-invaded bacteria, with reference to the SOS controls. Interacting with cancer cells, the bacteria are likely to evolve a pattern of surface proteins suitable for bacterial adhesion to and invasion of cancer cells (Fig. 11.1). The pattern is likely to be unique to the cancer adhesion-invasion phenotypes as compared with that of the SOS and non-invaded controls. These proteins are expected to be deficient in the *recA* mutant, in which SOS is precluded. They are unlikely to appear on the non-invaded bacterial control.

11.4.3.3 Identification of Bacterial Genes Encoding Cancer Adhesion and Invasion

It is plausible that knocking out the bacterial cancer adhesion-invasion genes would disrupt the adhesion-invasion phenotypes. If the proteomic analysis indicates increase in the levels of the proteins from the bacteria with the adhesion-invasion phenotypes, and if mutations inactivate the genes encoding these proteins, then the mutant bacteria will be unable to attach to and invade the cancer cells. However, over-production of these proteins is expected to enhance bacterial capture and invasion of the cancer cells but not the non cancer control. Then, it can be concluded that these genes are required for the cancer adhesion-invasion phenotypes.

Collectively, the high resolution of the 2-D based proteomic approach will allow us to identify the patterns of the cancer-inducible surface proteins on bacteria so that the encoding genes can be identified. These genes can be cloned and expressed in non-pathogenic bacteria that are safe to the hosts but selectively lethal to cancer cells.

11.4.4 One Implication of the Hypothesis

Our hypothesis regarding evolution of bacterial anticancer phenotypes implies that bacteria can evolve cancer cell-specific phenotypes when growing with cancer cells of certain types, for instance, the metastatic cells. While metastasis can be impaired by the antagonist bacterial biofilms (Weitao 2009b) or by the bacterial proteins (Kwan et al. 2009; Mahfouz et al. 2007; Yamada et al. 2005) during treatment with the anticancer drugs as proposed previously (Weitao 2009b), the underlying mechanism for bacterial recognition or tropism for cancer cells is not fully understood and thus addressed by this hypothesis. As bacterial tropism is still in its infancy, if this hypothesis is true, the outcomes may stimulate future research interest into development and evolution of the bacterial tropism for cancer cells, contributing to formulating novel bacterial anti-metastasis regimens.

11.5 Conclusion

Can bacteria be tamed to cure cancer? This question stems from the nineteenth century observation of regression of Sarcomas in patients who were infected with S*treptococcal* infections (Coley 1991; Nauts et al. 1946). Over centuries, administration of live bacteria as anticancer therapeutics has been abandoned because of the lethal side effect of infections. The idea to pursue further development has been neglected. The neglect was followed by a dark age in this area. Nevertheless, it seems to come into light that both bacterial biofilm formation and cancer development are evolutionary and both show analogous lifestyles. These thoughts have led to futuristic hypotheses as presented in this chapter. It can be imagined that these hypotheses may have profound impacts in the research arena against cancer. As scope of human imagination can not be limited, so our thinking, reasoning, and hypothesizing can move forward to cross the boundary that once confined our enthusiasm about the bacteria-cancer interactions.

Acknowledgements This work is supported by San Antonio Area Foundation, Barshop 2008 Seed grant program in the comparative biology of aging, and Collaborative Research Seed Grant Program (CRSGP 2010–2011). We are indebted to Peter Lentz, a graduate student working on his Master's thesis, for his proofreading of this manuscript.

References

Berdis AJ (2008) DNA polymerases as therapeutic targets. Biochemistry 47:8253–8260

Bridges BA, Woodgate R (1985) Mutagenic repair in *Escherichia coli*: products of the recA gene and of the umuD and umuC genes act at different steps in UV-induced mutagenesis. Proc Natl Acad Sci USA 82:4193–4197

Carterson AJ, Honer zu Bentrup K, Ott CM et al (2005) A549 lung epithelial cells grown as three-dimensional aggregates: alternative tissue culture model for *Pseudomonas aeruginosa* pathogenesis. Infect Immun 73:1129–1140

Chakrabarty AM (2003) Microorganisms and cancer: quest for a therapy. J Bacteriol 185:2683–2686

Cirz RT, O'Neill BM, Hammond JA et al (2006) Defining the *Pseudomonas aeruginosa* SOS response and its role in the global response to the antibiotic ciprofloxacin. J Bacteriol 188:7101–7110

Coley WB (1991) The treatment of malignant tumors by repeated inoculations of erysipelas. With a report of ten original cases. 1893. Clin Orthop Relat Res 262:3–13

Courcelle J, Khodursky A, Peter B et al (2001) Comparative gene expression profiles following UV exposure in wild-type and SOS-deficient *Escherichia coli*. Genetics 158:41–64

Cox MM (1998) A broadening view of recombinational DNA repair in bacteria. Genes Cells 3:65–78

Dallo SF, Weitao T (2010) Bacteria under SOS evolve anticancer phenotypes. Infect Agent Cancer 5:3

Fernandez de Henestrosa AR, Ogi T, Aoyagi S et al (2000) Identification of additional genes belonging to the LexA regulon in *Escherichia coli*. Mol Microbiol 35:1560–1572

Garcia-Medina R, Dunne WM, Singh PK et al (2005) *Pseudomonas aeruginosa* acquires biofilm-like properties within airway epithelial cells. Infect Immun 73:8298–8305

Gotoh H, Zhang Y, Dallo SF et al (2008) *Pseudomonas aeruginosa* under DNA replication inhibition tends to form biofilms via Arr. Res Microbiol 159:294–302

Gotoh H, Kasaraneni N, Devineni N et al (2010) SOS involvement in stress-inducible biofilm formation. Biofouling 26:603–611

Gupta GP, Massague J (2006) Cancer metastasis: building a framework. Cell 127:679–695

Haug IJ, Siebke EM, Grimstad IA et al (1993) Simultaneous assessment of migration and proliferation of murine fibrosarcoma cells, as affected by hydroxyurea, vinblastine, cytochalasin B, razoxane and interferon. Cell Prolif 26:251–261

Jain KK (2001) Use of bacteria as anticancer agents. Expert Opin Biol Ther 1:291–300

Khil PP, Camerini-Otero RD (2002) Over 1000 genes are involved in the DNA damage response of *Escherichia coli*. Mol Microbiol 44:89–105

Kitagawa Y, Akaboshi E, Shinagawa H et al (1985) Structural analysis of the umu operon required for inducible mutagenesis in *Escherichia coli*. Proc Natl Acad Sci USA 82:4336–4340

Kreft JU (2004) Biofilms promote altruism. Microbiology 150:2751–2760

Kwan JM, Fialho AM, Kundu M et al (2009) Bacterial proteins as potential drugs in the treatment of leukemia. Leuk Res 33:1392–1399

Lehrnbecher T, Koehl U, Wittekindt B et al (2008) Changes in host defence induced by malignancies and antineoplastic treatment: implication for immunotherapeutic strategies. Lancet Oncol 9:269–278

Little JW (1991) Mechanism of specific LexA cleavage: autodigestion and the role of RecA coprotease. Biochimie 73:411–421

Liu TC, Galanis E, Kirn D (2007) Clinical trial results with oncolytic virotherapy: a century of promise, a decade of progress. Nat Clin Pract Oncol 4:101–117

Loven D, Hardoff R, Bar Sever Z et al (2004) Non-resectable slow-growing meningiomas treated by Hydroxyurea. J Neurooncol 67:221–226

Mahfouz M, Hashimoto W, Das Gupta TK et al (2007) Bacterial proteins and CpG-rich extrachromosomal DNA in potential cancer therapy. Plasmid 57:4–17

Miyake K, Yamamoto S, Iijima S (1996) Blocking adhesion of cancer cells to endothelial cell types by *S. agalactiae* type-specific.polysaccharides. Cytotechnology 22:205–209

Nauts HC, Swift WE, Coley BL (1946) The treatment of malignant tumors by bacterial toxins as developed by the late William B. Coley, MD, reviewed in the light of modern research. Cancer Res 6:205–216

Nordman J, Skovgaard O, Wright A (2007) A novel class of mutations that affect DNA replication in *E. coli*. Mol Microbiol 64:125–138

Power AT, Bell JC (2007) Cell-based delivery of oncolytic viruses: a new strategic alliance for a biological strike against cancer. Mol Ther 15:660–665

Ries L, Melbert D, Krapcho M et al (2009) SEER cancer statistics review, 1975–2005. National Cancer Institute, Bethesda

Roland KL, Tinge SA, Killeen KP et al (2005) Recent advances in the development of live, attenuated bacterial vectors. Curr Opin Mol Ther 7:62–72

Russell SJ, Peng KW (2007) Viruses as anticancer drugs. Trends Pharmacol Sci 28:326–333

Sassanfar M, Roberts JW (1990) Nature of the SOS-inducing signal in *Escherichia coli*: the involvement of DNA replication. J Mol Biol 212:79–96

Shao J, Zhou B, Bernard C et al (2006) Ribonucleotide reductase inhibitors and future drug design. Curr Cancer Drug Targets 6:409–431

Sharma S, Doherty KM, Brosh RM Jr (2005) DNA helicases as targets for anti-cancer drugs. Curr Med Chem Anticancer Agents 5:183–199

Sherratt DJ (2003) Bacterial chromosome dynamics. Science 301:780–785

Teicher BA (2008) Next generation topoisomerase I inhibitors: rationale and biomarker strategies. Biochem Pharmacol 75:1262–1271

Walker GC (1984) Mutagenesis and inducible responses to deoxyribonucleic acid damage in *Escherichia coli*. Microbiol Mol Biol Rev 48:60–93

Weitao T (2009a) Multicellularity of a unicellular organism in response to DNA replication stress. Res Microbiol 160:87–88

Weitao T (2009b) Bacteria form biofilms against cancer metastasis. Med Hypothesis 72:477–478
Weitao T, Nordstrom K (2010) The cycle of *Escherichia coli* chromosomes and plasmids. VDM Publishing House Ltd/LAP Lambert Academic Publishing/AG & Co. KG, Saarbrucken
Yamada T, Fialho AM, Punj V et al (2005) Internalization of bacterial redox protein azurin in mammalian cells: entry domain and specificity. Cell Microbiol 7:1418–1431
Yu YA, Shabahang S, Timiryasova TM et al (2004) Visualization of tumors and metastases in live animals with bacteria and vaccinia virus encoding light-emitting proteins. Nat Biotechnol 22:313–320

Chapter 12
Management of Bacterial Infectious Complications in Cancer Patients

Kenneth V.I. Rolston

Abstract Serious bacterial infections occur frequently in patients with cancer, both as a result of their underlying disease and its treatment. Several factors increase the risk of bacterial infections, the foremost of which is neutropenia (absolute neutrophil count [ANC] ≤500 cells/mm^3). Neutropenia, however, seldom occurs alone, but is often superimposed on other risk factors, such as impaired humoral or cellular immunity, the presence of foreign medical devices, surgery, radiation, and poor nutrition. Other factors that influence the frequency and spectrum of bacterial infection include the use of antimicrobial prophylaxis, the intensity and nature of chemotherapy, and local institutional epidemiology. This chapter will focus on the management of bacterial infections in cancer patients, including infection prevention, treatment, infection control, and antimicrobial stewardship.

Keywords Bacterial infection • Neutropenia • Cancer patients • Antimicrobial prophylaxis • Infection control • Antimicrobial stewardship

Abbreviations

ANC	Absolute neutrophil count
ESBL	Extended-spectrum ß-Lactamases
KPC	*Klebsiella pneumoniae* carbapenemase
MASCC	Multinational Association for Supportive Care in Cancer
MRSA	Methicillin-resistant *Staphylococcus aureus*
VRE	Vancomycin-resistant enterococci

K.V.I. Rolston (✉)
Department of Infectious Diseases, Infection Control and Employee Health,
The University of Texas MD Anderson Cancer Center, Houston, TX 77030, USA
e-mail: krolston@mdanderson.org

A.A. Khan (ed.), *Bacteria and Cancer*, DOI 10.1007/978-94-007-2585-0_12,

12.1 Bacterial Infections in Neutropenic Patients

Neutrophils provide protection against many bacterial pathogens. Neutropenia from any cause increases the frequency and severity of bacterial infection. Bodey and colleagues were the first to describe the association between neutropenia and infection (Bodey et al. 1966). They demonstrated that the frequency and severity of infection was directly related to the degree and duration of neutropenia, once the ANC fell below 1,000/mm^3. The currently accepted definition of neutropenia is an ANC of ≤500/mm^3 (Freifeld et al. 2011). Until the late 1980s it was traditional to admit all febrile neutropenic patients to the hospital for close monitoring and the administration of broad-spectrum, parenteral antibiotic therapy for the entire duration of the febrile episode (Hughes et al. 1997). Our understanding of the syndrome of febrile neutropenia has improved substantially during the ensuing years. The availability of truly broad-spectrum antimicrobial agents (carbapenems, the extended spectrum cephalosporins) has enabled clinicians to administer monotherapy to many febrile neutropenic patients instead of always using two or three drug combinations (Bodey et al. 1990). The development of accurate risk-prediction rules, major improvements in infusion therapy and supportive care, and the increasing role played by home healthcare agencies has enabled clinicians to shift the site of care in low-risk patients from the hospital to the ambulatory clinic/home environment (Kern 2006; Rolston 2003). The development of oral agents such as the fluoroquinolones, with activity against important gram-negative pathogens including *Pseudomonas aeruginosa* has improved the efficacy of antimicrobial prophylaxis for high-risk neutropenic patients. On the other hand, the frequent use (misuse?) of antimicrobial agents in this setting has led to reduced susceptibility or overt resistance among common bacterial pathogens. All these factors add to the complexity of the management of bacterial infections in neutropenic patients.

12.2 Spectrum of Infection

The epidemiology/spectrum of bacterial infection in neutropenic patients undergoes periodic changes. Additionally, geographic and local institutional differences do exist; consequently, it is advisable to conduct frequent surveillance studies, particularly in institutions dealing with large numbers of neutropenic patients (e.g., Comprehensive Cancer Centers) in order to detect these changes in a timely manner (Yadegarynia et al. 2003). Bacteria frequently isolated from neutropenic patients are listed in Table 12.1. Current surveillance studies demonstrate a predominance of gram-positive bacteria over gram-negative and anaerobic bacteria (Wisplinghoff et al. 2003; Zinner 1999). Unfortunately, some of these surveys provide details only about monomicrobial bacteremias, and ignore other sites of infection such as the respiratory tract, urinary tract, and gastrointestinal tract, as well as polymicrobial infections. This paints an incomplete picture since gram-positive organisms are the predominant cause of bacteremia, whereas many other sites of infection have gram-negative organisms as the predominant pathogens (Rolston and Bodey 1993;

Table 12.1 Common bacterial pathogens isolated from neutropenic patients[ab]

Gram-positive
Coagulase-negative staphylococci
Staphylococcus aureus (including MRSA[a])
Enterococcus species (including VRE[c])
Viridans group streptococci
Beta-haemolytic streptococci
Gram-negative
Escherichia coli (including ESBL[b] producers)
Klebsiella species (including ESBL[b] producers)
Pseudomonas aeruginosa
Stenotrophomonas maltophilia

[a]*MRSA* methicillin-resistant *S. aureus*
[b]*ESBL* extended spectrum beta-lactamase
[c]*VRE* vancomycin-resistant enterococci

Yadegarynia et al. 2003). Additionally, data from our institution document the fact that polymicrobial infections have more than doubled in frequency since the 1970s and currently account for ~30% of documented bacterial infections (Rolston et al. 2007). Our data also document that ~80% of polymicrobial infections have a gram-negative component, and ~30–35% are caused exclusively by multiple gram-negative species (Elting et al. 1986). All these issues should be taken into consideration when choosing empiric or targeted therapeutic regimens in neutropenic patients.

12.3 Initial Assessment of Neutropenic Patients

One of the cornerstones in the management of febrile neutropenic patients is to perform a quick but thorough evaluation before the administration of empiric antibiotic therapy (Freifeld et al. 2011). A complete history and physical examination is essential. Historical information of particular interest includes details of antineoplastic and immunosuppressive therapy, the use of antimicrobial prophylaxis, and previous episodes of infection (or colonization) with important pathogens and their treatment, recent surgical/dental procedures, recent or remote travel history, and potential exposure to sick contacts. Underlying co-morbid conditions such as diabetes mellitus, chronic lung disease, and cardiovascular, renal and hepatic problems should also be documented as they might impact the nature and severity of infection, the risk of complications, and the agents selected for empiric therapy.

Neutropenic patients may have a blunted inflammatory response which may result in a paucity of clinical signs/symptoms. Consequently, the physical examination should focus on the detection of subtle manifestations especially at frequently infected sites such as the skin, oro-pharynx, gastrointestinal tract, lungs, and perineum. Although fever is the most common sign of infection, some patients may develop a serious infection without mounting a febrile response, particularly if they are receiving corticosteroids or other immunosuppressive agents.

Standard laboratory investigations include blood and urine cultures, and cultures from other sites (e.g., respiratory specimens, CSF, wounds, biliary drainage) when indicated. In patients with diarrhea, stool cultures are not very informative, but stool specimens for the detection of *Clostridium difficile* toxins should be obtained. Patients with pulmonary symptoms or an infiltrate on imaging often require a diagnostic bronchoscopy. Nasal specimens are recommended for detecting the presence of community respiratory viruses (e.g., influenza A or B, RSV) especially in the winter season.

Routine chest imaging is not recommended and should be performed in patients with respiratory manifestations. Computerized tomography of the chest, paranasal sinuses, abdomen/pelvis, head and neck, should be obtained when clinically indicated, and is usually more informative (but more expensive) than routine radiography. Other standard laboratory tests include complete and differential blood cell counts, an electrolyte panel, and tests for renal and hepatic function. These tests should be repeated as clinically indicated.

12.4 Risk Assessment

Prior to the development of reliable risk assessment models, all febrile neutropenic patients received hospital-based, parenteral empiric antibiotic therapy for the entire duration of the episode of infection (Hughes et al. 1997). With a better understanding of the syndrome of "febrile neutropenia" several investigators have developed reliable risk-prediction rules. The most widely used method to identify low-risk patients is the risk-index devised by the Multinational Association for Supportive Care in Cancer (MASCC) (Klastersky et al. 2000). This method assigns integer weights to seven characteristics to develop an index score. A score of 21 identified low-risk patients with a positive predictive value of 91%. Higher scores impart greater specificity with a corresponding loss in sensitivity. Other investigators use simple clinical criteria to identify low-risk patients, without having to calculate a score (Freifeld et al. 1999; Kern et al. 1999; Rubenstein et al. 1993). This might be a more practical method of identifying low-risk patients in a busy clinical practice setting. There is a >95% concordance between the MASCC risk-index and clinical risk-prediction rules (Rolston et al. 2009).

12.5 Risk Based Therapy

There is uniform agreement that patients who are not identified as low-risk should be hospitalized for the administration of empiric antibiotic therapy and close monitoring (Lyman and Rolston 2010). Several different options for the treatment of low-risk patients have been recently evaluated. These include the route of administration of the regimen (parenteral, sequential, i.e., IV → PO, oral), and the setting of

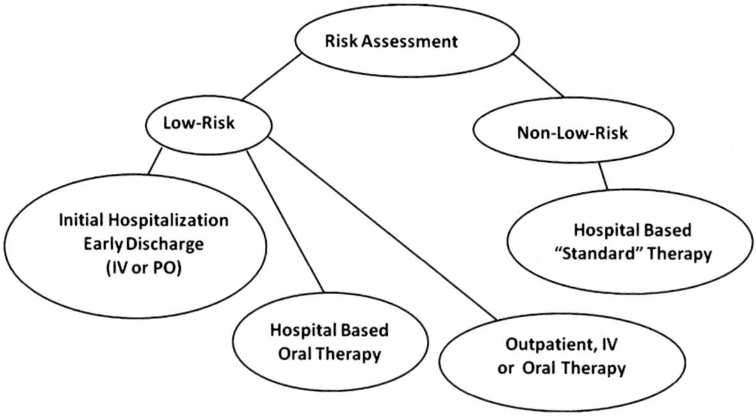

Fig. 12.1 Risk-based management of febrile neutropenic patients

therapy (initial hospitalization followed by early discharge; out-patient management of the entire febrile episode). These options constitute the entire spectrum of risk-based therapy (Fig. 12.1).

12.5.1 Empiric Antibiotic Therapy in Low Risk Patients

Most low-risk patients can safely be treated with oral antibiotics after a short period (4–12 h) of observation to ensure stability (Freifeld et al. 2011). Despite several prospective trials documenting the safety and efficacy of oral, out-patient therapy, some clinicians are still uncomfortable with this concept and prefer to admit low-risk patients for a longer (48–72 h) period before discharging them on oral (occasionally parenteral) regimens (Lyman and Rolston 2010; Vidal et al. 2004). The most commonly used regimens are listed in Table 12.2. Most oral regimens combine a fluoroquinolone with an agent with better gram-positive activity. Currently, fluoroquinolone monotherapy is not recommended routinely, although some pilot studies have successfully evaluated this option (Chamilos et al. 2005; Rolston et al. 2006, 2009). In patients who are not suitable candidates for oral therapy (e.g., mild to moderate nausea and/or mild mucositis) but are otherwise low-risk, parenteral, out-patient therapy is an option.

Out-patient management of low-risk neutropenic patients does require institutional support and infrastructure which may not always be feasible, especially in institutions that care for limited numbers of neutropenic patients (Table 12.3). Additionally, some medically low-risk patients may not have the psychosocial back-up and support to be candidates for out-patient therapy. It is prudent to treat such patients in the hospital.

Most studies of out-patient therapy in well selected, low-risk patients have shown a very low frequency of hospital admissions, or complications requiring intensive care.

Table 12.2 Antibiotic regimens in low-risk patients

Oral regimens
Fluoroquinolone[a] – amoxicillin/clavulanate or clindamycin, or azithromycin
Fluoroquinolone monotherapy[b]
Parenteral regimens
Aztreonam – clindamycin
Fluoroquinolone[a] + clindamycin
Ceftriaxone (±) amikacin
Ertapenem (±) amikacin
Ceftazidime or cefepime

[a]Ciprofloxacin has been used most often but other fluoroquinolones (levofloxacin, moxifloxacin) have also been used
[b]Limited data with fluoroquinolone monotherapy

Table 12.3 Requirements for a successful program of outpatient therapy in low-risk patients

Institutional support for necessary infrastructure (e.g., 24/7 emergency center)
Dedicated, multidisciplinary team of healthcare providers (physicians, nurses, pharmacists, infusion therapists, etc.)
Local and real-time epidemiologic/ microbiologic data including current susceptibility/resistance patterns
Adequate monitoring and follow-up (e.g., 24/7 'hot line' and access to healthcare team; febrile neutropenia clinic, etc.)
Adequate transport and communication for patients

The average duration of therapy is 5–6 days. Hopefully, emergence of resistance will not become a significant issue, as the pipeline for new drug development is relatively dry.

12.5.2 *Empiric Therapy for Patients That Are Not Low Risk*

The accepted standard of care for febrile neutropenic patients that do not fall into the low-risk category is the prompt administration of broad-spectrum empiric antibiotics (based on local susceptibility/resistance patterns) with close monitoring in the hospital for response, and the development of complications (Freifeld et al. 2011).

The various treatment options are listed in Table 12.4. They include combination antibiotic regimens (usually a combination of an anti-pseudomonal beta-lactam and an aminoglycoside or an agent with enhanced gram-positive activity e.g., vancomycin or linezolid); or monotherapy with a single broad-spectrum, anti-pseudomonal beta-lactam.

The most recently updated guidelines published by the Infectious Diseases Society of America (IDSA) recommend monotherapy for most febrile neutropenic patients, with the addition of other antimicrobial agents to the initial regimen for the management of complications (e.g., hypotension or pneumonia) or if antimicrobial resistance is suspected (Freifeld et al. 2011). Vancomycin or other agents (linezolid, daptomycin) with enhanced gram-positive activity are not recommended as a

Table 12.4 Common empiric regimens in neutropenic patients not classified as low-risk

Monotherapy
Cefepime or ceftazidime[a]
Imipenem or meropenem[b]
Piperacillin/tazobactam
Combination regimens without vancomycin
Aminoglycoside + cefepime or ceftazidime[a]
or imipenem or meropenem[b]
or piperacillin/tazobactam
or quinolone[c]
Combination regimens with vancomycin[d]
Vancomycin + cefepime or ceftazidime[a]
or imipenem or meropenem[b]
or piperacillin/tazobactam
or aztreonam[e]
or quinolone[c]

[a]Ceftazidime not optimal in some institutions due to declining susceptibility
[b]No clinical experience with doripenem
[c]Quinolones should not be used if patients have been receiving quinolone prophylaxis
[d]Teicoplanin used in some countries. Vancomycin occasionally replaced by linezolid
[e]Often used in patients with severe beta-lactam allergy

standard part of the initial antibiotic regimen for fever and neutropenia but should be considered for conditions such as catheter-related infections, skin and skin structure infections, pneumonia, or hemodynamic instability. The recommendations for specific bacterial pathogens include:

- MRSA – vancomycin, linezolid, or daptomycin
- VRE – linezolid or daptomycin
- ESBL producing gram-negative bacilli – carbapenems
- KPC producing organisms – polymyxin/colistin or tigecycline

In certain institutions, *Stenotrophomonas maltophilia* has emerged as a frequent pathogen in this patient population (Safdar and Rolston 2007). Most isolates still remain susceptible to trimethoprim/sulfamethoxazole, although declining susceptibility rates are being reported. Other agents with variable activity against these organisms include tigecycline, ticaricillin/clavulanate, moxifloxacin, minocycline, and ceftazidime. Therapy based on individualized susceptibility of the isolates is recommended. Combination therapy might be necessary in patients refractory to monotherapy.

12.6 Evaluation of Response

The median time to defervescence in low-risk patients is 2 days, and is approximately 5 days in moderate to high-risk patients (Corey and Boeckh 2002; Elting et al. 2000). Persistence of fever for 3–5 days in otherwise stable patients does not necessarily indicate failure of the initial regimen. Approximately 70–80% of patients will respond to the initial empiric regimen during this period (Freifeld et al. 2011). Persistence of fever beyond 5 days should lead to a full re-evaluation of the patient including a search for a drainable focus (abscess) or removable focus (infected medical device), or the development of a super-infection. A change in the initial regimen is recommended at this stage, based on specific clinical and/or microbiological findings.

In patients who remain febrile, imaging of various sites (paranasal sinuses, chest, abdomen and pelvis, head and neck), Doppler or venous flow studies, and various serologic studies might provide diagnostic information. Occasionally, more invasive procedures (biopsy of specific tissues or organs) might be necessary, but are often deferred as many neutropenic patients are also severely thrombocytopenic. A small proportion of patients (~5%) will have a non-infectious cause of fever such as tumour fever or drug fever.

12.7 Duration of Therapy

The duration of therapy continues to be a subject of debate. In patients with clinically or microbiologically documented infections, the duration of therapy will depend upon the particular organism(s) isolated and the site of infection (Elting et al. 1997; Freifeld et al. 2011). Appropriate antibiotics based on susceptibility data should be continued until resolution of neutropenia (ANC >500/mm^3) or longer, if clinically indicated.

In patients with unexplained fever, two schools of thought exist. One is to continue the initial regimen until signs of marrow recovery. The other is to discontinue therapy if all signs and symptoms of infection have resolved, even if the patient is still neutropenic. Some experts recommend reinstituting antimicrobial prophylaxis in such patients. The former approach may result in needless administration of antibiotics to many patients, potentially increasing healthcare costs, toxicity, and the development of bacterial and fungal superinfection, in addition to selection of resistant microorganisms. The latter approach requires careful observation of the patient after discontinuation of therapy. The ultimate decision as to when to stop therapy often needs to be individualized based on (1) the patient's risk group; (2) the presence and site of a documented (e.g. bacteremia, pneumonia, enterocolitis, etc.); (3) the underlying malignancy (solid tumour or hematologic malignancy); (4) the need for chemotherapy and/or immunosuppressive therapy and (5) the persistence of neutropenia. Some patients with documented infections and persistent neutropenia might benefit from the administration of hematopoietic growth factors (G-CSF, GM-CSF) and/or granulocyte transfusions, but their use remains unconventional (Hübel et al. 2002; Smith et al. 2006).

12.8 Antimicrobial Prophylaxis

The use of routine antimicrobial prophylaxis in neutropenic patients also continues to be a subject of debate. Most studies have demonstrated that the use of antibacterial prophylaxis results in a reduction in the number of febrile episodes and documented infections, particularly those caused by gram-negative bacteria. A recent meta-analysis showed increased survival in patients receiving quinolone prophylaxis (Gafter-Gvili et al. 2005). The main drawback of antimicrobial prophylaxis even when clinically justified is the emergence of resistant micro-organisms (Kern et al. 1994). Fluoroquinolone prophylaxis should be considered for patients with expected durations of neutropenia that exceed 7–10 days (i.e., ANC $\leq$ 100 cells/mm^3 for >7–10 days). Real time microbiological monitoring for the emergence of resistant organisms is recommended in institutions where prophylaxis is used commonly (Baden 2005). The addition of an agent with enhanced gram-positive activity (e.g., vancomycin, linezolid) is not recommended.

12.9 Infections in Cancer Patients Without Neutropenia

Prolonged and profound neutropenia is seen most often in patients with hematologic malignancies and in recipients of hematopoietic stem cell transplantation, and much less so in patients with solid tumours. However, solid tumours account for the vast majority of cancers in adults. Data published by the American Cancer Society indicate that approximately 1.5 million new cases of solid tumours are diagnosed each year in the USA (American Cancer Society 2010). The spectrum, clinical features, diagnosis, and management of infection in these patients is substantially different to those encountered in neutropenic patients and treatment strategies specific for these patients needs to be developed. The predominant sites of infection in solid tumour patients are listed in Table 12.5.

12.10 Predominant Sites of Infection

The predominant sites of infection depend upon the location and size of the primary tumour and/or metastatic lesions, and the site and nature of medical devices and surgical procedures. Surgical wound infections are not uncommon regardless of the site of tumour. Patients with CNS infections often have partial or complete loss of the gag reflex, predisposing them to aspiration pneumonia. Impaired micturation and urinary retention as a result of neurological impairment lead to urinary tract infection. Following surgery for tumour resection and/or the placement of shunts, surgical wound infections, epidural or subdural infections, cerebral abscesses, meningitis, and shunt related infection can occur.

In the United States, approximately 200,000 breast cancer surgical procedures are performed annually (Penel et al. 2007). The frequency of infection following such procedures is estimated to be between 4% and 8%, which translates into an annual figure of 8,000–16,000 cases. These include surgical wound infections, cellulitis, and lymphangitis secondary to axillary lymph node dissection, mastitis, breast abscess, and breast tissue expanders associated infections.

Table 12.5 Predominant sites of infection in cancer patients with various solid tumours

Tumour	Common sites of infection
Brain (CNS)	Wound infection; epidural and/or subdural infection; brain abscess; meningitis/ventriculitis; shunt-related infection; aspiration pneumonia; urinary tract infection
Breast	Wound infection; cellulitis/lymphangitis following axillary lymph node dissection; mastitis; breast abscess
Bone/joints/cartilage	Wound infection; septic arthritis; osteomyelitis,; bursitis; synovitis; infected prosthesis
Genitourinary/prostate	Wound infection; cystitis; prostatitis; pyelonephritis; catheter-related complication, urinary tract infection
Hepato-biliary/pancreatic	Wound infection; peritonitis; ascending cholangitis ± bacteremia; hepatic, pancreatic, or sub diaphragmatic abscess
Upper gastrointestinal	Wound infection; mediastinitis; trachea-esophageal fistula; gastric perforation and abscess; PEG tube-related infections
Head and neck	Wound infections; cellulitis; aspiration pneumonia; PEG tube-related infection; mastoiditis; sinusitis; cavernous (or other) sinus thrombosis; meningitis; brain abscess; retropharyngeal and paravertebral abscess; osteomyelitis ± osteoradionecrosis
Lower gastrointestinal	Wound infection; peritonitis; abdominal/pelvic abscess; necrotizing fasciitis; enterocolitis; perianal or perirectal infection; urinary tract infections
Respiratory (lung)	Wound infection; pneumonia (post-obstructive); empyema; broncho-pleural fistula; aspiration
Gynecologic	Wound infection; complicated urinary tract infection; abdominal/pelvic abscess, tubo-ovarian abscess; pyometra; fistula formation and associated infections

Infection of the upper respiratory tract such as sinusitis, pneumonia (including aspiration pneumonia and ventilator-associated pneumonia), and local cellulitis and necrotizing infections following surgical excision and reconstruction, are the most common sites in patients with head and neck tumours. These patients also frequently need PEG tube placement for alimentation, and develop PEG tube associated infections (local cellulitis, abscesses, perforation and peritonitis) as well (Walton 1999).

Patients with carcinoma of the lung develop pulmonary infections such as post-obstructive and/or necrotizing pneumonia, lung abscess, empyema, and surgical wound infections. Localized infections may lead to the development of bacteremia or disseminated infections.

Cholangitis with or without bacteremia, solitary or multiple hepatic abscesses, and peritonitis are not infrequent in patients who have hepato-biliary/pancreatic tumours (Rolston et al. 1995). Abscesses in the pancreatic bed and subdiaphragmatic abscesses can occur following extensive surgical resection. Patients receiving intra-arterial chemotherapy for hepatic tumours are also at risk for such infections. Osteomyelitis, osteoradionecrosis, and infected prosthetic devices with adjacent bone, joint, or soft tissue infections predominate in patients with osteosarcoma and other bone neoplasms.

Local obstruction caused by tumour, tumour necrosis, and therapeutic modalities (chemoradiation, surgery) all contribute to infections in patients with gynecologic malignancies. Tumour-related infections depend on the site and size of the tumour. For example, infections complicating stage I cervical cancer generally involve the surfaces of the tumour and are usually limited to the vagina (Brooker et al. 1987). As tumours enlarge, obstruction to various organs results in the development of urinary tract infections, tubo-ovarian abscesses and pyometra. Rupture of tubo-ovarian abscesses or pyometra can lead to the development of acute peritonitis (Barton et al. 1993; Imachi et al. 1993). These complications are rare since most gynecological cancers are detected and treated at an earlier stage. Complications of radiation include bowel obstruction/stricture, perforation, and fistula formation. These complications are often difficult to deal with, particularly due to impaired healing in previously radiated areas.

In contrast to neutropenic patients with hematologic malignancies, patients with solid tumours represent an extremely heterogeneous group. Consequently, the treatment of infections occurring in these patients is usually site and organism specific. A detailed discussion of all the infections listed in Table 12.5 is beyond the scope of this chapter. Nevertheless, the general principles relating to appropriate initial evaluation and the prompt administration of appropriate antimicrobial therapy, based on local epidemiology and susceptibility/resistance patterns are applicable in these patients as well. Surgical intervention (or other approaches such as the placement of stents) is occasionally required to remove devitalized tissue and overcome obstruction. In order to better understand the diversity of infections seen and to develop management strategies specific for the different tumour groups (CNS, lung, gastrointestinal, breast, gynecologic, etc.) carefully designed studies focusing on the predisposing factors, changing epidemiology, clinical manifestations, diagnosis, and treatment of these infections need to be conducted (Rolston 2001). Such studies will provide the information needed to appropriately manage infections in patients with solid tumours, rather than applying management strategies that have been developed for and are more pertinent in patients with hematologic malignancies.

12.11 Infection Control and Antimicrobial Stewardship

Infection control is an extremely important aspect of the management of infections in neutropenic and non-neutropenic cancer patients. Strict adherence to infection control policies and procedures is mandatory. These policies (a detailed discussion is outside of the scope of this chapter) are not only critical in the investigation and disruption of outbreaks, but also in the day to day setting as they limit the spread of resistant microorganisms.

Antimicrobial agents are used with greater frequency and for a larger number of indications (prophylaxis, pre-emptive therapy, empiric therapy, targeted or specific therapy of a documented infection, maintenance/suppressive therapy) in cancer patients than in most other patient populations (Freifeld et al. 2011). Although

Table 12.6 Recommendations for antimicrobial stewardship

Baseline data/infrastructure

- Determine local epidemiology and resistance patterns
- Know institutional formulary and prescribing habits
- Develop multidisciplinary antimicrobial stewardship team (MAST)

Recommendations for antimicrobial usage

- Limit antibacterial prophylaxis
- Encourage targeted/specific therapy
- Consider formulary restriction and/or pre-authorization
- Create guidelines and clinical pathways
- Consider antimicrobial heterogeneity
- Consider de-escalation (streamlining) of empiric regimen
- Dose optimization
- Parenteral to oral conversion
- Optimization of duration of therapy

Other strategies

- Prospective audits of antimicrobial usage with feedback to prescribers
- Educational activities (Grand Rounds, in-services)
- Strict adherence to infection control policies

justified, this has created pressures leading to the emergence of resistant organisms (Rolston 2005). Traditionally, the development of novel antimicrobial agents has been an important tool in battling the problems caused by resistant organisms. However, the development of novel agents is at an all time low, mandating the judicious use of currently available agents – i.e. antimicrobial stewardship. The various strategies for an antimicrobial stewardship program are listed in Table 12.6, and include a multidisciplinary antibiotic stewardship team (MAST), institutional pathways/guidelines, formulary restrictions or pre-approval requirements for certain agents, and de-escalation or streamlining of therapy when appropriate (Dellit et al. 2007). Antibiotic stewardship programs have been successfully implemented at several institutions (including ours) and, in the opinion of this investigator, will soon become mandatory at most institutions (Agwu et al. 2008; Metjian et al. 2008; Mulanovich et al. 2009).

12.12 Conclusion

Infection remains the most common complication of cancer and its therapy. Bacterial infections predominate, although fungal and viral infections are increasing in frequency. The epidemiology of bacterial infections is different in patients with hematologic malignancy when compared to those with solid tumours, and undergoes periodic changes. It is important to monitor local epidemiology and susceptibility/resistance patterns in order to determine the most appropriate empiric regimens. The emergence of resistance among common bacterial pathogens has posed new therapeutic challenges, especially since new drug development is practically at a standstill. Consequently, antimicrobial stewardship and infection control are vital aspects in the management of infections in cancer patients.

References

Agwu A, Lee CKK, Jain SK et al (2008) A worldwide web-based antimicrobial stewardship program improves efficiency, communication, and user satisfaction and reduces cost in a tertiary care pediatric medical center. Clin Infect Dis 47:747–753

Baden LR (2005) Prophylactic antimicrobial agents and the importance of fitness. N Engl J Med 353:1052–1054

Barton DPJ, Fiorica JV, Hoffman MS et al (1993) Cervical cancer and tubo-ovarian abscesses: a report of three cases. J Reprod Med 38:562–564

Bodey GP, Buckley M, Sathe YS et al (1966) Quantitative relationships between circulating leukocytes and infection in patients with acute leukemia. Ann Intern Med 64:328–340

Bodey GP, Fainstein V, Elting LS et al (1990) Beta-lactam regimens for the febrile neutropenic patient. Cancer 65:9–16

Brooker DE, Savage JE, Twiggs LB et al (1987) Infectious morbidity in gynecologic cancer. Am J Obstet Gynecol 156:515–520

Chamilos G, Bamias A, Efstathiou E et al (2005) Outpatient treatment of low-risk neutropenic fever in cancer patients using oral moxifloxacin. Cancer 103:2629–2635

Corey L, Boeckh M (2002) Persistent fever in patients with neutropenia. N Engl J Med 346:222–224

Dellit TH, Owens RC, McGowan JE et al (2007) Infectious Diseases Society of America and the Society for Healthcare Epidemiology of America Guidelines for developing an institutional program to enhance antimicrobial stewardship. Clin Infect Dis 44:159–177

Elting LS, Bodey GP, Fainstein V (1986) Polymicrobial septicemia in the cancer patient. Medicine 65:218–225

Elting LS, Rubenstein EB, Rolston KV et al (1997) Outcomes of bacteremia in patients with cancer and neutropenia: observations from two decades of epidemiological and clinical trials. Clin Infect Dis 25(2):247–259

Elting LS, Rubenstein EB, Rolston K et al (2000) Time to clinical response: an outcome of antibiotic therapy of febrile neutropenia with implications for quality and cost of care. J Clin Oncol 8:3699–3706

Freifeld A, Marchigiani D, Walsh T et al (1999) A double-blind comparison of empirical oral and intravenous antibiotic therapy for low-risk febrile patients with neutropenia during cancer chemotherapy. N Engl J Med 341:305–311

Freifeld AG, Bow EJ, Sepkowitz KA et al (2011) Clinical practice guideline for the use of antimicrobial agents in neutropenic patients with cancer: 2010 update by the Infectious Diseases Society of America. Clin Infect Dis 52(4):e56–e93

Gafter-Gvili A, Fraser A, Paul M et al (2005) Meta-analysis: antibiotic prophylaxis reduces mortality in neutropenic patients. Ann Intern Med 142:979–995

Hübel K, Carter RA, Liles WC et al (2002) Granulocyte transfusion therapy for infections in candidates and recipients of HPC transplantation: a comparative analysis of feasibility and outcome for community donors versus related donors. Transfusion 42:1414–1421

Hughes WT, Armstrong D, Bodey GP et al (1997) Guidelines for the use of antimicrobial agents in neutropenic patients with unexplained fever. Infectious Diseases Society of America. Clin Infect Dis 25(3):551–573

Imachi M, Tanaka S, Ishikawa S et al (1993) Spontaneous perforation of pyometra presenting as generalized peritonitis in a patient with cervical cancer. Gynecol Oncol 50:384–388

Kern KV (2006) Risk assessment and treatment of low-risk patients with febrile neutropenia. Clin Infect Dis 15:533–540

Kern WV, Andriof E, Oethinger M et al (1994) Emergence of fluoroquinolone-resistant *Escherichia coli* at a cancer center. Antimicrob Agents Chemother 38:681–687

Kern WV, Cometta A, De Bock R et al (1999) Oral versus intravenous empirical antimicrobial therapy for fever in patients with granulocytopenia who are receiving cancer chemotherapy. International Antimicrobial Therapy Cooperative Group of the European Organization for Research and Treatment of Cancer. N Engl J Med 341:312–318

Klastersky J, Paesmans M, Rubenstein E et al (2000) The MASCC risk index: a multinational scoring system to predict low-risk febrile neutropenic cancer patients. J Clin Oncol 18:3038–3051

Lyman GH, Rolston KV (2010) How we treat febrile neutropenia in patients receiving cancer chemotherapy. J Oncol Pract 6(3):149–152

Metjian TA, Prasad PA, Kogon A et al (2008) Evaluation of an antimicrobial stewardship program at a pediatric teaching hospital. Pediatr Infect Dis J 27:106–111

Mulanovich V, Chemaly R, Mihu C et al (2009) Antimicrobial stewardship program in The Critical Care Unit (CCU) of a comprehensive cancer center (Abst. 09–070). In: Multinational Association for supportive care in cancer 2009 international MASCC/ISOO symposium, Rome

Penel N, Yazdanpanah Y, Chauvet MP et al (2007) Prevention of surgical site infection after breast cancer surgery by targeted prophylaxis antibiotic in patients at high risk of surgical site infection. J Surg Oncol 96(2):124–129

Rolston KVI (2001) Infections in patients with solid tumors. In: Rolston KVI, Rubenstein E (eds) Textbook of febrile neutropenia. Martin Dunitz Ltd, London

Rolston KVI (2003) Oral antibiotic administration and early hospital discharge is a safe and effective alternative for treatment of low-risk neutropenic fever. Cancer Treat Rev 29:551–554

Rolston KV (2005) Challenges in the treatment of infections caused by gram-positive and gram-negative bacteria in patients with cancer and neutropenia. Clin Infect Dis 40(4):246–252

Rolston KVI, Bodey GP (1993) Diagnosis and management of perianal and perirectal infection in the granulocytopenic patient. In: Remington J, Swartz MN (eds) Current clinical topics in infectious diseases. Blackwell Scientific Publications, Boston

Rolston KVI, Dholakia N, Rodriguez S et al (1995) Nature and outcome of febrile episodes in patients with pancreatic and hepato-biliary cancer. Support Care Cancer 3:414–417

Rolston KVI, Manzullo EF, Elting LS et al (2006) Once daily, oral, outpatient quinolone monotherapy for low-risk cancer patients with fever and neutropenia. Cancer 106:2489–2494

Rolston KVI, Bodey GP, Safdar A (2007) Polymicrobial infection in patients with cancer: an underappreciated and underreported entity. Clin Infect Dis 45:228–233

Rolston KVI, Frisbee-Hume SE, Patel S et al (2009) Oral moxifloxacin for outpatient treatment of low-risk, febrile neutropenic patients. Support Care Cancer 18(1):89–94

Rubenstein EB, Rolston K, Benjamin RS et al (1993) Outpatient treatment of febrile episodes in low risk neutropenic cancer patients. Cancer 71:3640–3646

Safdar A, Rolston KV (2007) *Stenotrophomonas maltophilia*: changing spectrum of a serious bacterial pathogen in patients with cancer. Clin Infect Dis 45:1602–1609

Smith TH, Khatcheressian J, Lyman GH et al (2006) Update of recommendations for the use of white blood cell growth factors: an evidence-based clinical practice guideline. J Clin Oncol 24:3187–3205

Statistics published by the American Cancer Society (2010) Cancer facts figures. www.cancer.org/Research/Cancer

Vidal L, Paul M, Ben-Dor I et al (2004) Oral versus intravenous antibiotic treatment for febrile neutropenia in cancer patients: a systematic review and meta-analysis of randomized trials. J Antimicrob Chemother 54:29–37

Walton GM (1999) Complications of percutaneous gastronomy in patients with head and neck cancer – an analysis of 42 consecutive patients. Ann R Coll Surg Engl 81:272–276

Wisplinghoff H, Seifert H, Wenzel RP et al (2003) Current trends in the epidemiology of nosocomial bloodstream infections in patients with hematological malignancies and solid neoplasms in hospitals in the United States. Clin Infect Dis 36:1103–1110

Yadegarynia D, Tarrand J, Raad I (2003) Current spectrum of bacterial infections in cancer patients. Clin Infect Dis 37:1144–1145

Zinner SH (1999) Changing epidemiology of infections in patients with neutropenia and cancer: emphasis on gram-positive and resistant bacteria. Clin Infect Dis 29:490–494

Index

A

Acute lymphoblastic leukaemia (ALL), 200, 225–239
Adenoma, 11, 43, 44, 64, 66, 69, 72, 128, 172, 179
ALL. *See* Acute lymphoblastic leukaemia
Amerindians, 128, 130, 133
Antibiotic therapy, 49, 84, 85, 94–96, 105, 154, 156–158, 260–264
Anticancer drug, 189, 197, 203, 246, 247, 249–255
Anticancer therapy, 188, 195, 203, 246, 256
Antimicrobial prophylaxis, 260, 261, 266–267
Antimicrobial stewardship, 269–271
Antineoplastic, 225–239, 261
Apoptosis, 27, 32, 33, 45–50, 84, 86–87, 146, 147, 168–170, 172, 175, 178, 180, 193, 194
Asparaginase, 225–239

B

Bacteremia, 63–65, 72, 121, 177, 178, 260, 266, 268
Bacteria, 1–17, 27, 62, 80, 118, 146, 165–181, 185–203, 210, 225–239, 245–256, 259–271
Bacterial attachment, 42, 83, 146, 170, 172, 178, 250, 251, 254
Bacterial enzymes, 176, 189, 232
Bacterial gene-directed enzyme prodrug therapy, 189
Bacterial infection, 1–17, 37, 38, 69, 80, 102, 167, 168, 170, 171, 173, 181, 195, 199, 211, 214, 215, 217, 227, 250, 260, 261, 270
Bacterial invasion, 27, 84, 118, 120, 178, 198, 210, 211, 249, 252–255
Bacterial therapy, 196, 199, 211, 217, 219
Bacterial toxin, 2, 84, 169–170, 181, 189, 193–195, 246
B-cell help, 49, 50
Bifidobacterium, 190, 191, 196, 197, 211
Biliary tract cancer, 13, 64, 131
Biofilms, 247–251, 254–256
Biomarker, 134
Bladder, 15–16, 190, 192, 202

C

Cancer
- causes, 1, 2, 6, 15, 17, 27, 28, 118, 129–131, 134, 167, 169, 171, 180, 181, 191, 196, 199, 211, 226, 227, 231, 233, 246, 266, 269
- patients, 14, 65, 69, 177, 178, 190, 191, 193, 197, 199, 202, 203, 211, 247, 259–271
- prevention, 8, 52, 133, 190, 193, 228, 251
- therapy, 2, 185–203, 211, 229

Carcinogenesis, 4, 10, 12, 15, 28–35, 39, 41–46, 51, 52, 63, 65–70, 72, 73, 80, 81, 84, 88, 90, 118, 167–174, 178–181, 188, 193, 247
Carcinoma, 2–7, 11, 27, 28, 30, 31, 39, 40, 42, 69, 72, 81, 86, 88, 108, 126, 127, 132, 147, 170, 171, 174–181, 188, 192, 195, 196, 198, 250, 253, 268
Cell culture, 45, 86, 104, 109, 177, 200, 201
Chemotherapy, 2, 27, 104, 106, 144, 145, 188, 189, 191–194, 197, 199, 201, 211, 220, 227–230, 236, 239, 249, 266, 268

A.A. Khan (ed.), *Bacteria and Cancer*, DOI 10.1007/978-94-007-2585-0,

Chlamydia
C. pneumoniae, 17, 81, 82, 84–90, 93, 96–98, 100–102, 108, 147, 148, 150, 152, 154, 156, 167, 179–180
C. psittaci, 81, 82, 86, 87, 90, 92–102, 108, 148–156
C. trachomatis, 81, 82, 85–87, 90, 93, 97–108, 147, 150, 152, 154, 156
Chlamydophila
C. pneumoniae, 81, 146, 154
C. psittaci, 81, 139–158
Cholesterol gallstone, 124, 133
Chronic carrier, 13, 80, 118, 120, 122–125, 131–133, 176
Chronic infection, 11–14, 17, 29, 37, 48, 79–109, 118, 123, 149, 168, 180
Chronic inflammation, 7, 28, 29, 31, 38, 39, 42, 62, 89, 108, 126, 128, 141, 148, 168–169, 174, 176, 180, 181
Clostridium, 169, 189, 191, 194, 195, 197, 211, 251, 262
Colon, 11, 12, 16, 63, 67, 69, 70, 72, 154, 167, 177–179, 190, 194, 197, 199, 202, 213
Colorectal cancer, 11, 12, 61–73, 179, 190
Cytosine deaminase (CD), 197, 200, 201, 203
Cytotoxicity, 47–49, 51, 192, 194, 227, 231

D
Diagnosis, 9, 42, 69, 92, 94, 102, 105–107, 109, 126, 143, 151, 156, 157, 227, 246, 267, 269
Digestive cancer, 64
Diptheria, 194
DNA
damage, 35–38, 51, 88, 124, 168, 169, 171, 175, 177, 180, 248, 252, 253
replication inhibitors, 203, 247–251
Doxycycline, 96, 104–108, 153, 156–158

E
Elementary bodies (EB), 82–85, 94, 146, 151
Endocarditis, 63, 64, 70, 72, 73, 177–179
Enteric fever, 125
Enteric infections, 125
Environmental exposure, 12
Epidemiology, 1–17, 102, 118, 125–126, 132, 148, 260, 269, 270
Esophagus, 3–6

F
Fas ligand, 47–49, 51
Fluorescence, 89, 156, 213, 216, 217, 219
5-Fluorocytosine (5FC), 197
5-Fluorouracil (5FU) gene therapy, 197, 211
Fusarium, 234

G
Gallbladder, 14, 80, 181
cancer, 12–13, 117–134, 167, 176–177
Gallstones (GS), 12, 118, 123, 124, 126, 128–133
Gastric cancer, 2, 5–10, 17, 25–52, 63, 71, 80, 167, 173–176, 181, 194
Gastric lymphoma, 2, 47, 52, 94, 154, 174, 176
Genetically engineered bacteria, 196
β-Glucuronidase, 197
Green fluorescent protein (GFP), 212–215, 217, 219

H
Heat shock proteins (Hsp), 41, 47, 49, 83–86, 88, 99, 100, 103–107, 147, 149, 155, 180
Helicobacter pylori (HP), 2, 4–8, 14, 15, 17, 25–52, 62, 63, 69, 71, 80, 91, 92, 94, 95, 102, 141, 149, 150, 154, 167–176, 178, 181
Hsp. *See* Heat shock proteins
Hypoxia, 190, 191, 197, 199, 211

I
Imaging, 143, 197, 201, 212, 213, 215–217, 219, 251, 262, 266
Infection, 1–17, 27, 62, 79–109, 118, 139–158, 167, 188, 211, 226, 246, 259–271
control, 269–271
Inflammation, 4, 7, 12, 27–31, 34, 36–38, 40–42, 51, 62, 63, 81, 83, 86, 89, 91, 126, 128, 143, 168–169, 171, 173, 174, 176, 180, 181, 190, 199, 210
Intestine, 63, 67, 120, 179, 246

L
Leucine-arginine auxotrophs, 211
Leukemia, 167, 192, 201, 225–239, 253
Lung, 14, 89–91, 134, 152, 155, 196, 217, 250, 253, 261, 268, 269
cancer, 16–17, 88, 108, 147, 167, 179–180, 192, 194, 218
Lymphoma genesis, 49, 50, 52, 90–93, 108, 149

M
Macrolides, 86, 107
MALT. *See* Mucosa-associated lymphoid tissue
Marginal zone B-cell lymphoma (MZL), 91, 94, 105, 107, 140–158
Metastasis, 63, 69, 168, 199, 203, 216–218, 246, 247, 250–251, 255
Mice, 29, 30, 34, 36, 37, 39–41, 46, 48–50, 63, 172, 190, 191, 193–196, 198, 199, 203, 211–221, 230, 233, 235
Microorganisms, 73, 88, 92, 94, 95, 108, 124, 234, 266, 269
Molecular evolution, 253
Mucosa-associated lymphoid tissue (MALT), 47, 81, 139, 140, 143, 146, 148–151, 153, 156–158
  MALT lymphoma (MALToma), 27, 28, 48–51, 90–108, 141, 142, 144, 145, 147, 152, 154, 155, 167, 176
Mucosal immunity, 42
MZL. *See* Marginal zone B-cell lymphoma

N
Neutropenia, 260, 262, 264–267
Nitroreductase (NR), 187, 197
Non-Hodgkin's lymphoma (NHL), 2, 81, 91, 92, 107, 108, 155, 229
Nude mice, 193, 195, 212–219

O
Ocular adnexal lymphoma (OAL), 81, 86, 90, 92–97, 100–102, 104–108, 139–158

P
Pancreas, 13–15, 64, 216, 235, 237
PBMC. *See* Peripheral blood mononuclear cell
Pegaspargase, 237, 238
Perforin, 47–51
Peripheral blood mononuclear cell (PBMC), 88–90, 94, 97–101, 104–107, 149–151, 154
Polymerase chain reaction (PCR), 68, 71, 90, 94, 96–101, 104–107, 123, 141, 147, 148, 152–156, 176, 219
Polyp, 44, 64–67, 69, 70, 73, 178, 179, 190
Proteins, 6, 30–37, 41, 43–48, 51, 63, 67, 72, 82–85, 87, 93, 102, 146, 147, 149–151, 155, 169–175, 178, 180, 188, 193–198, 201, 212, 219, 229–233, 235–238, 250–255
Proteomics, 251, 254, 255
Pseudomonas, 193–195, 203, 233–235, 248–250, 252–254, 260, 261

Q
Quinolones, 86, 107, 265, 267

R
Radiotherapy, 96, 144–145, 188, 191, 194, 211
Rectum, 11
Red fluorescent protein (RFP), 212, 213, 215–219
Reticulate-bodies (RB), 82–86, 146, 147, 151
Reverse transcriptase PCR (RT-PCR), 100, 101, 104–107, 154
Review, 5, 8, 10, 11, 16, 28, 68, 118, 121, 124, 128, 130, 131, 211, 251, 253

S
Salmonella
  S. enterica, 118, 119
  S. typhi, 12, 13, 80, 117–134, 167, 169, 176–177, 181, 193
  S. typhimurium, 187, 191, 196–198, 209–221
Serology, 68–69, 102, 122, 123, 154, 176
SOS, 203, 247, 248, 251–255
Spores, 189, 191, 195–196, 199, 211
Stomach, 1, 2, 4, 6–10, 12, 14, 27–30, 32, 34, 39, 40, 42, 43, 48–50, 80, 91, 175, 176
Streptococcus
  S. bovis (SB), 11, 61–73, 167, 177–179
  S. gallolyticus, 63, 64, 71–73, 179

T
T cells, 32, 33, 44, 47–51, 83, 87, 88, 108, 141, 147, 172, 175–177, 198, 231, 233
T helper 1 (Th1), 40–42, 83, 172, 177
T helper 2 (Th2), 172, 180
T helper, 49
Time-release polymerase chain reaction (TETR-PCR), 96–101, 147, 152, 153, 155, 156
Toxin, 2, 6, 32–34, 50, 69, 70, 81, 84, 124, 169–170, 174–177, 181, 188, 189, 191, 193–195, 199–201, 203, 211, 220, 246, 262

Tumour targeting, 212–215, 220
Tumour, 2–4, 17, 26, 29, 34–36, 38–50, 63–65, 68–73, 81, 91–93, 97, 104, 107, 126, 140, 141, 143–147, 150, 157, 167, 168, 170, 171, 173, 178–180, 188–203, 210–221, 226, 230, 231, 246, 247, 249, 251, 253, 254, 266–270
Typhoid fever, 13–15, 17, 119–125, 130–134, 176, 246

V

Vaccines, 8, 119, 125, 188, 189, 198
Vector, 37, 120, 189, 196–199, 203, 211
Vi antibodies, 123, 132, 134

MIX
Papier aus verantwortungsvollen Quellen
Paper from responsible sources
FSC® C105338

Printed by Books on Demand, Germany